ANALYTICAL PHOTOC
AND PHOTOCHEMICAL

Solids, Solutions, and Polymers

Analytical Photochemistry and Photochemical Analysis

SOLIDS, SOLUTIONS, AND POLYMERS

Edited by J. M. Fitzgerald

DEPARTMENT OF CHEMISTRY
UNIVERSITY OF HOUSTON
HOUSTON, TEXAS

MARCEL DEKKER, INC., New York 1971

MARCEL DEKKER, INC.
95 Madison Avenue, New York, New York 10016

LIBRARY OF CONGRESS CATALOG CARD NUMBER 78-142894

ISBN 0-8247-1191-2

PRINTED IN THE UNITED STATES OF AMERICA

DEDICATION

To Professor Clark E. Bricker

PREFACE

The last few years have seen a tremendous growth in the amount of research and the number of publications in photochemistry. Much of the rapid growth of the field is due to new analytical instrumentation and techniques. However, as the title indicates, one can either be concerned with the analysis of a photochemical system or with the use of photochemistry for an analytical system. Analytical chemists have been involved with the determinations and apparatus used in photochemical research, but only a small portion of this total research effort has been directed towards developing photochemistry as an analytical tool.

Many analytical chemists are quite familiar with the theory and practice of electronic spectroscopy, and yet working in the closely related field of photochemistry has been viewed as a "forbidden transition." The original aim in assembling this book was to collect information that would be useful to analytical chemists. In addition, it is now hoped that the monograph that finally developed will also prove useful to those "pure" photochemists without the training or time to critically choose from among the many analytical methods presently used in photochemical research. In general, books already available have concentrated on photochemical reaction mechanisms, basic theory, and synthetic applications. Any discussion of the interrelation of the photochemical reaction and the analytical method used is treated secondarily, if at all. The emphasis in the present volume remains, as originally planned, on the analytical methods.

In order to facilitate use of this book by both practicing photochemists and analytical chemists, the material has been divided into three sections. Part I, "Photochemical Equipment and Actinometers," covers the major experimental apparatus and the measurement of intensity with actinometry. This section provides essential background information for any person working in the field. The second part, "Use of Photochemistry in Analytical Determinations," is concerned with methods in which the photochemical reaction is used as the means to some other end. The final part, "Analytical Methods for the Study of Photochemistry," reviews methods that have already been found useful in the study of photochemical reactions. Gas phase reactions are not treated in this section; it was felt that this area is relatively specialized and is already adequately reviewed in several well-known books. The important field of polymer photodegradation was worthy of an integrated review from the analytical point of view.

The reader will find that, wherever possible, important analytical criteria, such as the selectivity, sensitivity, precision, and concentration range of a method, are summarized in the latter two sections. It was originally planned that enough experimental detail would be included so that the book could be used for direct application to new problems as they arose in the laboratory. However, it soon became clear to all of those involved that such detail would result in an unreasonably long and tedious book. Therefore extensive references to original papers are included, and an Author Index has been included so that the full range of application for a given reference can be traced back to each of the chapters in which it is cited.

Each author was given a free hand to develop the specific topic as he saw fit, and in general a surprisingly small amount of overlap between chapters was found during the editorial process. Wherever possible, this overlap has been eliminated and an appropriate cross reference to another chapter has been substituted. Some techniques, e.g., flash photolysis, are covered from different viewpoints in two or more chapters; many of these sections have been allowed to stand as is. The careful reader will note, however, that there are differences of opinion between authors as to the utility or suitability of a given experimental technique. This is to be expected in an experimental science. No effort was made to force all the chapters into one rigid style or order of presentation. The material covered is quite divergent. It is also felt that the personality of the writers can serve as an aid to the reader; since the material discussed in most of the chapters has been selected from a wider literature, the personal attitudes and judgements of the writer must enter in.

Since the editor's primary research interest is in the use of photochemical reactions to obtain analytical data, it was somewhat disappointing when this part of the book turned out to be the shortest of the three. However, the field is still in its infancy and it is hoped that new applications will soon be reported. It is tempting to draw an analogy between the early days of electroanalytical chemistry and the present status of photoanalytical chemistry; if the analogy holds, there should be many more advances in the near future.

The special help of several persons at various stages in the development of this book should be acknowledged. Franklin Fry and William Riggs provided many useful suggestions during the preparation of the book. Hugh A. Taylor has been a co-worker in the field for some years now and was, as always, a great help with the present project. Finally, the pleasant and constant diversion provided by many associates in the laboratory at the University of Houston is acknowledged. Without their constant companionship, this book would have been completed much earlier, but the task would have been much more tedious.

Houston, Texas JERRY M. FITZGERALD

CONTRIBUTORS

H. D. Drew, Faculty of Chemistry, Southern Illinois University, Edwardsville, Illinois

J. M. Fitzgerald, Department of Chemistry, University of Houston, Houston, Texas

Franklin H. Fry, American Can Company, Neenah, Wisconsin

S. P. Perone, Department of Chemistry, Purdue University, Lafayette, Indiana

William M. Riggs, E. I. du Pont de Nemours and Company, Central Research Department, Wilmington, Delaware

Norma Z. Searle, American Cyanamid Company, Central Research Division, Stamford, Connecticut

Edward J. Smith, St. Joseph's College, Philadelphia, Pennsylvania

Hugh A. Taylor, Chemstrand Research Center, Durham, North Carolina

Earl L. Wehry, Department of Chemistry, University of Tennessee, Knoxville, Tennessee

W. W. Wendlandt, Department of Chemistry, University of Houston, Houston, Texas

CONTENTS

PART II

USE OF PHOTOCHEMISTRY IN ANALYTICAL DETERMINATIONS

PART III

ANALYTICAL METHODS FOR THE STUDY OF PHOTOCHEMISTRY

Part I

PHOTOCHEMICAL EQUIPMENT AND ACTINOMETERS

CHAPTER 1

APPARATUS FOR PHOTOCHEMICAL EXPERIMENTATION

Edward J. Smith

ST. JOSEPH'S COLLEGE
PHILADELPHIA, PENNSYLVANIA

I. Introduction

In this chapter, sources of photolysis radiation with which the experimental photochemist should be familiar are discussed. Specialized sources, such as flash lamps and lasers, are not discussed here since these areas are treated in other chapters. Methods and apparatus for wavelength and intensity selection discussed here will emphasize commercial equipment or other easily available approaches. The selection of optical materials used in the constructions of photolysis cells, filter bottles, etc., is also discussed. Some commercially available grades of glass and quartz and the transmittance at photochemically important wavelengths are tabulated.

In recent years reaction vessels specially constructed for photolysis studies have become abundantly available. Both lamp manufacturers and laboratory supply houses have entered the field, and vessels of varied shape, size, and material are now available from stock. Many manufacturers also will construct specialized apparatus to fit the needs of a particular problem. Wherever possible in this chapter commercial sources are cited. However, such information is by no means complete and the field is rapidly changing at this time.

Before beginning consideration of equipment, a discussion of units and relationships commonly encountered in expressing the properties of filters and sources will be of use in later sections.

II. Fundamental Units and Relationships

Electromagnetic radiation (EM) has a dual nature and may be considered as a wave or as a particle. When discussing physical optics, one would find it difficult to consider EM anything but a wave. On the other hand, when discussing the photoelectric effect it is convenient to regard EM as a particle or packet of energy.

Electromagnetic radiation has four parameters associated with it: amplitude, velocity, wavelength, and frequency. Of the four, only frequency is independent of the medium through which the EM is transmitted:

$$\text{Frequency} = \nu = v/\lambda \tag{1}$$

where v is the velocity of EM in centimeters per second and λ is the wavelength in centimeters. Both the velocity and wavelength vary with the refractive index of the medium.

$$\text{Velocity in medium} = \frac{\text{Velocity of EM in a vacuum}}{\text{Refractive index of medium}}$$

$$v = \frac{c}{n} \tag{2}$$

The velocity of light in a vacuum is given the symbol *c* and has a value of 2.9979 × 10^{10} cm/sec (*1*).

Analytical chemists generally use units of wavelength to specify portions of the EM spectrum where many pure spectroscopists use units of frequency, ν, or most often, units of wave number, $\bar{\nu}$.

$$\bar{\nu} = \frac{1}{\lambda} = \frac{\nu}{c} \tag{3}$$

Common units of EM are summarized in Table 1.

TABLE 1
COMMON UNITS OF ELECTROMAGNETIC RADIATION

Name	Symbol	Equalities
Angstrom	Å	1 Å = 10^{-8} cm = 10^{-10} m
Millimicron	mμ	1 mμ = 10 Å = 10^{-7} cm
Nanometer	nm	1 nm = 1 mμ
Micron	μ	1 μ = 10^{-4} cm = 10^{-6} m
Wave number	$\bar{\nu}$	$\bar{\nu}$ (in cm^{-1}) = $1/\lambda$ (in cm)
Frequency	ν	$\nu \text{ (in sec}^{-1}) = \dfrac{c \text{ (in cm/sec)}}{\lambda \text{ (in cm)}}$

The quantum theory introduces a relationship between the wave and particle natures of EM, this relationship being

$$E = h\nu = hc/\lambda = hc\bar{\nu} \tag{4}$$

The constant *h* is called Planck's constant (*1*) and is a fundamental constant of nature. The accepted value of *h* is 6.6256 × 10^{-27} erg-sec/quanta. Equation (4) relates the energy *E* (in ergs) of one quanta or photon to its frequency ν (in sec^{-1}), velocity *c* (in cm/sec), and wavelength λ (in cm). The energy in ergs of 1 mole of photons, commonly referred to as an Einstein, may be calculated by multiplying the energy of a single photon by *N*, Avogadro's number, 6.023 × 10^{23} (*1*).

TABLE 2
COMMON UNITS OF ENERGY AND POWER AND RELATIONSHIPS BETWEEN THEM

1 erg	$= 10^{-7}$ J
	$= 2.390 \times 10^{-8}$ cal
	$= 6.242 \times 10^{11}$ eV
1 eV	$= 1.602 \times 10^{-12}$ erg
1 cal	$= 4.184$ J
1W (power)	$= 10^{7}$ ergs/sec
	$= 1$ J/sec
1 erg/quanta	$= 1.440 \times 10^{13}$ kcal/Einstein
	$= 3.759 \times 10^{34}$ eV/Einstein
	$= 5.035 \times 10^{15}$ cm^{-1}
	$= 6.023 \times 10^{16}$ J/Einstein
E (in kcal/Einstein)	$= 9.536 \times 10^{-14} \times \nu$ (in sec^{-1})
	$= 2.859 \times 10^{-13} \times \bar{\nu}$ (in cm^{-1})
	$= 2.859 \times 10^{5}/\lambda$ (in Å)

In terms of kilocalories, the energy of an Einstein of photons may be conveniently calculated by including a conversion factor from ergs to kilocalories.

$$E \text{ (in kcal/Einstein)} = \nu \times 9.537 \times 10^{-14} \qquad (5)$$

It is often necessary to convert from one energy unit to another during photochemical calculations. Common units and relationships developed as shown in the examples above are given in Table 2.

The photochemist is often confronted by spectra of EM sources published in terms of power (in watts) versus wavelength. What the photochemist is really concerned with is the energy of an individual quantum and the intensity or number of quanta of a particular energy emitted per unit time. The energy can be obtained easily by referral to Table 2.

Since power is energy per unit time and each photon has an energy associated with it, power may be easily converted to intensity, I. (See Table 3.) For example, a helium–argon laser with a power of 2×10^{-3} W

TABLE 3
FACTORS FOR INTRACONVERSION OF POWER AND INTENSITY

I (in quanta/sec)	$= 5.035 \times 10^{14} \times \lambda$ (in Å) $\times$ power (in watts)
I (in μEinsteins/sec)	$= 0.836 \times 10^{-3} \times \lambda$ (in Å) $\times$ power (in watts)
Power (in watts)	$= I$ (in quanta/sec) $\times 1.986 \times 10^{-15}/\lambda$ (in Å)
	$= I$ (in μEinsteins/sec) $\times 1.196 \times 10^{3}/\lambda$ (in Å)

at 6328 Å is emitting 6.37×10^{15} quanta/sec or 1.06×10^{-2} μEinsteins/sec at 6328 Å.

In the case of photochemical titrations it is convenient to compare them with coulometric titrations because of the many similarities. In coulometry the current–time integral is a measure of the electrons added to the chemical system.

$$1 \text{ A} = 6.242 \times 10^{18} \text{ electrons/sec} \tag{6}$$

In photochemical titrations quanta are added to the system. The quantum intensity equivalent of current is

$$\begin{aligned} 1 \text{ A} &\equiv 10.036\ \mu\text{Einsteins/sec} \\ &\equiv 6.242 \times 10^{18} \text{ quanta/sec} \end{aligned} \tag{7}$$

Thus photolysis at an intensity of 10.036 μEinsteins/sec is identical to electrolysis at 1 A of current in terms of the number of equivalents of electrons or photons introduced into the system.

Whether or not the same amount of electrolysis and photolysis will produce an equivalent amount of chemical change depends upon the particular chemical system; more specifically, the current efficiency, the quantum yield, and the molar absorptivity of all species must be considered.

III. Photolysis Region

Broadly speaking, photochemistry is the study of the effects of electromagnetic radiation on chemical systems. Most photochemistry, however, is initiated by irradiation with wavelengths lying in the visible and ultraviolet regions. Photolysis resulting from X-ray or gamma-ray irradiation is usually referred to as radiolysis. The chemical effects of irradiation by frequencies lying in the infrared, microwave, and radio regions result for the most part from induced thermal effects and are not usually referred to as photolytic effects.

The visible region includes all frequencies which the human eye can perceive as light. The longest and shortest perceived wavelengths are limited by the ability of the radiation to penetrate the eye and reach the retina. For the average child the visible region is between 3130 and about 7600 Å; while for the average adult the region is more limited, extending from 3800 to 7600 Å.

TABLE 4
THE ELECTROMAGNETIC SPECTRUM

Region	Wavelength range	Frequency range, cps or Hz	Energy range, kcal/Einstein
Gamma	Below 0.1 Å	Above 3.0×10^{19}	Above 2.9×10^{6}
X-ray	0.1–400 Å	3.0×10^{19}–7.5×10^{15}	2.9×10^{6}–720
Ultraviolet			
Extreme	40–100 nm	7.5×10^{15}–3.0×10^{15}	720–286
Far	100–200	3.0×10^{15}–1.5×10^{15}	286–143
Middle	200–300	1.5×10^{15}–1.0×10^{15}	143–95.3
Near	300–400	1.0×10^{15}–7.5×10^{14}	95.3–71.5
Visible			
Violet	400–450 nm	7.5×10^{14}–6.7×10^{14}	71.5–63.9
Blue	450–480	6.7×10^{14}–6.3×10^{14}	63.9–60.1
Green-blue	480–490	6.3×10^{14}–6.1×10^{14}	60.1–58.2
Blue-green	490–500	6.1×10^{14}–6.0×10^{14}	58.2–57.2
Green	500–560	6.0×10^{14}–5.4×10^{14}	57.2–51.5
Yellow-green	560–575	5.4×10^{14}–5.2×10^{14}	51.5–49.6
Yellow	575–590	5.2×10^{14}–5.1×10^{14}	49.6–48.6
Orange	590–625	5.1×10^{14}–4.8×10^{14}	48.6–45.8
Red	625–760	4.8×10^{14}–4.0×10^{14}	45.8–38.1
Infrared			
Near	0.76–2.5 μ	4.0×10^{14}–1.2×10^{14}	38.1–11.4
Fundamental	2.5–25	1.2×10^{14}–1.2×10^{13}	11.4–0.11
Far	25–500	1.2×10^{13}–6.0×10^{11}	0.11–0.06
Microwave	0.05–100 cm	6.0×10^{11}–3.0×10^{8}	0.06–2.9×10^{-6}
Radio	1–1000 m	3×10^{8}–3.0×10^{4}	2.9×10^{-6}–2.9×10^{-9}

The ultraviolet region lies between the X-ray and visible regions, extending from about 40 to 4000 Å. In terms of frequency range the ultraviolet region is about two hundred times as wide as the visible. Table 4 lists the regions of the electromagnetic spectrum and their associated energies and wavelengths.

IV. Sources of Photolysis Radiation

The requirements for a source of radiation to be used by a spectroscopist or a photochemist are usually quite different. For spectrophotometric measurements a continuous source of relatively low intensity is desirable while for most photochemical work a source which emits a few intense frequencies is more desirable. In both cases the intensity of the source (number of photons emitted per second) should be stable with time.

Moreover, the source itself should have a long life, be inexpensive, and be easy to replace or repair.

By far the most important consideration in choosing a source for photolysis is the absorption spectrum of the compound to be photolyzed. Photochemical reaction can take place only if the compound absorbs at the frequencies emitted by the source. Another consideration is the rate of photolysis. Since the rate of most photochemical reaction is directly proportional to the source intensity, a source intensity must be available to bring about a desirable conversion of reactants to products in a convenient time interval. The technique for recording the emission spectrum of a source is discussed in Chapter 4.

A. *Arcs*

By far the most popular source for photolytic application is the mercury vapor arc. Mercury arcs are classified according to the pressure within the arc envelope: low, medium, or high pressure. Each type emits a characteristic spectrum and intensity and has specific applications for the photochemist.

B. *Low-Pressure Mercury Arcs*

Low-pressure arcs in which the vapor pressure of mercury is about 10^{-3} Torr emit radiation primarily at two frequencies, 1849 and 2537 Å. These are the "resonance frequencies" for mercury and are required to initiate mercury-photosensitized reactions. Medium- and high-pressure arcs emit "reversed" radiation due to self-absorption and are not suitable as sources for mercury-photosensitized reactions. The 2537-Å line from a low-pressure arc is usually the more intense of the two lines. The intensity of the 1849-Å line depends on the conditions of excitation, the transmission characteristics of the arc envelope, and the length of the air path between the arc and the absorber.

Three types of low-pressure arcs are available. The heated-cathode arc operates at low ac or dc voltage and the cold-cathode lamp which must be operated at higher voltages to induce electron emission at the cold cathode. The electrodeless discharge lamp can also be used.

One of the features of the low-pressure arc that photochemists find most valuable is its long life and dependability. Cold-cathode lamps are generally rated at 10,000 hr of life or more. Considerably less heat is produced by a cold-cathode lamp than by a hot-cathode lamp since the former does not require heating of the electrodes to maintain the arc. The cold cathode also does not require electron emissive coatings which are partially consumed with each start up of the lamp. Low-pressure mercury arcs are often marketed under the name of germicidal or Sterilamps (registered trademark of Westinghouse Electric Corp.). Germicidal lamps are hot-cathode lamps which are usually cylindrical or U-shaped. They have the same operating parameters as ordinary visible emitting fluorescent lamps; i.e., the cathode must be heated before the arc can be started. Sterilamps are cold-cathode, low-pressure lamps which operate at high voltage (*2*). The lifetime of cold-cathode lamps is limited by the ultraviolet transmission of the glass rather than the electrode life or frequency of starts. The Westinghouse 782H-10 is rated at 17,500 hr of life (*3*). The life of hot-cathode lamps, on the other hand, is governed by the life of the electrodes, which is governed by the frequency of starts and operating current.

Cold-cathode lamps can be shaped into any form and size to provide maximum use of their output. Lamps are available commercially from Englehard–Hanovia, Ultra-Violet Products, Westinghouse, and Oriel Optics Corp. The C-13 Series miniature lamps marketed by Oriel are extremely stable. The output intensity varies only 2% when the power supply input voltage varies between 110 and 130 V (*4*).

Resonance radiation from electrodeless mercury discharge lamps containing ^{198}Hg are available commercially from Hilger and Watts, Ltd. The radiation band from these lamps is so narrow that it selectively excites only ^{198}Hg in mercury vapor also containing other isotopes of mercury.

The proper operation of low-pressure arcs is a topic which deserves some mention. Operating instructions for each arc should be followed as directed by the manufacturer in order to receive maximum stability and life. Manufacturers are very helpful in designing equipment utilizing their arc and will provide information concerning the power and cooling requirements and the safety precautions that should be followed. Some general comments on low-pressure arcs as enumerated by Calvert and Pitts (*5a*) and Noyes and Leighton (*6a*) are summarized here.

1. All arcs possess a negative volt–ampere characteristic. As the current through the lamp increases, the voltage decreases. If a conventional low-impedance power source is used, the arc will destroy itself. A power source that places a sufficiently large ballast resistance (or reactance in the case of ac arcs) in series with the arc is necessary to achieve a stable arc intensity.

2. The intensity emitted by an arc is sensitive to the current through the arc and the temperature of the arc electrodes. At low currents, 15 mA, the intensity at 2537 Å is directly proportional to the current, while at higher currents, 130 mA, the rate of change in intensity with current increases rapidly with increasing current.

If the electrode portion of an arc is held at constant temperature, the arc tube may be heated as high as 600°C with little variation in arc intensity. Thus, an arc can be placed in an oven along with the sample as long as the electrodes remain at constant temperature outside the oven.

If either the arc tube or electrodes are cooled below room temperature, however, the intensity of the arc will decrease.

Since the intensity of an arc will change with aging, it is recommended that an elapsed-time meter, a voltmeter, and an ammeter be included in the lamp circuit.

3. If the rate of the photoreaction is not directly proportional to the light intensity, then a 60-cycle ac arc may give anomalous results. The intensity of such an arc drops to almost zero 120 times a second. A dc arc should, therefore, be used in cases which may be subject to the effects of intermittent light (*6b*).

C. *Medium-Pressure Mercury Arcs*

As the pressure of the arc is increased from low ($\sim 10^{-3}$ Torr) to medium ($\sim$1 atm), two effects can be observed. The intensity of the arc increases with increasing pressure, and also the number of different frequencies emitted increases. The spectra of two popular low- and medium-pressure lamps are compared in Table 5.

The Hanovia Utility Model 30620 has found particular application as the source in photochemical titrations because of the many frequencies emitted and because of stability and reproducibility of its intensity (*7*, *8*).

A few points to note concerning the operation of medium-pressure arcs are:

1. The intensity of the arc is determined by the pressure, current, and voltage. Fixing any two of these variables fixes the third. At constant current, the intensity is directly proportional to the pressure and the square of the voltage. At constant pressure the intensity is proportional to the square of the current.

TABLE 5
RELATIVE INTENSITY DISTRIBUTION IN LOW- AND MEDIUM-PRESSURE MERCURY ARCS

	Relative intensity	
Wavelength, Å	Low pressure[a]	Medium pressure[b]
2537	1.000	0.037
2965	0.006	0.099
3022	0.011	0.150
3126–3132	0.008	0.223
3650–3663	0.012	0.457
4047	0.009	0.200
4358	0.017	0.420
5461	0.012	0.740
5770–5790	0.004	1.000

[a] Ultra-Violet Products, Inc., Model 11 SC-1C.
[b] Englehard–Hanovia, Inc., Model 30620 Utility.

2. For a given power input in watts per centimeter of arc length, the maximum intensity is obtained at high pressures, i.e., with high voltage, low current, and a minimum of cooling.

3. As with the low-pressure arc, a ballast resistance must be placed in series with the lamp. Commercial power sources supplied with the lamps have the proper resistances built in.

4. Higher voltage can be applied to the arc power source to increase the intensity, but at the cost of a shorter life. A Variac or a feedback-stabilized voltage regulator can be used to select and maintain a desired arc intensity during the course of an extended photochemical investigation.

5. Arcs should be aged before being used in quantitative studies. The intensity of a new arc is usually much greater than the average lifetime intensity. About 24 hr of intermittent use is usually required to "age" the Hanovia Utility 30620. Ozone production is quite evident in the first hours of operation but falls off as the arc-envelope transmittance characteristics change.

6. A warm-up time of about 15 min for the Hanovia Utility 30620 was found necessary to reach a steady intensity (*7*). This parameter should be determined for each model lamp and the cooling conditions employed.

Arcs may be cooled by a stream of clean, compressed air or nitrogen, but this must be done reproducibly if constant intensity is desired.

7. Restarting of a high- or medium-pressure mercury arc may be difficult if the arc is still hot. This is due to the fact that the voltage required to

start the arc at the operating pressure and temperature is higher than that required to start a cold lamp (*2*). Restarting time may be reduced by passing air over the arc to accelerate cooling.

D. *High-Pressure Mercury Arcs*

The most intense of all arcs is the high-pressure arc which operates typically at internal pressures of several hundred atmospheres and temperatures of 5500 to 8000°K. The high-pressure emission spectrum is characterized by broad bands instead of the discrete line spectrum of low- and medium-pressure arcs. The intensity of the arc is due to asymmetric broadening of the characteristic mercury emission lines and the addition of continuous radiation due to thermal emission of the hot arc.

A high-pressure arc is used in studies where intensity is of primary importance even at the sacrifice of stability, reproducibility, and monochromaticity. Per unit length of arc, the output of a high-pressure arc is about ten times that of a medium-pressure arc and 1000 times the intensity of a low-pressure arc.

The use of high-pressure arcs presents two dangers to the photochemist if not used properly. Since the internal pressure of the arc envelope may be very high during arc operation, failure of the envelope could present a serious hazard if a safe lamp housing is not employed. Also, the intense short-wavelength radiation from the arc will cause severe discomfort if allowed to strike the naked eyeball. The discomfort is not long lasting and no permanent eye damage results. A conveniently designed housing is described by Calvert and Pitts (*5b*). Similar housing may be obtained from Oriel Optics Corp., Schoeffel Instrument Corp., and the arc manufacturers themselves. A typical housing for point-source arcs is shown in Fig. 1-1.

Two general types of high-pressure arcs are on the market, the capillary arc and the point-source arc. Capillary arcs generally range in size from 1–2 mm in diameter and 1–12 in. in length. The average life of such a lamp ranges from 10 to 80 hr, depending on cooling method and number of starts. Capillary arcs are very intense sources; the AH6 2, 2-kW arc (PEK, Inc.) is about 80 times more intense in terms of luminous flux than an ordinary 100-W tungsten bulb. Capillary arcs operate at very high pressures and temperatures, and they must be cooled with forced air or water to prevent envelope failure. Water must be circulated at very high velocities around the arc to prevent formation of steam bubbles and also should be of high purity for maximum transmittance and minimum electrical conductivity.

Point-source, short arc lamps consisting of a high-powered discharge in mercury, mercury–xenon mixtures, or xenon, compressed in a small

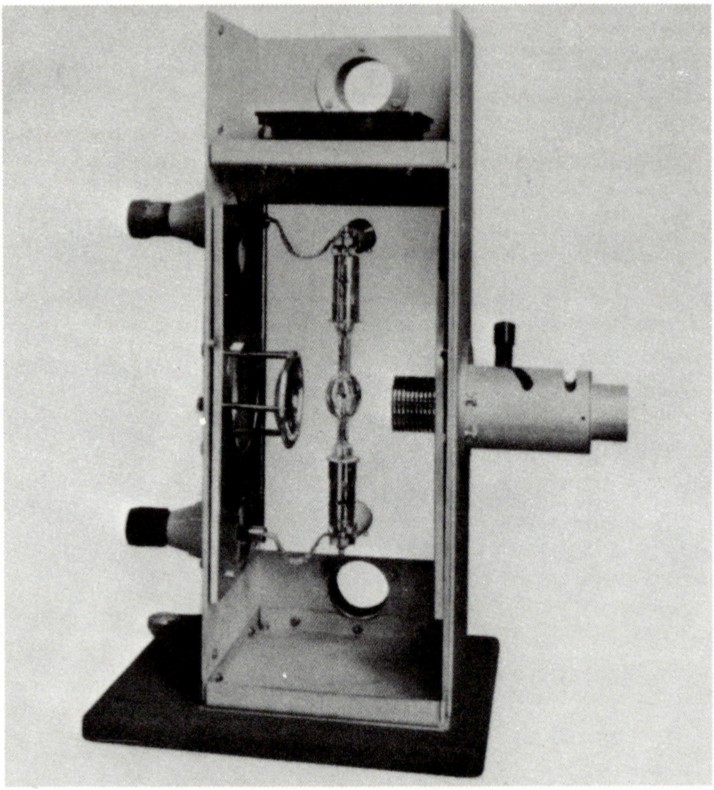

FIG. 1-1. Oriel Optics Model C-60-50 universal lamp housing for lamps up to 1000 W, showing lamp, rear reflector, beam port, and reflective thermal shielding (Oriel Optics Corp.).

volume of bulblike design, are simpler to operate than the capillary arcs. Usually they need no more cooling than forced air. Point-source lamps are currently available from PEK, Inc.; OSRAM Lamps; George W. Gates and Co., Inc.; Illumination Industries, Inc.; Hanovia; and others, ranging in power from 35 to 2500 W. The expected lifetime of point-source arcs is greater than that for capillary arcs and ranges from 200 to 1500 hr or longer.

The operating voltage of mercury and mercury–xenon point-source arcs range from 20 to 80 V and currents of 1 to 50 A. However, a pulse of from 50 to 30 kV is necessary to start the arc. Since the power supplies for these arcs obviously require some special components, it is easier to order them directly from commercial sources than to try to construct one. Mercury lamps containing some xenon or pure xenon lamps emit more energy in the visible and near ir than lamps filled with mercury only.

E. *Other Line Sources for Photochemical Studies*

Medium-pressure, rare-gas lamps of the type available from Oriel Optics Corp. provide a useful source of closely spaced frequencies in the ultraviolet and short-wavelength visible regions that may be useful for low-intensity, low-energy photolysis.

TABLE 6
WAVELENGTHS AVAILABLE FROM RARE GAS LAMPS

Rare gas	Potentially useful range of wavelengths (nm) available
Argon	390–440
Helium	380–450
Krypton	430–450
Neon	340–360
Xenon	460–480

Table 6 lists the range of useful wavelengths that may find use in photolytic applications. Most metal-vapor resonance lamps emit frequencies of such small energies that they have not been used for photolysis. However, cadmium, thallium, zinc, and mercury–thallium lamps emit at high enough frequencies to find some use. Cadmium emits in the 325-to-365-nm region, thallium at 350 to 385 nm, and zinc at 305 to 335 nm.

High-intensity, hollow-cathode lamps, such as those used in atomic flame fluorescence, coupled with a monochromator also offer photochemists a monochromatic source of low-intensity radiation (*9*). Lasers as sources are considered in the following chapter by F. H. Fry.

F. *Other Continuous Sources for Photochemical Studies*

The high-pressure xenon lamp is useful as a source of continuous radiation in the ultraviolet and visible regions emitting proportionally more of its energy in the visible and infrared regions than a mercury arc. Point-source arc lamps, similar in design to mercury arcs, are commercially available (e.g., PEK, Inc.).

Hydrogen and deuterium lamps with Suprasil quartz windows emit intense radiation at frequencies useful for photolysis. High-pressure, water-cooled deuterium lamps and power supplies are convenient to work with and are commercially available (e.g., Scientific Cell Co., Inc.). The zirconium arc emits a continuum of radiation extending from 2000 Å into the infrared region. It consists of a tantalum cathode with a zirconium oxide core and an anode sealed into a bulb in an atmosphere of argon or other inert gas (*2*). The spectrum shows zirconium and argon lines superimposed upon a broad continuum.

Carbon arcs are another possible source of low-wavelength, continuous radiation. The open carbon arc lasts only a few hours, however, due to consumption of the carbon electrodes during operation. Also, in order to maintain constant output, the arc gap must be maintained constant either manually or automatically.

Incandescent solids are excellent sources throughout the visible and infrared regions, but are generally not intense enough to be useful as ultraviolet sources. The spectral distribution of radiation from an incandescent solid at a given temperature will usually approximate that from a blackbody at a somewhat lower temperature. Auto headlights and projection lamps have a color temperature of about 3400°K and can be used as sources of radiation between 300–400 nm. When wavelengths below 300 nm are desired, color temperatures above 3500°K are necessary.

Many photochemical studies require simulation of natural sunlight; for example, solar cell and panel testing, material degradation studies, and nutrient growth tests. The principal reduction in energy of sunlight as it passes through the atmosphere comes from absorption in the ultraviolet end of the spectrum below 0.3 μ by the ozone and absorption at various frequencies in the infrared, primarily by water vapor. The amount of solar ultraviolet radiation reaching the earth's surface depends on many factors, including time of day, month, year, latitude, elevations, thickness of the ozone layer, and concentration and nature of airborne pollutants. One commercial solar simulator, i.e., Spectrolab, uses a high-pressure xenon arc as the source filtered by ultraviolet-absorbing filter and distilled water. Simulators for sunlight at various altitudes are available. An excellent review of lamps is given by McLanahan (*24*).

V. Methods and Apparatus for Wavelength and Intensity Selection

In many cases lamps available to the experimenter have spectral properties different from that desired for a particular application. If the sources available do not emit at the proper wavelengths or if the intensity is too low, then the only recourse is to purchase or construct another source. However, if the source emits excess wavelengths and the intensity is more than adequate, the experimenter can obtain the spectrum he desires by proper attenuation.

A. *Wavelength Selection*

In general there are three types of wavelength selectors available: absorption filters, interference filters, and monochromators. Absorption filters are constructed in three different types: liquid or chemical, gelatin, and glass.

Liquid filters consist of chemical solutions of appropriate concentration and path length through which the beam of radiation must pass before reaching the system to be irradiated. Quartz filter cells with several compartments of various path lengths are usually needed to isolate narrow portions of the spectrum. Each compartment contains a solution which is designed to remove a portion of the spectrum. Several recipes of liquid filters and their stabilities are discussed in Calvert and Pitts (*5c*) and in Noyes and Leighton (*6c*).

Gelatin filters are usually thin (0.1 mm) and are liable to damage by moisture and heat unless protected (*10*). These objections can be removed by placing the gelatin between glass or quartz plates and sealing the edges to keep out moisture. Like solution filters, gelatin filters change transmittance characteristics due to repeated exposure to ultraviolet radiation.

Because of their more permanent nature, glass filters are preferred to gelatin or solution filters. The radiation that is not transmitted by the filter is absorbed by the glass. If infrared radiation is absorbed, the filter will heat up and may crack. For this reason special heat-absorbing filters should be placed between the source and the absorption filter.

Absorption filters are usually divided into three types, depending on their transmittance spectrum: short-wavelength-transmitting filters, long-wavelength-transmitting filters, and bandpass filters.

Short- and long-wavelength-transmitting filters are also known as cutoff filters. Cutoff filters come in a variety of cutoff sharpnesses and wavelengths. A 600-nm short-wavelength cutoff filter with an 8% cutoff sharpness is one that is designed to transmit ($> 80\%\ T$) wavelengths longer than 600 nm and reject ($< 1\%\ T$) wavelengths shorter than this value. At 600 nm the transmittance would be 50%. A "sharpness" of 8% means that the percentage transmittance drops from above 80% to below

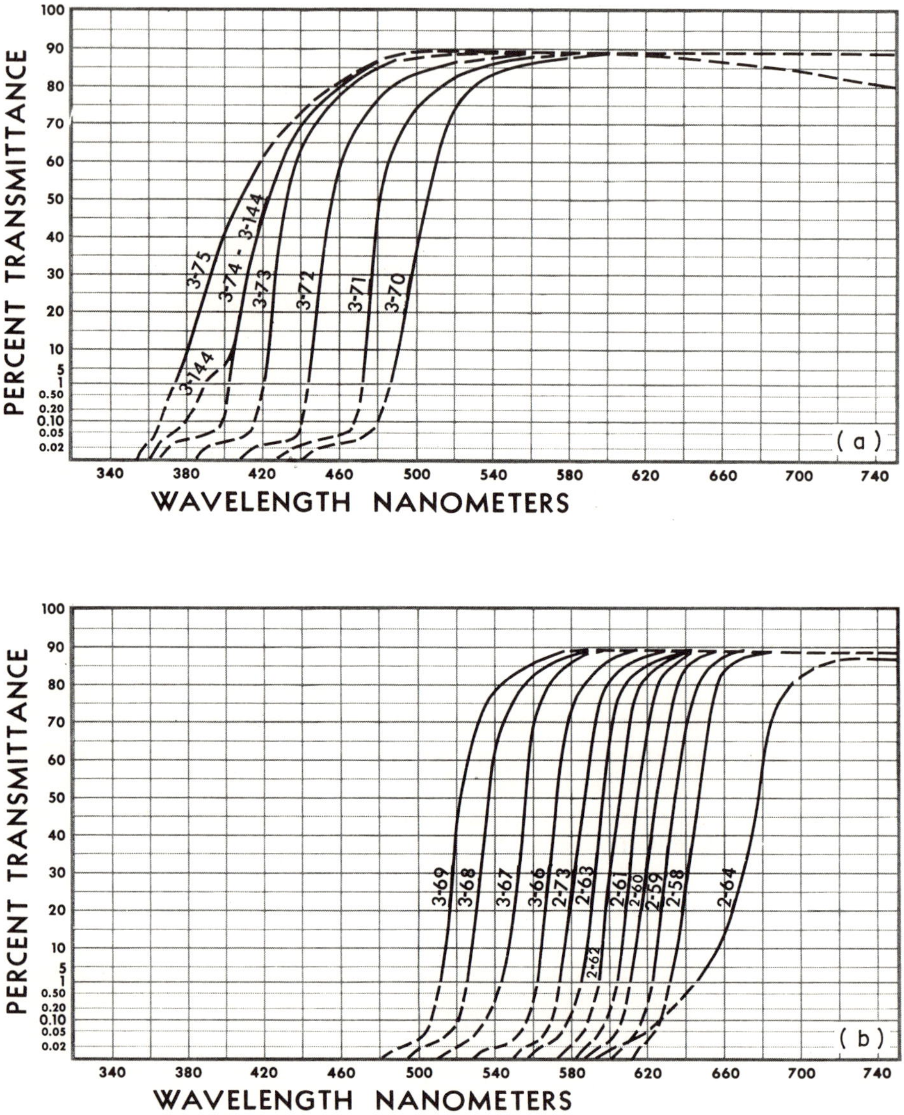

FIG. 1-2. Transmission spectra of cutoff filters available from Corning (Corning Glass Works).

1% in a 0.08 × 600-nm or a 48-nm increment centered about a wavelength of 600 nm.

Spectra of several long-wavelength-transmitting filters available from Corning Glass Works are shown in Fig. 1-2. The filters also have cutoff points in the infrared due to the absorbance by the glass itself. Long-wavelength-transmitting filters are available with various cutoff points ranging from approximately 240 to 670 nm. Short-wavelength-transmitting filters, those which will transmit only ultraviolet and not visible, are not as readily available.

Band filters are designed to transmit a specified portion of the spectrum. Filters are usually designated by their wavelengths of peak transmission and their half-bandwidth or percent half-bandwidth. The half-bandwidth is the width of the band, in wavelength units, between points which the intensity is one-half the peak intensity. The percentage half-bandwidth is the bandwidth divided by the wavelength of maximum transmission multiplied by 100. The narrower the half-bandwidth the purer the transmitted radiation; but, in general, intensity must be sacrificed to gain spectral purity, especially in the ultraviolet region. Narrow bands of radiation may be obtained by filtering the radiation through both a short-wavelength- and a long-wavelength-transmitting filter. Radiation will pass through both filters only where their transmission spectra overlap.

The third type of filter, the interference filter, works not by absorption of unwanted radiation but by destructive interference. Interference filters of the Fabry–Perot type are constructed of thin dielectric films vacuum-deposited on substrates of quartz or optical glass. The dielectric is usually sandwiched by another sheet of quartz or glass. The equation for determining a maximum for the transmission of a spectral band when light is normally incident upon the filter is

$$m\lambda = 2dn \tag{8}$$

where d is the thickness of the dielectric film of refractive index n, and m is the order and λ is the wavelength transmitted. For a filter of a particular d, there will be m wavelengths passed

$$\lambda = \frac{2dn}{m} \tag{9}$$

That is, a filter transmitting 400 nm in the first order ($m = 1$) will transmit 400/2 or 200 nm in the second order, and so on. If only one order is desired, other absorption filters must be used in conjunction with the interference filters.

The half-bandwidth of interference filters in the ultraviolet and visible regions normally varies 2.5 to 50 Å, depending on price. Filters with peak response from 200 to 1600 nm are available from several sources such as International Light, Inc., Bausch and Lomb, Inc., and Oriel Optics, Corp.

Interference filters generally transmit 40–60% of the selected wavelength. An inherent property of interference filters is the decrease in wavelength position of a bandpass when the filter is tilted. Large changes result if the tilt exceeds 15 to 20° from the normal. Temperatures also affect the bandpass position by changing the thickness of the dielectric film. Operating temperatures of greater than 80°C should be avoided.

A few points should be noted in the selection of the proper filters for a given photochemical experiment.

1. Cutoff filters are an inexpensive way of selecting large regions of the spectrum. Cutoff filters passing only ultraviolet radiation and no visible are not readily available.
2. Absorption band filters are available with a variety of peak transmissions and half-bandwidths in both the visible and ultraviolet regions.
3. Interference filters give the highest spectral purity, but at an increased dollar cost and at the cost of intensity. Interference filters are available for most of the emission lines of common arc lamps.

The use of monochromators in photochemistry is usually limited to applications where the monochromaticity of the radiation is of greater importance than high intensity. Even so, monochromators commonly used in spectroscopy are unsuitable for photochemical application because of the very low intensities transmitted. The requirements of a monochromator used in photochemistry are usually met by choosing a monochromator with a large aperture and wide slits. The aperture should be *f*/3 to *f*/5 whereas the aperture in most spectrophotometer monochromators is *f*/7 to *f*/14.

The aperture number, or *f* number, is a measure of the light-gathering ability of a monochromator and is defined as the ratio of the focal length of the collimating mirror to the diameter of a circle with an area equal to the cross section of the beam at the dispersing device.

Early studies employed prism monochromators almost exclusively (see Ref. *5* for a thorough discussion), but with the advent of inexpensive replica gratings more photochemists are utilizing grating monochromators.

For most photochemical purposes a high resolving power in a monochromator is not necessary. If a line source is employed, the slit should be made as wide as possible without overlap from other intense lines. The line-source/monochromator combination gives an ideal source of monochromatic radiation.

High-intensity monochromators designed for photochemical applications can be obtained from several distributors including Bausch and Lomb, Inc., Schoeffel Instrument Corp., and Rudolph Inst. Engineering Co., Inc. A typical high-intensity monochromator is the Bausch and Lomb Model No. 33-86-25-01 which has an aperture of *f*/3.5 and adjustable slits with a maximum width of 6.0 mm.

B. *Intensity Selection*

Several methods of attenuating the intensity of source radiation are available to the photochemist.

The lamp voltage, or current, may be varied to increase or decrease the intensity. The use of higher voltage than recommended by the manufacturer is, however, not recommended since this will result in decreased lamp life. When the voltage is changed, a lamp must be allowed to reach thermal equilibrium with its environment before the intensity will stabilize. When reproducible intensities are desired, it is not recommended that variation of lamp voltage be the intensity-attenuation method used.

By far the most popular method of intensity selection is by the use of neutral density filters. A neutral density filter has the unique property of being capable of reducing the transmission of light over a wide wavelength range while at the same time maintaining the spectral integrity of the attenuated beam. Two main types are available. One type employs colloidal carbon dispersed in gelatin sandwiched between glass plates. Kodak's Wratten neutral density filters No. 96 are of this type; they are useful only in the visible where the glass and gelatin do not absorb.

The second, and most popular, type of neutral density filter is made by depositing a thin film of a metal such as aluminum, gold, silver, copper, or nickel on a suitable substrate such as glass or quartz. Attenuation is both by reflection and absorption. Depositions on quartz extend use in the ultraviolet region. Neutral density filters of this type are unaffected by temperature, humidity, aging, and ultraviolet irradiation. Filters may be obtained having absorbances ranging from 0.1 to 6.0 from sources such as Corion Inst. Corp., The Ealing Corp., Oriel Optics Corp., and Bausch and Lomb, Inc.

Inexpensive neutral density filters can be made from ordinary wire screens. Screens should be as rigid and flat as possible to avoid changes in transmittance with handling. A useful relationship for estimating the percentage transmittance of wire screens is

$$\%T = 100[1 - MS(2 - MS)] \tag{10}$$

where M is the mesh number, or number of wire strands per inch, and S is the wire diameter in inches (*11*).

Another method of radiation attenuation is by the use of contrarotating polarizing filters which allow the light intensity to be varied continuously from maximum transmission down to complete extinction. Maximum transmission occurs when the two filters which are placed together between the lamp and the material to be irradiated have their polarizing directions parallel. As one filter is rotated with respect to the other the transmission decreases until complete extinction is reached when the polarizing directions of the two filters are 90°. The effect of ultraviolet light on the polarizing and transmission properties of these filters should be obtained from the manufacturer (e.g., Polacoat, Inc.) before use. Filters are available for use down to 200 nm.

VI. Optical Materials for Construction of Photochemical Apparatus

The proper choice of optical materials is a prime consideration in the construction of any photochemical apparatus. Important parameters to consider are: (1) percentage transmittance at the wavelengths to be used for photolysis; (2) ability to resist cracking due to heat; and (3) chemical resistance.

Table 7 lists the percentage transmittance for various optical materials at the important mercury emission lines. The exact percentage transmittance of glasses and quartz varies from batch to batch and should be checked if the exact value must be known.

The apparent transmittance of optical materials also varies with the refractive index of the medium in contact with them. For monochromatic radiation incident upon the material at 90° to the surface, the fraction of light reflected, f, can be calculated from the following relationship (*12*):

$$f = \frac{(n_1 - n_2)^2}{(n_1 + n_2)^2} \tag{11}$$

In this relationship n_1 represents the refractive index of the optical

TABLE 7

PERCENTAGE TRANSMITTANCE FOR VARIOUS THICKNESSES OF COMMON OPTICAL MATERIALS AT SOME MERCURY EMISSION LINES

Material	Thickness, mm	%T[a] at wavelengths (nm) indicated			
		185	254	313	366
Borosilicate crown glass	10	<1	<1	40	90
Corning lime glass[b]	1	<1	<1	25	86
Corning Pyrex glass	1	<1	<1	63	87
Corning Corex D glass	1	<1	<1	80	90
Corning No. 9720 glass	1	<1	45	87	92
Corning No. 9741 glass	1	<1	80	92	92
Corning No. 7910 glass	1	<1	86	92	92
Corning 101 crystalline quartz	1	30	86	95	>95
Corning 151 fused silica (silica glass)	1	85	95	95	>95
Amersil Suprasil[c]	10	79	92	92	92
Amersil Infrasil	10	—	90	92	92
Amersil Ultrasil	10	—	92	92	92
Amersil Optosil	10	—	75	92	92
Q and S Purasil[d]	1	—	82	92	—
Q and S Germisil	1	<1	15	92	—
Q and S Heliosil II	1	<1	1	55	—
CaF_2[e]	3.6	85	90	93	93
LiF	10	87	92	93	93
MgF_2	1.5	88	93	94	94
KBr	6.4	1	79	84	87
KCl	10	1	80	85	87

[a] Reflective losses included.
[b] Corning Glass Works.
[c] Amersil, Inc.
[d] Quartz and Silice.
[e] Harshaw Chemical Co.

material and n_2 the refractive index of the medium in contact with the optical material. Thus 3.64% of the 5893-Å radiation incident upon a borosilicate glass ($n = 1.47$)–air interface ($n = 1.00$) will be reflected. When possible, the transmittance of optical ware should be measured containing a solution of the same refractive index as the solution to be photolyzed. Since the refractive index is also a function of temperature, the transmittance of optical materials will also be a function of temperature. The temperature dependence of the refractive index ($\Delta n/\Delta t$) of Tetrasil fused silica has an average value of 10^{-5}/°C.

Some optical materials also undergo "solarization," that is, changes in their transmittance characteristics upon prolonged irradiation at certain wavelengths. For example, Amersil's T14-Optosil III darkens upon relatively short exposure to wavelengths below 2000 Å, while other materials such as Amersil's T19-Suprasil I does not undergo solarization (*13*). The original transmittance characteristics of many materials may be restored by proper heat treatments. Specific information for various materials are available from Corning Glass Works.

Most glasses fluoresce to some extent when irradiated by ultraviolet radiation. This can cause anomalous results in actinometric studies if it is not realized. Corning 0160 glass exhibits a blue fluorescence while 3750 glass exhibits a yellow-green fluorescence (*14*).

The thermal durability of optical materials is of great importance in constructing photochemical apparatus. High-pressure arc lamps dissipate considerable heat and may shatter carefully constructed (and costly) apparatus if improper optical materials are used. Heat resistance generally decreases as the coefficient of thermal expansion increases. The following values will serve as a basis for comparison: fused silica, 5.5; Vycor No. 7913, 8.0; borosilicate glass (Corning No. 7740), 32.5; window glass, 92. Values are given in units of cm × 10^{-7}/cm-°C and are valid between 0 and 300°C (*14*).

Borosilicate and silica glass compositions are outstanding for chemical services. Certain limitations, however, must be recognized. Fused quartz or fused silica is not attacked by acids below 150°C. Above this temperature, hydrofluoric and phosphoric acids attack the material slowly. Other glasses are attacked more readily by these acids. Alkalies attack slowly even at room temperature.

Devitrification of silica glass may take place above 1000°C if heated in contact with impurities. Devitrification brings out small fractures which tend to lower the transmittance of the silica. Finger marks are especially detrimental. As far as practicable all silica ware should be thoroughly clean before heating. The envelope of arc lamps should never be touched with the bare hand.

Further information on construction materials for photochemical apparatus can be found in the treatises by Calvert and Pitts (*5d*) and Koller (*2*). Manufacturers and distributors of optical materials are also quite willing to provide information concerning any property of their materials and should be consulted for specific information.

VII. Photolysis Reaction Vessels and Associated Apparatus

Reaction vessels are usually designed either for synthetic work or for quantitative studies.

A. *Vessels for Synthetic Applications*

For synthetic application the most important consideration is a design that will allow a large amount of sample to be irradiated in an efficient way. Thus, a high-intensity lamp surrounded by the sample in such a way that all of the intensity emitted by the lamp is absorbed by the sample is the obviously best geometrical arrangement. An example of this approach is shown in Fig. 1-3. Most vessels of the type shown in Fig. 1-3 contain a water-jacketed immersion well in which the lamp is placed. The immersion well must pass all wavelengths of the lamp that will be needed for photolysis and also resist cracking due to the heat of the lamp. Quartz, which has high transmittance in the ultraviolet and a low coefficient of thermal expansion, is the most commonly used material. The reaction vessel itself is usually constructed of borosilicate glass which is inexpensive and further serves to protect the operator from the ultraviolet radiation. Vessels varying in volume from 50 ml to 12 liters and employing lamps with powers from 10 to 2800 W are available from commercial sources such as Hanovia Lamp Division and Ace Glass, Inc. Special Dewar vessels designed for the study of photochemical reactions at low temperatures are marketed by Pope Scientific, Inc. Immersion-type equipment for large industrial applications is also available from Hanovia.

The photochemical chamber reaction is also a popular design. In this type a bank of lamps surrounds the sample vessel to be irradiated. In the Rayonet RPR-204 shown in Fig. 1-4, a bank of four mercury lamps is utilized for photolysis. Lamps may be changed for wavelength selection. Four lamps emitting primarily either 2537, 3000, or 3500 Å can be used alone or in combination. Using 2537-Å lamps allows one to irradiate samples with 60 W of radiation. The reactors will accommodate vessels with capacities up to 1.5 liters and are cooled by forced air. Larger scale reactors are also available.

B. *Vessels for Quantitative Studies*

Quantitative studies usually consist in the determination of the quantum yield of a reaction or the effects of temperature, pressure, photolysis intensity, and other variables on the quantum yield.

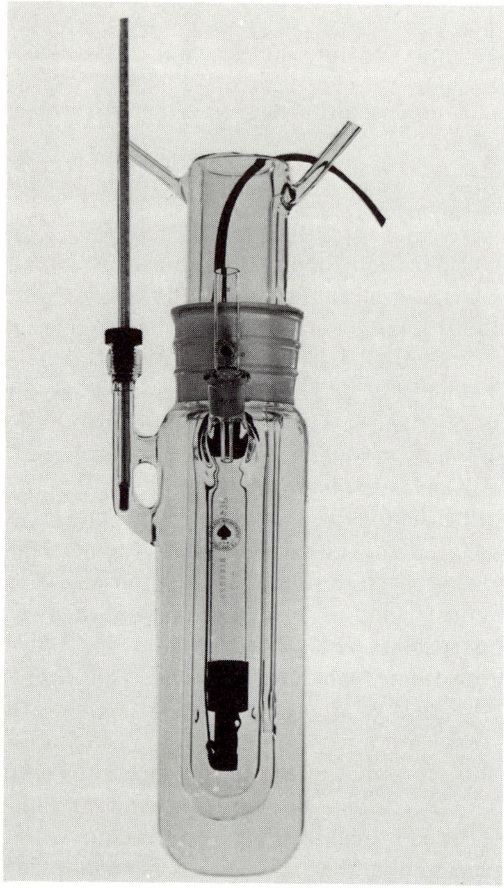

FIG. 1-3. Immersion-type photochemical reaction vessel. Arc lamp is shown in double-walled well through which cooling water may be circulated. Solution to be photolyzed is placed in outermost area protected from the heat of the arc (Ace Glass, Inc.).

In quantitative studies it is necessary to know the number of quanta absorbed. This is best done by designing a photolysis vessel with a known, and uniform, path length. If the concentration and molar absorptivities at the photolytic wavelengths are known, it is then possible to calculate the fraction of incident radiation being absorbed using Beer's law.

For gas studies, a cylindrical glass cell with fused-quartz, optically flat windows at the bases serves as a convenient cell. The window area should be large enough to intercept all of the incident beam and must be flat to avoid reflective losses. Reflection losses become significant when the angle between the beam and the window is less than 60°. Glass and quartz tubing, both cylindrical and square, can be obtained commercially (Amersil, Inc., Quartz and Silice, Thermal American Fused Quartz Co.). Optically flat windows of several materials are also available (General Electric Co., Lamp Glass Dept.; Kaufman Glass Co.). If quartz windows are used with glass cells, one must be aware that the two materials have

FIG. 1-4. The Rayonet RPR-204 preparative photochemical reactor. One lamp module is removed and shown at right (The Southern New England Ultraviolet Co.).

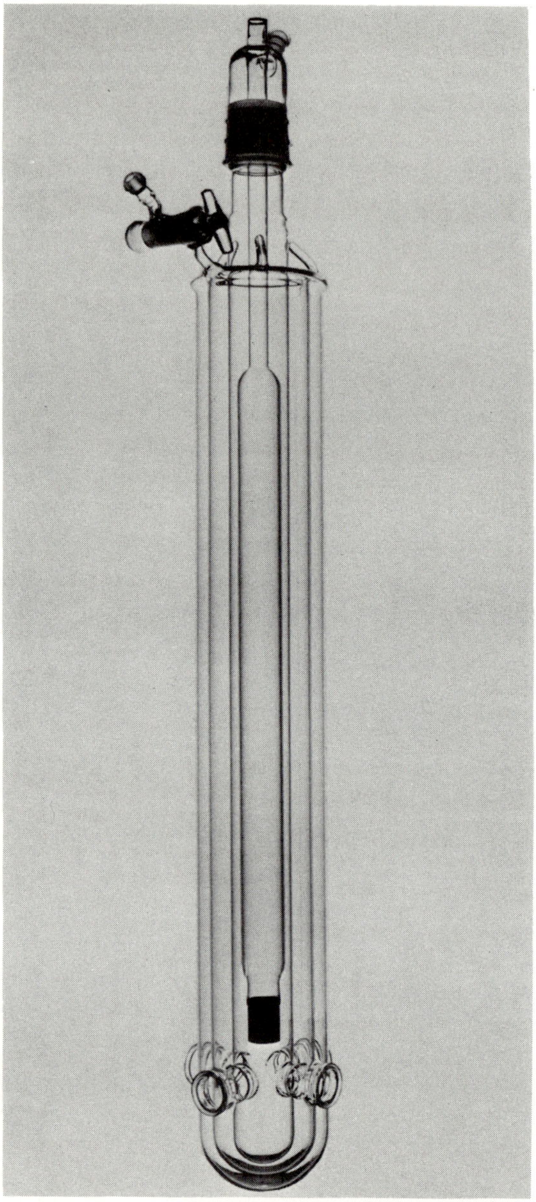

FIG. 1-5. Optical Dewar flask for photolysis at low temperatures, shown before silvering was applied. Observations can be made through windows at bottom (Pope Scientific, Inc.).

different coefficients of thermal expansion and the cell therefore cannot be exposed to sudden temperature changes without the risk of cracked joints. All-quartz construction avoids this problem but at additional expense. Graded seals should be used when Pyrex tubing must be attached to the cell body.

Ordinary quartz or silica spectrophotometer cells can be utilized for the photolysis of liquid samples. Cylindrical and rectangular cells of path lengths from 5 mm to 10 cm are found in manufacturers' stock, and special dimensions are easily manufactured. Anaerobic and water-jacketed cells of all shapes and dimensions are also available. Rectangular cells with four quartz windows are especially convenient since the photolysis reaction may be continuously monitored spectrophotometrically 90° to the direction of photolysis.

The photolysis of compounds in rigid media at low temperatures requires the use of an insulated cell, Dewar type, to avoid fusion of the solid. Photolysis may be carried out on the neat compound or in solvents such as EPA (a mixture of ether, isopentane, and ethyl alcohol, 5:5:2 ratio) which form glasses at liquid-nitrogen temperature (77.3°K).

An example of an optical Dewar flask used for photolysis studies at low temperature is shown in Fig. 1-5.

Other cells especially designed for the photolysis of solids are described in detail in Calvert and Pitts (*5e*).

Specific methods for the determination of the quantum yield of a reaction are discussed in other chapters in this book, but two cell designs employing chemical actinometers as references from which one may calculate the quantum yields of other compounds deserve mention here. Figure 1-6 is a four-compartment cell designed specifically for the

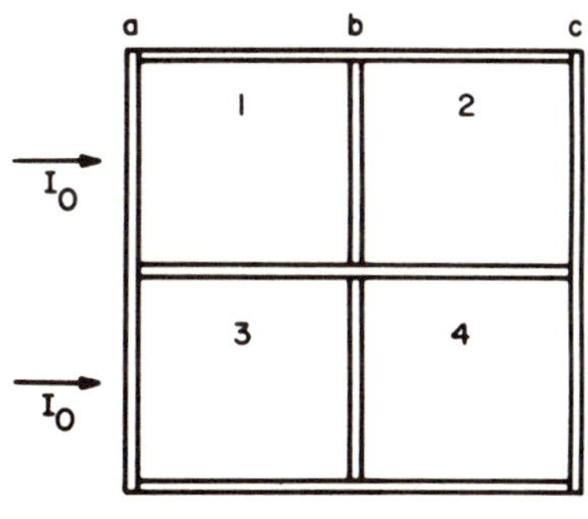

FIG. 1-6. Compound cell for actinometry in which absorption is not complete. From Noyes and Leighton (*6*); reprinted through permission of the publisher.

FIG. 1-7. Rayonet Merry-Go-Round designed to assure equal irradiation of several solutions. The cell rotator is shown above the reaction chamber (The Southern New England Ultraviolet Co.).

determination of quantum yield of compounds in which absorption is not complete. The system consists of four cells of identical dimensions with quartz windows at *a* and *b* and a nonreflecting rear surface *c*. Actinometer solution is placed in cells 1, 2, and 4, and the solution of unknown quantum yield in cell 3. The system is then exposed to a beam of radiation which is uniform in intensity over the entire surface of the front window *a*. If Φ_A is the quantum yield for the actinometer, Φ_U the quantum yield of the solution in cell 3, and m_1, m_2, m_3, m_4 are the respective moles transformed in cells 1, 2, 3, and 4, then

$$\Phi_U = \Phi_A \frac{m_3}{m_1}\left(\frac{m_1 - m_2}{m_1 - m_4}\right) \tag{12}$$

The recently marketed Rayonet Merry-Go-Round is designed to ensure equal irradiation of sample and actinometer solutions (Fig. 1-7). Eight samples may be irradiated in the reactor. Samples are rotated past the light sources at a speed of 5 rpm to ensure equal irradiation regardless of lamp position or intensity fluctuations. Quartz and glass cells are available for the Merry-Go-Round.

Sample and actinometer solutions must be prepared so that an equal fraction of the incident light is absorbed by both. If this condition exists, then

$$\Phi_U = \Phi_A \frac{m_u}{m_a} \tag{13}$$

where Φ_U and Φ_A are the quantum yields of unknown and actinometer solutions, respectively, and m_u and m_a are the number of moles transformed in unknown and actinometer solutions during the irradiation period.

VIII. Apparatus for Photochemical Titrations

Photochemical titration is a method in which photons are used either directly as a titrant or to generate a suitable chemical titrant. Photochemical titrations are similar in concept to coulometric titrations and thus they are sometimes referred to as photonometric titrations, a term first used by Bricker and Schonberg (*7*).

There are two general types of photochemical titrations. In the direct type photons, $h\nu$, are used as the titrant in the titration of an analyte A to form products P.

$$A + h\nu \rightarrow P$$

In the indirect type photons are absorbed by a suitable absorber X, which leads to the formation of a chemical titrant T.

$$X + h\nu \rightarrow T$$

$$T + A \rightarrow P$$

As long as photons are absorbed at a constant rate, the analyte concentration will decrease at a constant rate and the photolysis time required to titrate the analyte will be proportional to the original amount of analyte present.

A detailed discussion of the theory and chemical reactions used in photochemical titrations is found in the chapter on that subject. Apparatus for photochemical titrations consists of a constant-intensity source of quanta, a filter system to allow wavelength and intensity selection, a shutter to facilitate accurate photolysis time measurement, and a cell in which photolysis takes place. All components must be arranged in a fixed geometry to assure a constant incident intensity on the photolysis cell. Some end-point method must also be available which will not interfere with nor be affected by the photolysis. Several design approaches to photochemical apparatus have been reported and these will be discussed here.

In the first photochemical titration reported, Bricker and Schonberg utilized a medium-pressure mercury vapor lamp as the source. The lamp, a Hanovia General Utility lamp, Type 30600, 115 V, 140 W, was kept at constant intensity by a Sola, Type CVH-1, 1000-W, constant-voltage transformer. The intensity of the lamp was found to increase 3% for every 1% increase in applied voltage. Since intensity control is an important factor in determining the precision of photochemical titrations, voltage regulation is of primary importance.

The intensity of the Hanovia lamp used decreased about 4% during the first week of lamp operation, then remained constant during the subsequent four months of use. A warm-up time of about 15 min was found necessary for the lamp to reach a constant intensity. The spectrum of the Hanovia lamp of the type employed by Bricker and Schonberg is shown in Fig. 1-8 (*15*).

A manually operated shutter connected to an electric microtimer allowed measurement of the photolysis time to $\pm$ 0.01 min. A photometric end point was used for all titrations. The photolysis cell, a 250-ml beaker, was situated in the cell compartment of a Beckman Model B spectrophotometer. Because of the construction of the Model B, photometric

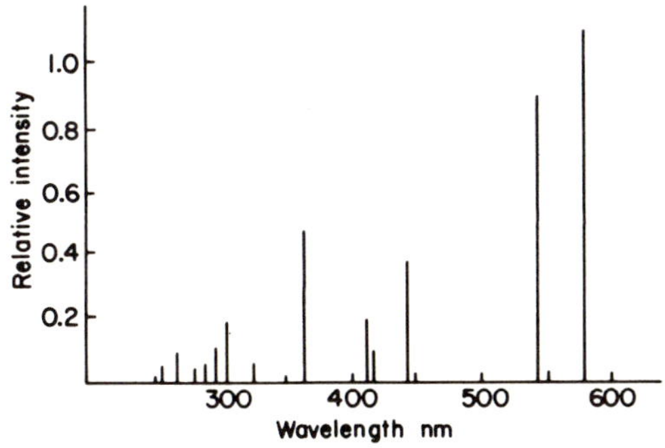

FIG. 1-8. Spectrum of Hanovia Type 30620 mercury vapor lamp corrected for detector response. From Smith (*15*).

measurements could not be made while photolysis was in progress without subjecting the photometric detector to the intense photolytic source. A diagram of the apparatus is shown in Fig. 1-9.

Bricker and Schonberg investigated the effect of geometry changes on the precision of photochemical titrations. As might be expected, small changes in the distance between the lamp and the surface of the solution being photolyzed change the titration time required.

To keep the solution as homogeneous as possible, stirring via a magnetic bar was employed. Stirring rate did not affect photolysis time to the end point unless the rate was sufficient to cause a vortex.

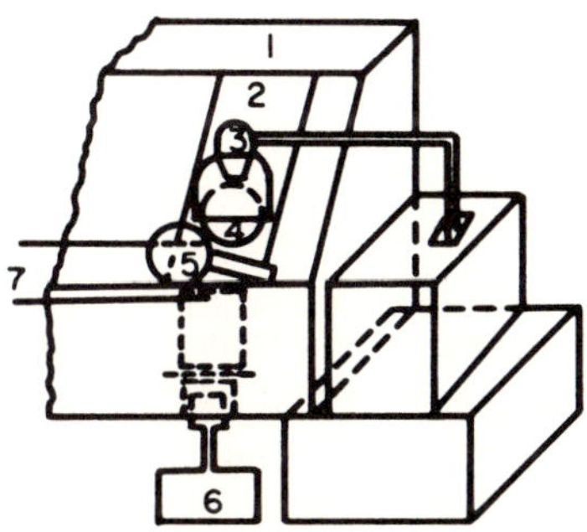

FIG. 1-9 Diagram of apparatus used by Bricker and Schonberg (*7*). (Reprinted through permission of the copyright owner.) 1, Beckman Model B spectrophotometer; 2, Bakelite cover for cell compartment; 3, mercury vapor lamp; 4, hole in Bakelite cover; 5, brass shutter; 6, magnetic stirrer; 7, leads for electric timer.

The apparatus used by Kuwana in the photochemical titration of oxygen and copper(II) in methanol is shown in Fig. 1-10 (*16*).

A PEK 109, 100-W, high-pressure mercury arc lamp was used in this study. The EM from the lamp was rendered monochromatic at 3130 Å by a series of filters. A stabilized dc power supply maintained constant voltage to the arc. Ten minutes for lamp warm-up was required. The

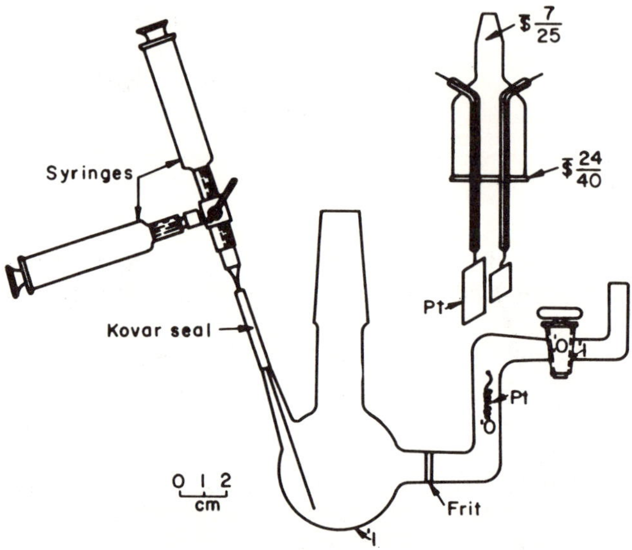

FIG. 1-10. Diagram of apparatus used by Kuwana (*16*) showing syringes for introduction and removal of samples. Platinum electrodes used for amperometric end points are shown at top right. (Reprinted through permission of the copyright owner.)

lamp intensity was monitored by a phototube whose photocurrent was assumed proportional to the lamp intensity. The lamp supply voltage was adjusted day to day to produce the same photocurrent in the monitoring system. This method is more direct than that employed by Bricker and Schonberg who maintained the lamp voltage constant and assumed a reproducible intensity. Since the intensity of mercury arcs does vary with age, it is advisable to monitor the intensity by either chemical actinometry or by the use of a phototube. A thermopile could also be used where precise intensity control is necessary.

Kuwana used a mechanical shutter to control the time of photolysis. The photolysis cell was constructed of borosilicate glass. Provisions for degassing the solution with nitrogen was made through a fritted side arm. Samples were introduced into the cell via syringes.

Three end-point methods were employed using this cell: potentiometric, amperometric, or photometric; and the titration curves could be recorded continuously during photolysis. Continuous recording represents a significant advance over the incremental methods used by Bricker and Schonberg (*7*).

Potentiometric measurements were made by conventional methods. Figure 1-10 shows the two platinum electrodes used for amperometric measurement. Photometric end points were obtained by recording the output from a photomultiplier tube which was positioned to monitor the light transmitted through the cell parallel to the photolysis beam. A tungsten bulb served as the photometric source. The phototube was protected from the high-intensity photolysis beam by an interference filter which did not pass 3130 Å, the only photolytic wavelength incident upon the cell.

Drew and Fitzgerald used a Hanovia Utility mercury vapor lamp in their study of turbidimetric titrations of ferricyanide (*8*). As in the original work of Bricker and Schonberg, the lamp intensity was regulated by maintaining constant lamp voltage.

The equipment used was simple. The lamp was mounted over a 250-ml photolysis beaker. Precautions were taken so that the beaker was situated under the lamp in a reproducible manner. A magnetic stirrer mixed the solution during photolysis. Turbidimetric end-point techniques were used and the titration was performed incrementally. At appropriate intervals the photolysis shutter was closed, the timer stopped, and the sample beaker slid out from under the lamp. A part of the solution was transferred to a cuvette for photometric measurement in a Spectronic 20 spectrophotometer. The aliquot was then returned to the beaker and photolysis carried out for another 0.5-to-1.0-min photolysis interval. Concentrations of ferricyanide down to 0.1 mM were titrated. The path length used for photometric measurements limited the sensitivity of the titration.

Smith and Fitzgerald have developed a single-source apparatus for photochemical titrations employing photometric end points. The design approach investigated is shown in Fig. 1-11. A mercury lamp (Hanovia Utility Model, Type 30620) mounted over the titration cell served as the source. Selection of wavelengths and intensities of the titrating radiation was accomplished by using appropriate filters and neutral-density wire screens in position *F*1. A 2.5-in.-diameter, electrically operated shutter

(Harvard Electric Shutter, Burke and James, Inc.) was mounted between the filter holder and lamp and served to control the time of photolysis. A stream of cool air was passed over the filters to prevent cracking due to heat. Heat-absorbing, neutral density filters are also available (Englehard–Hanovia No. 6540) which, when placed between the arc and other filters, will prevent cracking but of course decrease the available intensity.

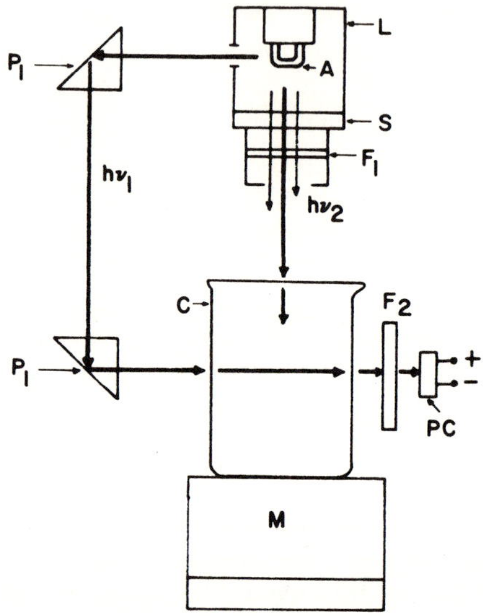

FIG. 1-11. Apparatus for photochemical titrations employing a single source for both photolysis and photometric measurements. *A*, mercury arc; *L*, lamp housing; *P*1, *P*2, reflecting prisms; *S*, electrically operated shutter; *F*1, photolysis radiation filter; *F*2, photometric radiation filter; *PC*, photocell; *C*, photolysis cell; $h\nu_1$, photometric radiation; $h\nu_2$, photolysis radiation; *M*, magnetic stirrer. From Smith (*17*).

In order for the mercury lamp to act as the source in the photometric circuit, a fraction of the total radiation is routed via reflecting prisms through the titration cell, a precalibrated 250-ml beaker, perpendicular to the direction of the photolysis beam.

Because glass optics limited the wavelengths and intensities passed through the titration cell in the photometric path to low-energy, low-intensity radiation, photolysis due to photometric radiation was negligible. Of the various mercury emission lines passing through the titration cell, one may be selected via an interference filter in position *F*2. Changes in solution transmission at this wavelength were measured with a photovoltaic cell, although a phototube detector could undoubtedly be substituted. The photocell output was converted to absorbance and recorded using a logarithmic potentiometric recorder. The shutter relay was coupled to the drive mechanism of the strip-chart recorder, allowing one to start photolysis simultaneously with the start of the recorder (*18*).

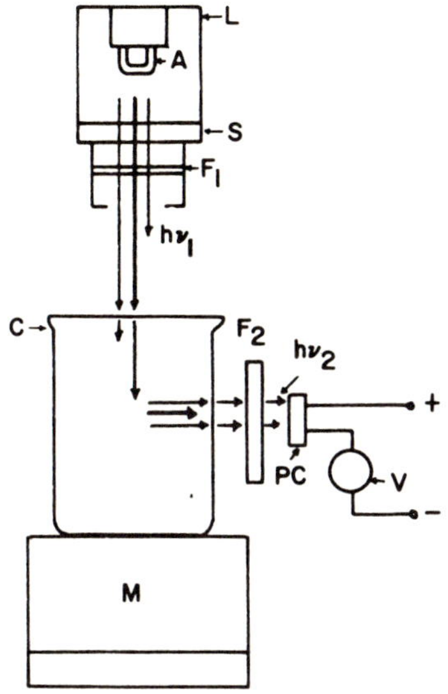

FIG. 1-12. Apparatus for photochemical titrations employing nephelometric end points. *A*, mercury lamp; *L*, lamp housing; *S*, shutter; *C*, photolysis cell; *F*1, photolysis filter; *F*2, photometric filter; *PC*, photocell; *M*, magnetic stirrer; *V*, offset voltage source for initial balance; $h\nu_1$, photolysis radiation; $h\nu_2$, visible light scattered by precipitate. From Smith (*19*).

A magnetic stirrer served to gently mix the contents of the titration cell assuring sharp end points. This apparatus was used in photometric titrations of chromium(VI), vanadium(V), and ferricyanide with photogenerated iron(II) (*17*), and in the turbidimetric titration of silver(I) with photogenerated chloride ion (*19*).

The apparatus as modified for nephelometric end-point titrations of silver(I) with chloride is shown in Fig. 1-12 (*19*). The nephelometric mode of operation was limited in sensitivity by the selenium photocell used as the photometric detector. The use of a phototube or photomultiplier tube would undoubtedly increase sensitivity significantly.

Lukasiewicz and Fitzgerald (*18*) used a modification of the apparatus of

Smith and Fitzgerald (*17*) for photochemical titrations employing a fluorescence end point. Using the full radiation from a medium-pressure mercury lamp, the short-wavelength mercury lines were used for direct photolytic decomposition of quinine while the longer-wavelength lines were used to excite visible fluorescence of the quinine remaining at any time during photolysis. Plots of fluorescence intensity versus photolysis time yielded a suitable end-point time which was related to the initial

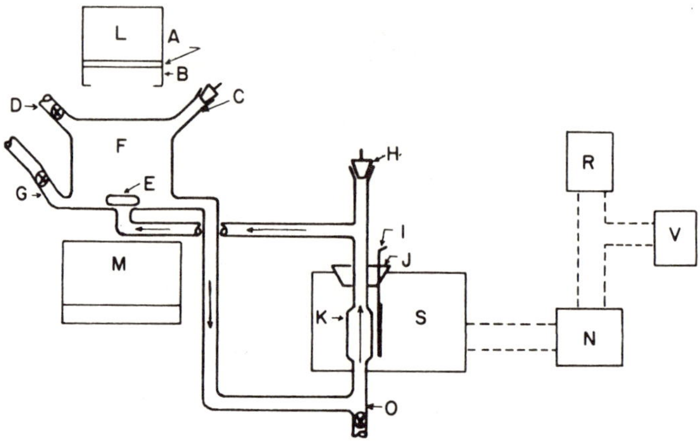

FIG. 1-13. Flow system for photometric end points in photochemical titrations. *A*, shutter; *B*, filter holder; *C*, large stoppered inlet; *D*, gas outlet; *E*, stirring bar; *F*, photolysis cell; *G*, gas inlet with frit; *H*, solution inlet; *I*, shutter; *J*, rubber-stopper cell holder; *K*, absorbance cell; *L*, mercury vapor lamp; *M*, magnetic stirrer; *N*, meter recorder switching circuit; *O*, solution outlet; *R*, recorder; *S*, Spectronic 20 spectrophotometer; *V*, backing voltage source; ↑, direction of solution flow; ⊗, stopcock. From Smith (*15*).

concentration of quinine. A relatively insensitive B-2M selenium photovoltaic cell (International Rectifier Corp.) was used as the detector. Titration of quinine in the parts-per-billion range were carried out with errors of about 4%.

A flow system for photometric end-point determinations in photochemical titrations has been employed by Smith and Fitzgerald (*20*). The system is shown schematically in Fig. 1-13 and is similar to others used for photometric titrations (*21*).

Solution photolyzed in the photolysis cell is pumped via a Teflon-coated stirring bar through an absorption cell located in the sample compartment of a Spectronic 20 spectrophotometer. Modification of the Spectronic 20 for direct recording of absorbance was made following a modified procedure of Frings and Pardue (*22*).

In order to increase the photometric end-point sensitivity in many of the photochemical titrations, scale-expansion techniques were employed (*20*). With the solution to be photolyzed and a 90% transmittance filter in the beam, the recorder was set to read at the zero end of the scale. A small millivolt source was required for this procedure. Upon removal of the filter, the recorder was set to read at the other extreme end of the scale. Photolysis was then initiated and the percentage transmittance between 100 and 90 was recorded versus photolysis time. Straight lines were obtained

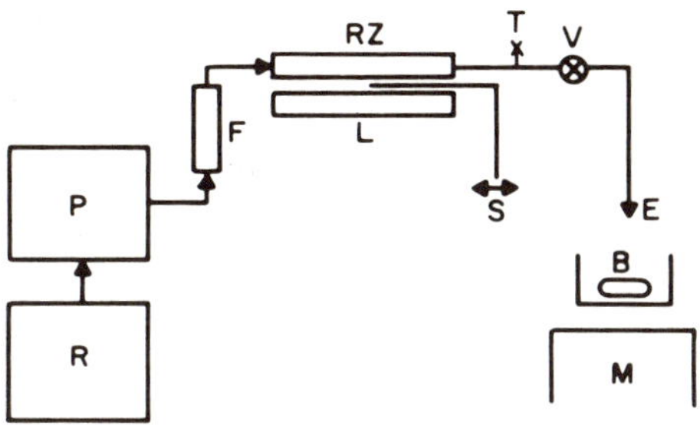

FIG. 1-14. Apparatus for external photochemical titrant generation. *R*, solution reservoir; *P*, pump; *F*, flowmeter; *R.Z.*, quartz tubing (reaction zone); *L*, germicidal lamp; *S*, sliding shutter; *T*, bubble trap; *V*, Teflon (DuPont) leak valve; *E*, titrant exit; *B*, titration vessel; *M*, magnetic stirrer. From Drew and Fitzgerald (*23*); reprinted through permission of the copyright owner.

since percentage transmittance is a linear function of concentration at high values of percentage transmittance.

Apparatus was developed by Drew and Fitzgerald which can be used for photochemical titration of species which cannot withstand direct irradiation by the photolysis lamp (*23*). The titrant-generating species is pumped past a photolysis source where the titrant is formed. The titrant then flows into a vessel containing the analyte. The method is similar to the method of external coulometric titrant generation. A schematic diagram of the external photochemical titrant generator of Drew and Fitzgerald is shown in Fig. 1-14.

A 57-cm, low-pressure mercury discharge lamp (Hanovia Model 94A-1, Englehard–Hanovia, Inc.) was used for photolysis. Photogenerator solution was pumped passed the lamp through 6-mm i.d. quartz tubing and flow rates of 1–4 ml/min were used in various titrations.

The major disadvantage of this method is that some method is required to dispose of unwanted titrant and dilution of the sample is of course required.

APPENDIX

Names and Addresses of Manufacturers Cited

Ace Glass, Inc.
1430 N. West Blvd.
Vineland, N.J. 08360

Amersil, Inc.
685 Ramsey Ave.
Hillside, N.J. 07205

Bausch and Lomb, Inc.
6098 Bausch St.
Rochester, N.Y. 14602

Beckman Instruments, Inc.
2500 Harbor Blvd.
Fullerton, Calif. 92634

Burke and James, Inc.
333 W. Lake St.
Chicago, Ill. 60606

Corion Inst. Corp.
156 Prospect St.
Waltham, Mass. 02154

Corning Glass Works
Corning, N.Y. 14830

The Ealing Corp.
2225 Massachusetts Ave.
Cambridge, Mass, 02140

Eastman Kodak Co.
343 State St.
Rochester, N.Y. 14650

Englehard–Hanovia, Inc.
Hanovia Lamp Division
100 Chestnut St.
Newark, N.J. 07105

George W. Gates and Co., Inc.
Tpke. and Lucille Ave.
Franklin Square
New York, N.Y. 11010

General Electric Co.
Lamp Glass Dept.
24400 Highland Rd.
Richmond Heights, Ohio 44143

The Harshaw Chemical Co.
1945 E. 97th St.
Cleveland, Ohio 44106

Hilger and Watts, Ltd.
98 St. Pancras Way
Camden Rd.
London NW 1, Great Britain

Illumination Industries, Inc.
610 Vagueros St.
Sunnyvale, Calif. 94086

International Light, Inc.
Dexter Industrial Green
Newburgport, Mass. 01950

International Rectifier Corp.
El Segundo, Calif.

Kaufman Glass Co.
1215 French St.
Wilmington, Del. 19899

Oriel Optics Corp.
1 Market St.
Stamford, Conn. 06902

OSRAM Lamps
Epic, Inc.
150 Nassau St.
New York, N.Y. 10038

PEK, Inc.
825 E. Evelyn Ave.
Sunnyvale, Calif. 94086

Polacoat, Inc.
9750 Conklin Rd.
Blue Ash, Ohio 45242

Pope Scientific, Inc.
13600 W. Reichert Ave.
Menomonee Falls, Wis. 53051

Quartz and Silice
8 Rue d'Anjou
75—Paris, France

Rudolph Inst. Engineering Co., Inc.
61 Stevens Ave.
Little Falls, N.J. 07424

Schoeffel Instrument Corp.
15 Douglas St.
Westwood, N.J. 07675

Scientific Cell Co., Inc.
118–21 Queens Blvd.
Forest Hills, N.Y. 11375

The Southern New England Ultraviolet Co.
954 Newfield St.
Middletown, Conn. 06457

Spectrolab
12484 Gladstone Ave.
Sylmar, Calif. 91342

Thermal American Fused Quartz Co.
Rt. 202 and Change Bridge Rd.
Montville, N.J. 07045

Ultra-Violet Products, Inc.
5114 Walnut Grove Ave.
San Gabriel, Calif. 91778.

Westinghouse Electric Crop
P.O. Box 868
Pittsburgh, Penna. 15230

REFERENCES

1. J. C. Davis, Jr., *Advanced Physical Chemistry*, Ronald, New York, 1965, pp. 615–617.
2. L. R. Koller, *Ultraviolet Radiation*, 2nd edition, Wiley, New York, 1965.
3. Westinghouse Electric Corp., Pittsburgh, Penna., Sterilamp Bulletin ASC-144, 1969.
4. Oriel Optics Corp., Stamford, Conn., Bulletin, 1969.
5. J. G. Calvert and J. N. Pitts, *Photochemistry*, Wiley, New York, 1966: (*a*) p. 693; (*b*) p. 706; (*c*) pp. 728–747; (*d*) pp. 747–750; (*e*) p. 756.
6. W. A. Noyes, Jr., and P. A. Leighton, *The Photochemistry of Gases*, Dover, New York, 1966: (*a*) p. 38; (*b*) p. 202; (*c*) p. 69.
7. C. E. Bricker and S. S. Schonberg, *Anal. Chem.*, **30,** 922 (1958).
8. H. D. Drew and J. M. Fitzgerald, *ibid.*, **38,** 778 (1966).
9. T. S. West, in *Trace Characterization* (W. W. Meinke and B. F. Scribner, eds.), NBS Monograph 100, National Bureau of Standards, Washington, D.C., 1967, p. 291.
10. R. H. Freere, *Laboratory Practice*, **16,** 1110 (1967).
11. B. Jaselskis, in *Treatise on Analytical Chemistry* (I. M. Kolthoff and P. J. Elving, eds.), Wiley-Interscience, New York, 1964, Vol. 5, Pt. 1, p. 3027.
12. H. A. Stobel, *Chemical Instrumentation*, Addison-Wesley, Reading, Mass., 1960, p. 71.
13. Amersil, Inc., Hillside, N.J., Bulletin Al-1056-3, 1968.
14. Corning Glass Works, Bulletin CF-3, 1965.
15. E. J. Smith, Ph.D. Thesis, Seton Hall Univ., S. Orange, N.J., 1968.
16. T. Kuwana, *Anal. Chem.*, **35,** 1398 (1963).
17. E. J. Smith and J. M. Fitzgerald, Abstracts, 153rd National Meeting, American Chemical Society, Miami Beach, Florida, April 1967, No. B40.
18. R. J. Lukasiewicz and J. M. Fitzgerald, *Anal. Letters*, **1,** 455 (1968).
19. J. M. Fitzgerald and E. J. Smith, Abstracts, 154th National Meeting, American Chemical Society, Chicago, Illinois, Sept. 1967, No. B46.
20. E. J. Smith and J. M. Fitzgerald, Abstracts, 156th National Meeting, American Chemical Society, Atlantic City, N.J., Sept. 1968, No. ANAL 74.
21. C. Rehm, J. I. Bodin, K. A. Connors, and T. Higuchi, *Anal. Chem.*, **31,** 483 (1959).
22. C. S. Frings and H. L. Pardue, *ibid.*, **36,** 2477 (1964).
23. H. D. Drew and J. M. Fitzgerald, *ibid.*, **41,** 974 (1969).
24. D. McLanahan, *Electro-Optical Systems Design*, **3,** 18 (1971).

CHAPTER 2

LASERS IN ANALYTICAL CHEMISTRY

Franklin H. Fry

AMERICAN CAN CO.
NEENAH, WISCONSIN

I. Introduction

Ever since the laser was discovered, great things have been predicted of it. It has been stated that the laser would find vast use as a weapon, as an industrial tool, and as a means of communication. All of these applications are, as of the present writing, the subject of intense scientific investigation. Most of the predictions were somewhat premature, however, and the widespread use of the laser in any of the above areas cannot be found yet.

The laser is in fact a very special kind of light source and has found immediate application in areas where it is an improvement over presently available light sources. In addition, its other unusual properties such as coherence and intensity have earned it a place where these characteristics are of value. The area of application in which the laser has found the widest and earliest broad use is scientific experimentation. Researchers in general have quickly learned to appreciate the laser as an important addition to their stock of available equipment.

In no specialty has the laser found a home quicker than in analytical chemistry. While its use in routine analysis is only beginning, the laser as a tool in analytical research already has an established place. The laser makes possible experiments in Raman spectroscopy and emission spectroscopy, for example, that were simply not possible before. Because the laser will be more and more important to analytical chemists, it is of interest to the professional analytical chemist to obtain a working knowledge of the laser and an idea of the breadth of scientific techniques in which this new and exciting device finds a use. Hopefully, this review of laser applications pertinent to analytical chemistry will encourage the specialists in that field to look further into the use of lasers in chemical analysis.

II. The Theory of Laser Operation

As is well known, atoms, ions, and molecules can exist only in fixed energy states. The lowest energy state is the ground state and any state of higher energy is known as an excited state. An atom, ion, or molecule can pass from an excited state to a lower energy state by loss of energy. Although energy can be lost from a system in numerous ways, the transition mechanism of interest here is radiation. An atom which drops from its first excited state to its ground state emits radiation with a frequency ν, where

$$E = h\nu \tag{1}$$

In Eq. (1), E is the energy difference between the excited and ground states and h is Planck's constant. The transitions discussed may occur from any excited state to any lower energy state providing the transition is allowed. A system in its ground state cannot emit radiation.

An aggregation of excited atoms will emit light in a random way. The fraction of the excited atoms that will emit radiation per unit time is A_{nm} where n designates the upper state and m is the lower state; A_{nm} is called the spontaneous transition probability or the Einstein coefficient of spontaneous emission. This radiation of light from excited atoms (or molecules or ions) without the external influence is called *spontaneous* emission. The radiation from this aggregate has no directionality or special phase and is therefore incoherent.

If an excited atom is bathed in radiation whose frequency is identical to that which it would emit spontaneously, it may emit its photon in phase with the influencing radiation. The light emission process dependent upon external radiation is called *stimulated* emission. The total probability of an excited atom emitting a photon in unit time is

$$P_{nm} = A_{nm} + U_{\nu}B_{nm} \tag{2}$$

where B_{nm} is a constant characteristic of the system and U_{ν} is the radiation density at the frequency of the transition; B_{mn} is the Einstein coefficient of stimulated or induced absorption. The second term in Eq. (2) is the probability of stimulated emission. The radiation produced in the process of stimulated emission has the same direction and phase as the incident radiation.

Equation (2) can be made to look more familiar by calculating the probability of an atom in lower state m absorbing a photon of frequency and being excited to state n.

$$P_{mn} = U_{\nu}B_{mn} \tag{3}$$

In Eq. (3) B_{mn} is the Einstein coefficient of induced absorption.

The connections between the various coefficients above are given by the Einstein relations:

$$B_{nm} = B_{mn} \tag{4a}$$

$$A_{nm} = \frac{8\pi h\nu^3 B_{nm}}{c^3} \tag{4b}$$

These equations hold only if the system has only one state at each energy level. If there are g_n states with energy E_n and g_m states with energy E_m, then the first Einstein relation takes the following form:

$$g_n B_{nm} = g_m B_{mn} \tag{5}$$

The terms g_n and g_m are called the *multiplicities* of levels n and m, respectively. The second Einstein relation Eq. (4b), remains unaffected by energy-level degeneracy.

It is well known that an ensemble of atoms in thermal equilibrium will have most of its members in the lowest energy state, a smaller number in the next higher state, and so on. The number of atoms, N_n, in an excited energy state n is given by the Boltzmann relation

$$N_n = N_m \exp\left(\frac{-\Delta E}{kT}\right) \tag{6}$$

In Eq. (6) N_m is the number of atoms in the lower state m, ΔE is the energy difference between the two states, k is the Boltzmann constant, and T is the absolute temperature.

Radiation of the proper frequency can either gain or lose intensity on passing through the collection of atoms. The intensity lost by absorption is proportional to $N_m U_\nu B_{mn}$. The intensity gained is $N_n(A_{nm} + U_\nu B_{nm})$. With the help of the Einstein relations it is possible to determine that the term A_{mn} is small relative to B_{nm} and may be neglected. Consequently, for a collection of atoms at thermal equilibrium, N_m is greater than N_n and radiation passing through the atoms decreases in intensity. This is the basis of light-absorption spectroscopy familiar to all analytical chemists.

If somehow the system can be so treated that the number of excited atoms exceeds the number of unexcited atoms, radiation of the right frequency passing through the atoms is easily seen to *gain* in intensity. Since the gain is by stimulated emission, the emitted radiation is in phase with the incident light. Much of laser technology is concerned with conditions which create and maximize this gain phenomenon. The situation in which there are more atoms in the upper state than in the lower state is called a *population inversion.*

As radiation passes through an unexcited gas, if its frequency corresponds to some transition of the atoms in the gas, the intensity of the radiation is decreased. The intensity decrease is described by the well-known Beer's law

$$I = I_0 \exp(-k_\nu x) \tag{7}$$

where I is the intensity of the light after passing through a distance x of the absorber and I_0 is the original radiation intensity. The factor k_ν of Eq. (7) is the absorption coefficient of the medium. The absorption coefficient k_ν is related to the usual analytical terms in Beer's law by the following relation

$$k_\nu = EC \tag{8}$$

where E is the molar absorptivity in liters per mole-centimeter and C is the concentration of the absorber in moles per liter. Equation (8) is relevant only if a solution absorber is concerned.

In the case of a system with an inverted population, the light wave striking the system will increase in intensity as it passes through the system. The intensity of the wave is expressed by an equation like (7) but with a positive term α replacing $-k_\nu$. The gain is represented by

$$I = I_0 \exp(\alpha x) \tag{9}$$

If the light wave and atomic system are placed between two mirrors, one highly reflecting and the other partially reflecting, the beam will grow in intensity as it is reflected from each mirror. Assume the beam starts at a partially reflecting mirror toward a second, completely reflecting, mirror; after it has traveled the distance between the mirrors L, the beam intensity is $I_0R_2 \exp(\alpha L)$, where R_2 is the fraction of light reflected by the totally reflecting mirror. The fraction of light reflected from the first, partially transmitting, mirror is R_1 and is called the reflectivity of that mirror. The intensity of light reflected from the first (partially reflecting) mirror is $I_0R_1R_2 \exp(2\alpha L)$.

The term $R_1R_2 \exp(2\alpha L)$ is the gain per pass or per round trip. Setting

$$R = (R_1R_2)^{1/2} \tag{10}$$

and

$$-\gamma = -\ln R \tag{11}$$

The fractional gain per pass can be written

$$F = \exp[(2(\alpha L - \gamma)] \tag{12}$$

If $\alpha L > \gamma$, the intensity of the wave between the mirrors will rapidly increase. The intensity of the beam transmitted by the partially reflecting or output mirror will also increase. If T is the transmission coefficient of the output mirror, the intensity of the beam after its first reflection from the output mirror is $I_0T \exp(2\alpha L)$. The intensity after the second pass is $I_0T \exp(2\alpha L)F$, and after n passes is

$$I_0T \exp(2\alpha LF^n) \tag{13}$$

Writing the fractional gain per pass out in full, the transmitted intensity I_T is obtained:

$$I_T = I_0 T \exp(2\alpha L)\exp(2nL\alpha - 2n\gamma) \tag{14}$$

It is apparent that the transmitted-intensity expression given in Eq. (14) is a very strong function of the gain coefficient. It has been implicitly assumed in the arguments above that α was not dependent on frequency, but in fact it is. The dependence of α (and k_ν) on frequency is described by a bell-shaped curve and can be approximated by a Lorentzian equation. The important thing to remember, however, is that the output or transmitted intensity of this hypothetical device falls off abruptly as the frequency scanned moves away from the frequency of maximum gain.

As already implied, in a laser two mirrors are set parallel to one another in order to gather and reflect back the light wave amplified by the medium with an inverted population. Because light waves can be reflected back and forth between the mirrors with little loss, the two mirrors are said to constitute a *resonant cavity*. The latter is a term borrowed from microwave technology.

Energy loss from the cavity is very slow only if the distance a photon travels making a complete round trip in the cavity is an integral multiple of whole wavelengths; that is, a cavity is resonant with respect to a given photon only if

$$\frac{q\lambda}{n} = 2L \tag{15}$$

where n is the refractive index of the material between the mirrors and q is an integer. The difference between two adjacent stable wavelengths in the cavity is given by the following equation:

$$\lambda_1 - \lambda_2 = \Delta\lambda = 2nL\left(\frac{1}{q} - \frac{1}{q+1}\right) \tag{16}$$

The width of an emission line of a fluorescent material can be quite narrow, as in the rare-earth compounds. It will be shown, however, that the emission-line width even in the case of the rare earths is much wider than necessary to permit oscillation in a laser.

In Eq. (16), for L of 10 cm, n of 1.76, and q of 500,000, the difference between wavelengths that are stable in the cavity is approximately 10^{-10} cm or 10^{-2} Å. Assuming that the emission-line width of the amplifying medium is 1 Å, it is clear that many wavelengths within the emission line are stable in the cavity defined by the mirrors. It can be shown that the widths of the lines that have low loss are of the order of 10^{-4} Å.

The latter is a property of the cavity alone. Consequently, the output from the laser is expected to be a group of very sharp lines spaced apart at distances of 10^{-2} Å. Depending on the gain in the laser as a function of wavelength, only those stable *resonances* in the cavity at highest amplification will be observed in the laser output. Consequently, the laser output will normally consist of only a few different wavelengths. In the infrared where the cavity resonance spacings are much greater, and in some cases in the visible, only one output wavelength will be observed.

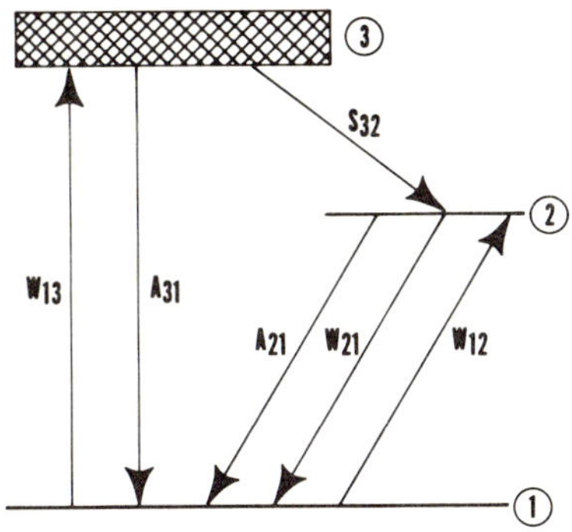

FIG. 2.1 Energy level diagram for a three-level laser.

Each stable wavelength or resonance in the cavity is called a *longitudinal mode*. The intensity of the laser output for a given longitudinal mode will vary as a function of the radial distance from the axis of the laser, Such radial-intensity variations are called *transverse* modes and specific, recurrent patterns of this type are identified by the letters TEM followed by subscripts (*1*).

It was assumed at the start of this discussion that the atomic system in which light amplification occurred had only two levels: the ground state and one excited state. While two-level systems are valuable for illustration purposes, they are not typical of real laser systems. As is very well known, atomic and molecular structures have complex energy schemes with many energy levels. For lasers it turns out that three- and four-level energy systems are the most common and in fact have practical advantages.

Figure 2-1 illustrates a three-level scheme that is frequently used to represent the energy levels of ruby, the first optical laser (*2*).

Level 1 is the ground or unexcited state; level 3 is the excited state that results on absorption of a visible photon in the visible absorption bands of ruby; level 2 is the emitting level for laser action. There are really two visible absorption bands and for simplicity the broadness of level 3 includes both excited states that may result from the absorption of a photon by one band or the other. The terms shown in Fig. 2-1 are as follows: W_{21} is the probability of the stimulated emission of a photon from level 2; A_{31} and A_{21} are the Einstein coefficients of spontaneous emission from the two excited levels; and S_{32} is the probability of transition for the non-radiative process of level 3 to level 2. The terms N_1, N_2, and N_3 are the respective level population densities and N_0 is the total chromium-ion density in the ruby crystal.

The equations describing the rate of change of populations for each of the levels are given below:

$$\frac{dN_3}{dt} = W_{13}N_1 - (W_{31} + A_{31} + S_{32})N_3 = 0 \tag{17}$$

$$\frac{dN_2}{dt} = W_{12}N_1 - (A_{21} + W_{21})N_2 + S_{32}N_3 = 0 \tag{18}$$

$$N_1 + N_2 + N_3 = 0 \tag{19}$$

In writing the equations the steady state has been assumed. One solution for the rate equations is

$$\frac{N_2}{N_1} = W_{13}\left(\frac{S_{32}}{W_{31} + A_{31} + S_{32}}\right) + \frac{W_{12}}{(A_{21} + W_{21})} \tag{20}$$

Assuming that the system under consideration has a fluorescence quantum yield, the emission from level 2 will be much greater than that from level 3, and $S_{32} \gg A_{31}$. Under this assumption it will also be true that the rate of the nonradiative transfer to the fluorescent level will be much greater than the rate of excitation from the ground level to level 3, or $W_{13} \ll S_{32}$. In such a case both N_1 and N_2 will be much greater than N_3 and Eq. (20) can be simplified to yield

$$\frac{N_2}{N_1} = \frac{(W_{13} + W_{12})}{(A_{21} + W_{21})} \tag{21}$$

Equation (21) can be rewritten as

$$\frac{(N_2 - N_1)}{N_0} = \frac{(W_{13} - A_{21})}{(W_{13} + A_{21} + 2W_{12})} \tag{22}$$

If stimulated emission is to occur, a population inversion must exist, and N_2 must be greater than N_1. From Eq. (22) this will only be true if the rate of excitation to level 3 is greater than the rate of spontaneous emission from level 2, or $W_{13} \gg A_{21}$. This is the minimum condition, and a further excess population of level 2 over level 1 is required for a useful power output from the device.

In the laser configuration described earlier, if a wave passes through the mirror once and is reflected from one mirror and gains in intensity, the laser will emit sufficient power to be useful. This situation is described as

$$R \exp(-\alpha_{12}L) \cong 1 \tag{23}$$

where R is obtained from Eq. (10) and α_{12} is the general absorption coefficient and may be positive or negative. Assuming that the absorption coefficient is small and using the infinite series approximation for the exponential, it can be shown that

$$-\alpha_{12}L = (1 - R) \tag{24}$$

The absorption coefficient for the transition from level 1 to level 2 is the product of the population density in level 1 and the absorption cross section per atom for the same transition, σ_{12}.

$$\alpha_{12} = (N_1 - N_2)\sigma_{12} \tag{25}$$

From Eqs. (24) and (25), the population inversion needed for stimulated emission can be obtained

$$(N_1 - N_2) = \frac{(1 - R)}{\sigma_{12}L} \tag{26}$$

Consequently, maximizing the reflectivity R and the length of active material, L, minimizes the population inversion necessary to obtain laser output.

When the excitation rate exceeds the spontaneous emission rate from the laser level, or the mirror reflectivity and cavity length are so arranged that the population inversion is just sufficient to overcome losses, the laser *threshold* has been reached. The onset of true laser oscillation is described as a threshold because experimentally the energy input to the device must be above a minimum value for a useful output to be obtained, and because the change in properties of the radiation emitted from a laser operating below threshold and in those of the radiation obtained above threshold is a marked change in kind. Laser radiation has special collective properties of monochromaticity, divergence, intensity, and coherence which nonlaser emission does not have. These will be discussed in more detail later.

A. *Time Variations of Laser Output*

The instantaneous power output–time plot of a laser depends on the type of laser and on the mode of excitation. The simplest case is the dc excited CW gas laser. In this case, the output is constant with time, perhaps fluctuating a few percent, depending on the filtering in the supply. The same is true of the optically pumped solid-state laser driven by a dc-powered lamp. An ac-powered gas laser will have an output that looks like rectified alternating current because power is transferred into the plasma on both halves of the ac signal. The output frequency will be twice the frequency of the ac input.

The optically pumped, pulsed solid-state laser gives the most complicated output–time structure. The flash-lamp pulse will last a few milliseconds. The laser emission may begin a fraction of a millisecond after the lamp fires and continue as long as the lamp provides adequate pumping, perhaps for half a millisecond. The structure of the output is not constant with time but is divided into a number of spikes (*3*). The reason for the spikes is qualitatively easy to explain: the pump light causes a population inversion which is increasing as stimulated emission begins. The induced emission is a much faster process than excitation and the first pulse of laser emission decreases the population level below the point of emission and the laser turns itself off. The flash lamp is still on, however, and excitation continues. A population inversion is reestablished and the cycle begins again. The pulses continue until the lamp output decreases to the point that the excitation rate is insufficient to form an inversion. Statz and deMars were the first to provide this explanation for laser pulsations (*4*).

B. *Q-Switching*

The gleaning of energy from excited-state atoms by induced emission depends upon the feedback of the mirrors in the resonant cavity. If one of the mirrors is blocked or effectively removed from the cavity, feedback is prevented and stimulated emission will not occur. (This is generally true. There are cases, however, in which the gain is so high that stimulated emission can be obtained from a long body of excited active material without mirrors at either end.) If stimulated emission is prevented, the population inversion caused by the excitation will be greater than normal. Now if the missing mirror is returned to its place very quickly, stimulated emission will occur at a greater rate than normal due to the higher inversion. The output–time relation in this case is one spike of tens of nanoseconds duration and much higher peak power than is obtained in normal operation. Instantaneous peak powers of over 500 MW have been observed by this technique.

Controlling the laser operation by varying the feedback of the resonant cavity is called *Q*-switching, because the feedback controls the quality or "*Q*" of the cavity. The methods used to *Q*-switch lasers are varied. They have one common property, however. The technique must alter the *Q* of the cavity from low to high in a period of nanoseconds. Kerr cells, spinning mirrors, crystals such as potassium dideuterophosphate, and bleachable absorbers have been used as *Q*-switchers (*5*, *6*). Bleachable absorbers are placed in the resonant cavity between the end of the laser rod and one mirror. The material absorbs the laser output until the radiation intensity becomes large, at which point the absorber becomes transparent to the laser radiation and the giant output pulse is formed. This method of *Q*-switching is termed "passive" because no input power to the switch is required for operation.

The energy obtained in one giant pulse from an optically pumped laser may be a small fraction of the energy emitted in a series of less intense pulses. The peak power in a giant pulse may be up to three orders of magnitude higher than that from an uncontrolled laser. The peak power that is produced in a giant pulse depends on the power input to the laser and on the lifetime of the emitting level. The lifetime of the excited state in some laser-active materials is so short as to prevent *Q*-switching. Some electrically pumped gas lasers, such as the carbon dioxide–nitrogen laser, can also be *Q*-switched.

III. Special Properties of Laser Radiation

The emission from a laser has characteristics which distinguish it from incoherent light sources in four properties: intensity, monochromaticity, directionality, and coherence. At present, no use for coherence has been found in analytical chemistry, but there undoubtedly will be some application of coherence in the future.

Probably the single most valuable property of laser emission at present for analytical chemistry is its intensity. Various analogies have been used to convey the extreme power densities available with lasers. For example, it has been shown that a blackbody light source must have a brightness temperature of 10^{18}–10^{19} °K to have a spectral brightness equivalent to that of a ruby or He–Ne laser (*7*). The sun has a brightness temperature of 5840°K. Several years ago peak powers of 500 million W had been reported, which if focused by a lens, would produce an optical electric field higher than the field binding the outer electrons in most atoms. The power density from such a laser is 10^{15} W/cm^2, and is sufficient to cause ionization or breakdown even in transparent substances (*8*). Higher output powers have been reported since then.

These high-intensity pulsed-light sources have been used with great success in emission spectroscopy and mass spectroscopy. Particularly, the laser has permitted analysis, by either of these techniques, of very small areas of a sample. The mass of material removed from the sample for analysis can now be so small as to represent essentially no loss. Materials very difficult to vaporize by other methods for emission-spectroscopic analysis can now be vaporized with ease. Laser sources in Raman spectroscopy have found ready acceptance because the source intensity allows strong emission lines from small samples (*9*).

The monochromaticity of lasers is a natural outgrowth of the process by which laser emission is obtained. This fact has already been discussed in the preceding section. The width of a laser emission line can be as narrow as 10^{-4} Å. Since this is much narrower than most absorption lines found in matter, the laser can be used as a source for very high resolution spectroscopy. Another use of the monochromaticity (sometimes called temporal coherence) is found in Raman spectroscopy. The narrowness of the laser lines emitted by the helium–neon and argon-ion lasers used as exciting sources removes the need for optical filtering of the source and permits observation of Raman emission lines closer in frequency to the exciting line than had been possible with other types of sources. Laser sources improve the resolution of Raman spectroscopy also and permit narrower instrumental slit widths and require less electronic signal amplification than the conventional mercury arc (*10*).

The emission from a laser is said to be highly directional or to have low divergence. This means that the beam does not spread rapidly as it emerges from the laser itself. Imagine the beam sent out by a flashlight. Close to the flashlight the beam diameter may be only a few inches but several feet from the source the beam diameter will be perhaps 10 ft wide. It is a characteristic of "thermal" light sources that they emit light in all directions. The direction of emission from a laser coincides with the axis of the cavity and the diameter of the beam increases only slightly with distance. The directional behavior of the beam is the result of the nearly constant phase of the radiation emerging from the end of the laser. The angle the edge of the beam makes with the axis of the cavity is called the *divergence* and for ruby has been observed to be less than 0.7° or 12 mrad (*11*). The lower limit of divergence for a perfect source is called the *diffraction limit* and the minimum angle of divergence is given by

$$\theta_m = \frac{1.22\lambda}{d} \tag{27}$$

In Eq. (27) d is the diameter of the source aperture and λ is the wavelength of the light. In the case of a laser, d is the diameter of the rod or cavity. For a 1-cm-diameter ruby rod, the minimum divergence or diffraction limit is about 0.085 mrad.

It should be noted that the directional properties of laser radiation do not appear until the energy input to the laser is above threshold. Buddenhagen and coworkers (*11*) clearly showed that the emission from a ruby laser optically pumped below threshold was nondirectional. Upon reaching laser threshold, the low-divergence behavior of laser emission appeared.

The coherence of laser radiation is also a property which is a direct consequence of the process of stimulated emission which creates the emission. In the earlier discussion of stimulated emission, it was stated that a photon produced by induced emission is in phase with the photons that induced the emission and have the same wavelength as the incoming radiation. The photons produced by spontaneous emission are emitted at random and consequently have no fixed phase relationship to any other photons produced by a noncoherent source. The measurement of coherence from a source is usually carried out by passing the beam through two slits and photographing the dark and light fringes of the resultant interference pattern. The densitometric trace of the interference pattern is then compared with the theoretical intensity pattern of interference from a perfectly coherent source (*12*). Although no use for coherence has appeared in analytical chemistry, great use of coherence is made in a fully three-dimensional photographic technique called holography (*13*).

IV. Methods of Establishing Population Inversion

There are two types of lasers which find application in analytical chemistry, the optically pumped laser and the electron discharge excited laser.

The discharge excited laser can be subclassified depending on the emitting medium. The active medium may be neutral atoms, molecules, or excited ions. An additional type of laser which does not appear to have been used at all in analytical applications is the semiconductor laser. This class of laser will not be discussed here.

A. *Optical Pumping: Solid- and Liquid-State Lasers*

Optical pumping was the method used to excite the first laser in Maiman's original experiments (*14*). The process of optical pumping is simply immersion of the laser medium, usually a solid laser rod, in intense radiation. The only radiation really useful in exciting the laser rod is that which the active material in the rod can absorb. Laser rods are dilutely doped crystals with doping levels of the order of 0.05 wt%. Only the impurity ion is involved in laser action and the bulk of the laser rod is a matrix for the dispersion of the ions. Since the light sources are usually flash lamps or filament lamps which emit broad-band radiation and because the matrix inevitably absorbs some of the exciting radiation, heating of the rod is frequently a problem. The frequency with which pulsed optically pumped lasers can be operated depends on how well the rod is cooled. In continuous optically pumped lasers, the rod is usually surrounded by an optically transparent sleeve containing a flowing coolant such as water.

Lamps used to optically pump lasers can be classified into several types depending on the relationship of their emission spectra to the absorption spectrum of the laser material. The selection of the lamp in each laser is made on the basis of two criteria: how well the lamp emission matches the absorption spectrum of the laser-active material, and whether the power output of the lamp is sufficient to cause a useful output from the laser. It is possible to excite the laser material with the noncoherent emission of the same material from a separate lamp. This has been done in the case of the cesium vapor infrared laser (*15*).

Often the emission lines of some noncontinuous source have the appropriate wavelength for absorption by a laser-active material of a totally different kind. Such a situation is called accidental coincidence. For example, the cesium vapor laser can be pumped by a strong emission line from a helium lamp (*16*, *17*). Other examples are the use of a high-pressure mercury arc to drive a ruby laser and a light-emitting semiconductor diode to pump a dysprosium-doped calcium fluoride laser (*18*, *19*).

If it is not feasible to excite the laser with a lamp of the active material or a line-emission lamp, a broad-band source must be used. These are either filaments or discharges and they approximate blackbody light sources in general. The original ruby laser was excited by a commercial helical xenon-filled quartz flash lamp. The lamp was driven by discharging a capacitor bank whose input was varied by changing the charging potential (*14*). The discharge was triggered by a single output from a pulse generator which also served to initiate the sweep on the output-sensing circuitry. The pulse was applied to a wire outside the flashtube. This has been the conventional procedure for flash-lamp pumping of lasers. The flash lamp may also be triggered internally by injecting a narrow high-voltage pulse in series with the power supply and flashtube. In a large-scale technique improvement, Markiewicz and Emmett have presented a simple design procedure for optimization of one type of flash-lamp driving circuits (*20*).

Flash lamps may be either helical or linear. The length of the lamp is 5 cm or greater and the bore diameters range from 4 to 30 mm. Helical lamps will withstand higher input voltages than linear units, but are less efficient as pumping sources since energy reflected from the cavity walls is reabsorbed in the lamp. The lifetime of modern flash lamps is upward of 100,000 shots if operated at less than 60% of rated maximum energy input. The latter is determined by the input required to explode the lamp. Hardway has discussed the factors involved in maximizing xenon-flash-lamp life (*21*).

Xenon flash lamps suffer from two major drawbacks: they are expensive and the maximum repetition rate for pulse operation is about 1 sec^{-1}. Roess and Zeidler suggest that the use of a mercury flash lamp may overcome both of these problems (*18*).

Considerable work has been carried out in the area of continuous broad-band light sources for pumping lasers. These are generally very inefficient sources and seriously restrict the shortest-wavelength laser output possible. For example, a tungsten filament lamp emits only 30% of its energy at wavelengths shorter than 1 μm. Other continuous broad-band sources are not much better (*22*). It appears that new exciting lamps will be required if optically pumped uv lasers are to become prevalent. So called "frequency doubling" of existing laser outputs is a possible uv pump source (*22*). The optical pump of maximum power output thus far is the vortex-stabilized arc. This device, which requires an input of 100 kW, has been used with a neodymium-doped yttrium aluminum garnet rod to produce a 200-W output continuously at 1.06 μm (*23*).

The simplest continuous broad-band pumping source is the quartz iodine tungsten filament lamp. The conventional filament configuration is either a single coil or coiled coil. These lamps can be driven as much as 50% over their rated power input for short periods of time. For maximum transfer efficiency, the filament should be no larger than the rod being pumped and preferably should be smaller. The choice is limited by commercial availability, however. Any laser pumped by a commercial quartz iodine lamp should have facility for moving the lamp after installation in the laser because the filaments sag with use and it is a distinct advantage to be able to optimize the position of the lamp relative to the laser rod. In addition, the position of the filament in the lamp envelope is not constant from lamp to lamp.

The mismatch between lamp emission spectrum and laser rod absorption spectrum can be reduced by doping the laser rod with an additional impurity which will absorb the pump radiation not otherwise absorbed and fluoresce at a wavelength absorbed by the laser-active dopant. This arrangement is called "cross pumping" (*24*, *25*).

1. Optical Pumping Geometries

To maximize the transfer of energy from the driving light source to the laser-active core, the core, usually a laser rod, and the lamp are surrounded by a mirror of some kind. The purpose of the mirror is to focus light from the source onto the active material. The mirror cavity configuration depends to some extent upon the shape of the light source. For example, if the optical pump is a helical flash lamp, the mirror around the lamp can only be a cylinder. This is the configuration used first by Maiman (*14*). It is usually referred to as *coaxial pumping*. Another form of coaxial pumping has been reported by Lesnick and Church (*26*). These authors used a hollow-walled quartz cylinder as the flashtube and placed the glass laser rod in the center of the cylinder. They report efficiencies of up to 5.1% for the glass laser rod they pumped.

If the light source is linear, such as a xenon flash lamp or tungsten filament, the choice of reflective housings is wider. One possibility is to place one or more of the sources as close as possible to the rod and wrap the arrangement with aluminum foil. A particularly efficient configuration when one linear source is used is to place the lamp at one focus of a cylindrical cavity of elliptical cross section with the active material at the other focus (*27*). The ideal source in this arrangement would be a line. Real lamps are only a poor approximation of the ideal, however. Bowness has considered the results of finite lamp and rod sizes on the efficiency of transfer of radiant energy in elliptical laser cavities (*28*). For increased maximum energy transfer to the laser rod, one or more elliptical cavities can be joined so that a number of lamps focus on one rod. While this configuration increases power output from the laser, it does so at the cost of decreased transfer efficiency.

Roess has discussed the advantages of cavities that are ellipsoidal rather than elliptical (*29*). The ellipsoidal geometry is obtained by rotating an ellipse about its major axis. In the laser, the pumping lamp and the laser rod are mounted in line on the major axis. Using the ellipsoidal configuration, Roess has shown that 1.8 W of average output power are obtainable from a small ruby rod pumped with a mercury discharge lamp (*30*). He has further shown that the ellipsoidal mirror permits lower input energy for threshold operation with a ruby rod.

Many commercially available high-power discharge lamps are effective point sources. Some workers have made use of such lamps in optical pumping arrangements which are less symmetrical than the geometries already discussed. One example is Young's use of an aluminized sphere to direct the emission from a lamp onto a glass laser rod (*31*). While setups of this sort permit use of pumping sources that could not be used otherwise, the distribution of absorbed energy cannot be as uniform as obtained with either coaxial or elliptical configurations.

The wavelength range obtainable from optically pumped solid-state lasers is 0.55 to 2.6 nm (*22*). The efficiency of operation is usually less than 1%, which makes solid-state lasers much less efficient than some types of gas lasers. Solid lasers are much more compact than high-powered gas lasers. Commercial optically pumped Nd:YAG lasers are contained in a package whose volume is less than 1 ft^3, exclusive of power supply.

The power easily available from cw lasers for applications is about 15 W maximum. This output is representative of a number of presently offered Nd:YAG lasers emitting at 1.06 μm. Continuous ruby lasers may also be obtained. These emit on the order of 1 W at 0.6943 μm. For most applications the Nd:YAG laser is probably the best unit because it can produce maximum power in a small, easy to operate package. The neodymium-laser output is in the near infrared but it can still be manipulated without large loss using standard visible light optical elements.

Multijoule, long-pulse lasers have been available now for some time. Here, too, the best devices for applications are the ruby and neodymium-doped solids. In the pulsed case, however, neodymium-doped glass is an attractive and inexpensive substitute for the neodymium-doped garnet rod.

Because the glass rod can be made much larger than the crystalline garnet rod, much higher energy storage is possible with glass. This permits greater energy outputs from long-pulsed glass rod lasers. For maximum peak power the optically pumped solid-state laser is *Q*-switched (*22*).

2. Liquid-State Lasers

It was recognized early in the history of lasers that liquid lasers would have enormous advantages over solid-state lasers. In principle, at least, liquid lasers can be used in active cores which are optically homogeneous and larger than is possible with solid lasers. The active material in a liquid laser can be recirculated to reduce heating problems, and the active material could be cheap.

Spurred by these advantages, research into liquid lasers was pressed at several laboratories in the United States in the middle 1960s. The sum of this research was the liquid-chelate laser, an organic solution of a rare-earth ion complexed by four β-diketone ligands, usually fluorinated. The liquid-chelate laser could be run at room temperatures in the pulsed mode and the active material could be recirculated without affecting laser action. Unfortunately, the organic ligand used to complex the rare-earth ion is such an intense absorber of the pumping light that only cells with a diameter of 1 mm or so could be uniformly optically pumped. Such small active cores prevented emission powers of any consequence from the liquid laser.

Subsequently, two new types of liquid lasers were discovered and work on them is continuing. The first is an inorganic solution of neodymium oxide. The dried oxide is dissolved in selenium oxychloride and tin tetrachloride and filtered (*32–34*). The solution is then used as the laser-active material pumped by standard discharge lamps. This solution has one of the highest gains of any known laser material. It has been observed to produce stimulated emission when flashed in a test tube without mirrors (*34*). The output wavelength of the material is 1.05 μm. Because of the toxicity of selenium oxychloride, handling is the major problem of this laser material.

The dye laser was discovered after interest in the liquid-chelate laser waned. Sorokin and Lankard first reported stimulated emission from an organic dye which was optically pumped by the output from a ruby laser (*35*). Later the same workers obtained laser action from organic dye solutions excited by a flash lamp (*36*). The typical bandwidth of the emission obtained was 25 Å, much wider than that observed from either solid or gas lasers. Replacement of one mirror in the cavity by a diffraction grating narrowed the bandwidth to 0.6 Å and permitted tuning of the output over a range of about 50 Å (*37*). The use of the dyes as laser-active materials has permitted laser action over the wavelength interval 10,600 to 4540 Å (*38*). The types of dyes used have been polymethine, rhodamine, and coumarin derivatives (*38*, *39*). It has also been shown that the dyes will produce stimulated emission when dispersed in solid plastics such as polymethyl methacrylate (*40*). The solvents used to make up the solutions are water, acetone, glycerin, and methanol. The peak output power obtained is a few megawatts with integrated energies of about 0.01 J.

The tunability of output, wide wavelength range, and economy of dye lasers begin to fulfill the original promise of liquid lasers. Much more is to be expected from them in the near future.

B. *Electron Discharge Excitation: Gas Lasers*

Gas lasers are probably the most common type of laser found in the laboratory today. Because of the convenience of the method of excitation and the lack of environmental influence on its operation, the gas laser finds widest use in scientific and technological applications.

An impurity atom in a crystal lattice can lose up to 1000 cm^{-1} of energy by photon emission. Consequently, no crystalline lasers are expected to emit quanta of energy less than 1000 cm^{-1}. No such restriction is found in gas lasers and numerous gas-laser lines have been reported in the far infrared. Few light sources have strong emission in the ultraviolet; the mercury arc is one clear exception to this rule. Consequently, optically pumped solid lasers in the ultraviolet are rare. Gas lasers are most commonly excited by electric discharge, by which means energy transfer equivalent to ultraviolet wavelengths is readily obtained. Gas lasers have been reported to operate well into the near ultraviolet and a prediction of vacuum ultraviolet lasers has been made (*41*). The gain in many gas lasers is lower than that obtained in solid lasers. As a result, the optical resonators used with gas lasers are much longer than those found in crystalline lasers. It is easy to make relatively long resonators with a gaseous active material. Crystalline laser rods more than a few centimeters long are very difficult to grow.

In addition to the wide range of wavelengths available, gas lasers have advantages which make them more popular than other lasers. Scientifically, gas lasers are attractive because they are essentially emitters of atomic or molecular line spectra. The wavelengths and the energy–level structures of the gaseous species are quite well known and are readily explainable.

The optical medium of gas lasers is homogeneous and has a low refractive index. These features make direct application of optical-resonator theory meaningful and valid. Finally, the continuous output power of the nitrogen–helium–carbon dioxide laser is the highest currently available from any laser.

Optically pumped gas lasers have been reported, but they are few in number. Absorption bands of atoms or molecules in the gas phase are narrow and the absorption constants are small. The latter is the result of the low pressure used in gas lasers. Because of these facts optical excitation of gas lasers is not a common means of excitation.

Gaseous atoms or molecules may also be excited by collision with an electron or with another excited atom. (Throughout the discussion on excitation of gas lasers where "atom" is used, it is understood that "ion" or "molecule" is also included implicitly.) Both forms of excitation are used to drive gas lasers, but when atom–atom collisions are used to excite the active species, electron–atom collisions are the means of excitation of the atom transferring its energy. If the population inversion is created by electron–atom interaction only, the excitation is said to be *direct*. *Indirect* excitation takes place when the first species is raised to a high energetic state by electron bombardment and the laser-active species is excited by the collision with the atoms originally excited.

The passage of an electric current through a gas creates electrons and ions. The ions are not important in exciting the rest of the gas. The electrons created are accelerated by the field and transfer this energy to the gas by colliding with the atoms in the medium. The electron–atom collisions may transfer any amount of energy, but the amount exchanged corresponds to an energy-level change in the atom. That is, the amount of energy gained or lost in a collision is the same quantity of energy absorbed or emitted in an interaction with a photon. The probability of energy transfer between atoms is proportional to $\exp(\Delta E/kT)$ where ΔE is the energy difference between the donating energy level in the excited atom and the accepting energy level in the unexcited atom. In the probability expression k is Boltzmann's constant and T is the absolute temperature of the gas. When two species can be found with energy levels that are approximately the same, it is easy to create a population inversion by this process. Exchange of energy between atoms with nearly identical energy levels is called *resonant transfer*. Two of the more common gas lasers, the helium–neon laser and the nitrogen–carbon-dioxide laser, operate in this manner. This fortuitous energetic arrangement is similar to the accidental coincidence described in the section on optical pumping.

The discharge used to drive continuous (cw for continuous wave) gas lasers can be either direct current or radiofrequency. Some gas lasers are driven in the pulsed mode by capacitive discharge. The voltage and current used in the discharge depend on the type of laser excited. Despite its advantages, the discharge mode of excitation is not without its drawbacks. Particularly in the high-current discharges, erosion of the electrodes by sputtering shortens the life of the laser and a great deal of research has been expended to overcome this problem.

There are three general types of gas lasers. The neutral-atom gas laser was the first type discovered and its most common example is the helium–neon laser. The ion-gas laser is the second class and is represented by argon and krypton lasers. The final class is the molecular-gas laser and a typical one is the nitrogen–carbon dioxide–helium laser already mentioned.

1. Neutral-Atom Gas Lasers

The wavelength region of this class of gas lasers is the infrared. The preponderance of lines emitted by these sources lies in the region of 1–25 μm. Lines have been reported further into the far infrared and neon lasers have visible lines, but these exceptions are rare. The power output of this class is about 100 mW maximum and from 20 to 80 W input is required to obtain this power delivery. The discharge in a He–Ne laser is frequently rf powered. The frequency of the input may be 23–40 Mc; the best results are obtained when the input is in the range 27–30 Mc (*42*).

Neutral-atom lasers are primarily excited by direct electron–atom collision. The most common example of this class, however, is excited by resonant transfer energy from helium to neon. The He–Ne laser emits lines at 0.6328, 1.15, and 3.39 μ.

The electron–atom collisions which excite the helium can also deexcite the neon atom at high current. As a result, the output power of the laser cannot be increased by increasing the current flow in the discharge. Consequently, there will be an optimum current beyond which the power output of the laser will fall off (*43*). The tube diameter in the He–Ne laser is very important in determining the output power of the device. It has been shown that the gain is increased in the He–Ne laser by decreasing the diameter of the tube containing the gas (*44*). Apparently altering the tube diameter changes the relative rates of excitation and deexcitation processes. The optimum tube diameter is 6–8 mm (*45*).

It is possible to select the wavelength of emission from a He–Ne laser by using mirrors whose reflectivities are maximum at the wavelength of interest. Another method of selecting output wavelength is to insert a prism in the cavity between the output window of the gas cell and the completely reflective mirror. In operation the prism is positioned so that only the desired wavelength is reflected from the mirror back into the cavity (*46*).

A great deal of the experimental technique for gas lasers was developed in research on He–Ne lasers. The original device had flat reflective mirrors exposed to the gas and the discharge (*47*). Internal mirrors were corroded by the discharge and a means was quickly devised to place the mirrors outside the gas discharge. In the new configuration, the tube itself was terminated by Brewster-angle windows, which have the property of passing one plane of polarization with minimum reflection loss (*48*). At low-energy output, due to the effect of the Brewster-angle window, the emission from the laser is plane polarized. At greater outputs the second polarization will also appear in the radiation (*49*). Early work on the He–Ne laser also demonstrated that spherical mirrors were much easier to align for laser action than flat mirrors (*48*). Further, the losses of a resonator using spherical mirrors are much lower than those made with plane mirrors (*50*). A number of firms make excellent dielectrically coated mirrors for use in lasers of all wavelengths.

The design of mirror mounts for a laser is no small feat. The mirror mounts must move the mirror through a very small angle uniformly and without backlash for at least two axes of rotation. Several solutions to these requirements of varying complexity have been reported (*51–53*). Acceptable commercial devices of reasonable cost are available. The mirror mounts are usually fastened to a rigid member such as an aluminum channel, at a separation of slightly less than the radius of curvature of the spherical mirrors.

The optimum ratio of helium to neon is 7:1 to 10:1 (*53*). Impurities of nitrogen, oxygen, or moisture will prevent laser action. As a result, it is necessary to clean the tube rigorously by evacuation to at least 10^{-5} Torr followed by purging with the discharge. Evacuation and refill cycles may also be required. The optimum final gas pressure in the tube in Torr is 3.6 divided by the inside diameter of the tube in millimeters (*53*).

While it is possible to build a He–Ne laser easily in a well-equipped laboratory, commercial small lasers are so low in price and so dependable that, except in special situations, purchase is recommended.

2. Ion-Gas Lasers

The wavelength range of ion-gas lasers is 0.26 to 0.8 μm. It is quite common for one active material to emit on a number of lines. The power output obtained may be as high as 10 W, and 100 W has been predicted (*41*). Efficiencies range from 0.03 to 0.3%.

It is believed that the excitation of ion-gas lasers occurs in more than one step since the gain varies as the square of the current density in the plasma tube. Presumably the first step is an atom–electron collision creating the ion, and the second step is an ion–electron collision resulting in excitation of the ion. The power input mechanism is usually a dc discharge created with the help of a heated cathode. The power input is much higher than that in neutral-atom gas lasers; argon, krypton, and xenon showed laser action in a discharge of 200 V and up to 50 A (*54*). The discharge may also be pulsed. In this case the input is obtained from a capacitor that is charged up to 10 kv with maximum currents of 40–300 A. The current density in the plasma tube can be increased by a longitudinal magnetic field (*56*). Another means employed to increase the current density is rf excitation of the discharge. This technique improves the current density, precludes the need for electrodes and permits the use of corrosive gases such as chlorine for the active medium (*57*). It was observed that the dc discharge caused "pumping" of the gas fill in the plasma tube to one end of the tube, and a return tube is required to establish equal pressure throughout the tube.

The tube carrying the plasma is usually water jacketed for cooling purposes. Cooling is required because the temperature of the plasma is approximately 3000°K and erosion of the plasma tube occurs very readily. Water cooling is usually not sufficient to sustain the plasma tube for very long, especially if it is made of quartz. The tube can be made of a high-temperature ceramic or of short metal sections spaced with ceramic disks. The latter construction is particularly successful and has been used in commercial ion-gas lasers.

3. Molecular-Gas Lasers

The output of this laser class is found in the infrared, primarily beyond 5 μm. Multiple frequencies from one active material is common. Specific output wavelengths are: the carbon dioxide laser, a family of lines around 10.6 μm; the water laser, 27.9 and 118.6 μm; and the nitrogen–carbon-disulfide device, a series of 10 wavelengths centered at 11.5 μm (*41*, *58*). One member of this variety of lasers, the nitrogen–carbon dioxide–helium laser, emits more cw power than any other currently known. The power output of the carbon dioxide laser is a function of the length of the tube containing the gas, and current output figures compute the laser emission power at 100 W per meter of discharge tube (*59*). In special configurations, large carbon dioxide lasers are reported to have emitted 2.5 kW on a continuous basis (*60*). In addition, due to the long upper-state lifetimes in the carbon dioxide laser, *Q*-switching of the device is possible by a number of methods. Other active materials used in molecule lasers are water, NO, N_2O, CO, SF_6, $SiCl_4$, O_2, Br_2, and Cs_2 (*58*, *61*, *62*). In several of these cases, emission is obtained from free atoms formed by dissociation of the electrical discharge. The efficiency of molecular-gas lasers is higher than that obtained from any other source. The carbon dioxide laser has been reported to produce efficiencies of 15% (*63*).

The mode of excitation of the active materials is either direct or indirect by electric discharge. In all the cases reported, direct excitation by electron collision is possible, but in the highest-output-power cases, another molecule is excited which transfers its energy by collision with the active molecule. Laser action takes place on a vibrational–rotational line of the neutral active molecule.

Carbon dioxide is a linear symmetric triatomic molecule with three normal modes of vibration. They are: a symmetric stretch along the molecular axis, designated ν_1; a doubly degenerate bending mode perpendicular to the molecular axis, designated ν_2; and a symmetric stretching mode along the axis, called ν_3. The energies of the first excited vibrational levels of the three normal modes are 1388, 667, and 2349 cm^{-1}, respectively. The normal modes may be approximated as independent harmonic oscillators. To describe the energetics of the molecule at any time it is necessary to give the energy level of each normal vibration. This is done by giving the number of excitations of each vibration. Since ν_2 is doubly degenerate it has a superscript which describes its degeneracy. The state of the molecule is then described by the four numbers (N_1, N_2^L, N_3). Laser action usually occurs between the (00^01) and (10^00) levels. The emission lines obtained in stimulated emission correspond to transitions in which the rotational quantum number of the terminal level is larger by one than the rotational quantum level of the emitting level. These transitions are known as the *P* branch. [For an excellent brief review of the molecular spectroscopy terminology especially relevant to the carbon dioxide laser, see the government report by Horrigan (*64*).]

Originally laser action was observed in pure CO_2 excited by electric discharge (*65*). Subsequently Patel demonstrated that nitrogen could be excited separately and then transfer its energy to the carbon dioxide molecule (*66*). The explanation for this highly efficient energy transfer lies in the energy level of nitrogen. As a homonuclear diatomic molecule, nitrogen cannot radiatively decay from its first excited vibrational level,

$N_2^*(V = 1)$. Once excited to this level by collision with an electron, loss of energy from this level is very slow unless the nitrogen interacts with another molecule with similar energy level. The energy level of the excited vibrational level of nitrogen is 2330.7 cm^{-1}, while the energy level of 00^01 state of carbon dioxide is 2349.16. The energy difference between the two levels is very small, about 18 cm^{-1}, which makes transfer of energy from the nitrogen to the carbon dioxide very probable (*67*). The energy exchange may be written as

$$N_2^*(V = 1) + CO_2(00^00) \rightarrow N_2(V = 0) + CO_2(00^01) - 18 \text{ cm}^{-1}$$

The vibrationally excited nitrogen molecule has proved useful as a means of exciting molecules other than carbon dioxide. Both nitrous oxide and carbon disulfide have excited levels that will accept energy from nitrogen and both have yielded stimulated emission when excited this way, although neither has produced the power of the CO_2 laser (*58*, *68*).

The presence of a third gas, helium, in the gas fed to the carbon dioxide laser has been shown to increase the power output of the laser by a factor of 10 (*69*). The role of the helium is to increase the number of excited nitrogen atoms by collision transfer of energy, and apparently to increase the population inversion in carbon dioxide by increasing the rate of de-excitation from the terminal state. Xenon, in addition to helium, also improves the efficiency of the carbon dioxide laser, but not the output power. Paananen ascribes the efficiency increase with xenon to the reduction in discharge voltage (*70*). The effects of hydrogen and water on the laser power output have also been investigated (*71*, *72*). Finally, Jacobs and Bowers have shown the wavelength range of the CO_2 laser can be expanded using isotopes and mixtures of isotopes (*73*).

The original nitrogen–carbon-dioxide laser was a continuous flowing system, exhausted by mechanical vacuum pumps (*74*). The flowing-gas fill requires vacuum pumps and gas bottles with a valved inlet system, all of which are quite cumbersome. Efforts have been made to operate carbon dioxide–nitrogen–helium lasers with static fills. It appears that scrupulous cleaning of the cavity is necessary to permit relatively long laser life without power falloff (*75*, *76*). The reason for power falloff in sealed carbon dioxide lasers is the formation of carbon monoxide which is deleterious to laser action, and to adsorption of the gas fill by the electrodes (*76*). Heating the cathode to free adsorbed gases and recombining CO and O on a hot platinum wire in the discharge tube improve the life of a sealed CO_2 laser (*76*, *87*).

Experimentally, the dynamic fill carbon dioxide laser is probably one of the easiest lasers to construct. The tube diameter is not critical, but a 1-in. tube will produce maximum power (*64*).

Because of the wavelength of the stimulated emission in the carbon dioxide laser, special transmissive optical materials are required. The alkali halides are available in large-size pieces and are quite inexpensive, but suffer from the drawback of corrosion by atmospheric moisture. Germanium, silicon, and a commercial material, Irtran-4, have been used with success. The major difficulty with optics lies in coupling out large amounts of power. All of the above materials increase in absorption at 10.6 μm as the power throughput increases (*78*). Commercial eyeglass lenses given a coating of evaporated gold have been used successfully as inexpensive mirrors (*79*). For the completely reflective mirror, it is possible to use coated stainless steel without cooling. The output mirror, if it is transmissive, must be cooled when power outputs exceed 100 W. Power may be coupled out of the laser through a hole in the output mirror (*74*). Hole coupling may require an external window beyond the mirror if the mirror is directly fastened to the discharge tube. Smiley has given one scheme for mounting output mirror and window in the same head (*80*). Experience has shown that the mirror will be heated at the edge of the hole and will crack at high powers. The output obtained by the hole coupling consists of high-order modes of high divergence and is generally not suitable for applications work (*64*, *81*). The power produced in a carbon dioxide laser can also be coupled out by use of a beam splitter between the end of the discharge tube and the mirror. The power transmitted out depends on the angle the beam splitter makes with the axis of the discharge tube (*64*).

The tube itself can be commercial Pyrex-glass tubing. Because cooling the discharge tube increases the output power, it is preferable to use a jacketed tube (*74*). Commercial glass-pipe heat-exchanger units work very well. It is important to keep the mirror mount and the discharge tube mechanically isolated so that heating of the discharge tube will not cause movement of the mirror mount. Bellows between the tube and the mirror mount in an internal-mirror laser will suffice. Desenberg and Merritt have discussed a multipath gas spectroscopy cell as a cavity for a carbon dioxide laser (*82*).

The power supply may be ac or dc. Neon sign transformers have been used (*79*). If a dc discharge is used, the maximum tube length is a function of the maximum voltage available. Split discharges are possible, and preferable if the maximum voltage from the power supply will not discharge the length of the tube. Split discharges permit lower voltages and are attractive from a safety viewpoint. The carbon dioxide laser is like the helium–neon laser in that there exists an optimum current flow for maximum output power. The reason for the optimum current is the same as that in the He–Ne case: Above a certain current, electron collisions with excited molecules begin to predominate as an energy-loss mechanism and thereby reduce the output power (*64*).

It is possible to obtain pulsed emission from a carbon dioxide laser and a number of techniques exist for doing so. A *Q*-switch method for obtaining pulses from a carbon dioxide laser was reported by Kovacs *et al.* in 1966 (*83*). These authors used a rotating flat mirror at the end of a laser terminated by a Brewster-angle window. The axis of mirror rotation was perpendicular to the axis of the discharge tube. Extending this method, Djeu, and co-workers substituted a diffraction grating for the mirror and found that the vibrational–rotational transitions in CO_2 could be *Q*-switched sequentially if the rotation rate of the grating was correct (*84*). The peak powers observed by these authors were 15–150 kW per pulse with an integrated energy per pulse of about 1.5 mJ. Finally, T. J. Bridges reported that moving the end mirror of the laser in and out along the resonator axis at velocities between 16 and 30 cm/sec produced uniform pulsing at repetition rates between 30,000 and 60,000 pulses/sec (*85*). The peak power in this case was 30 W. At these high repetition rates the average output power obtained in pulsed operation was the same as that obtained in CW mode.

Passive *Q*-switching techniques are also available for the carbon dioxide laser. A cell containing a gas that absorbs the wavelength of stimulated emission is inserted in the resonator after the discharge tube. The laser is turned on and when the absorber gas pressure is at the proper value, usually a few Torr, the CW output from the laser is transformed into a uniform sequence of pulses. Sulfur hexafluoride (*86*), formic acid, and carbon-dioxide–propylene mixtures (*87*) have been used. Pulse rates for the carbon dioxide–propylene mixture have attained 10^5 sec^{-1}. The mechanism of *Q*-switching for the absorber gases is the same as that of saturable absorbers in pulsed optically pumped lasers. The absorber delays the onset of stimulated emission by absorbing the initial output. When the population inversion is much greater than normal, the *Q*-switching gas absorbs so much light that it is saturated at that wavelength and becomes transparent to it. The laser pulse then forms as if the absorber were not present.

Pulsed emission from carbon dioxide lasers can be caused electrically as well as optically. Hill has shown that the dimensions of an electrically pulsed gas laser can be much larger than an optically switched device and that the energy delivered per pulse is correspondingly larger (*88*). Hill used total gas pressures in the discharge tube of up to 60 Torr and applied pulses of up to 1 MV to the tube for periods of 5–50 μsec. Output pulses of 200-kW peak power and 5-J integrated energy have been obtained by this method. The average output power of his 3-in.-diameter, 8-ft-long discharge tube was 210 W maximum. Repetition rates of over 100 pulses/sec were observed.

V. Analytical Applications of Lasers

A. *Raman Spectroscopy*

1. The Raman Effect

Photons striking molecules which are not absorbed may be scattered with the incident frequency or with a new frequency. The new frequency may be either greater or less than the original frequency. The process by which light is scattered with the original frequency is called *Rayleigh scattering* while light of some other frequency is due to the *Raman effect*.

Rayleigh scattering may be thought of as a perfectly elastic collision between a photon and a molecule, occurring with no energy transfer between the particules involved in the collision. In the Raman effect the collision involves a transfer of energy between the molecule and the photon.

The Raman effect is due to the excitation or deexcitation of some rotational or vibrational level of the molecule. Not all rotational and vibrational transitions are Raman active. Those conditions which must be met for a transition to be Raman active will be discussed later. In the scattering of light by molecules, the Rayleigh effect is always present and is always much stronger than the Raman scattering. Consequently, the Raman spectrum of a molecule always includes the frequency of the incident light.

Experimentally, Raman spectrum contains three types of information: the shift in frequency of the scattered light; the intensity of the frequency-shifted radiation; and the polarization of the scattered light. The range of frequency shift can be from 5 to 3000 cm^{-1}. The intensity of any band depends on the strength of the transition. The polarization of the scattered light is usually expressed as the ratio of the intensity of the light in the horizontal plane to that in the vertical plane. It can vary from zero to infinity.

The selection rules which determine whether a vibration or rotation are Raman active are as follows: A normal vibration is active if and only if the nuclear displacements change the polarizability of the molecule (*89*). Similarly a pure rotational transition of a molecule can be expected to appear if and only if the rotation of the molecule causes a change of the polarizability of the molecule. The rotational Raman effect is observable only in the light scattered from gases, and for analytical purposes this phenomenon is much less important than the vibrational Raman effect.

From a practical standpoint, there are several interesting points about Raman-active transitions. They are independent of light absorption by the molecule. The selection rules for light absorption are different from those for Raman transitions and the absorption spectra and Raman spectrum of a molecule are often complementary and nonredundant. Organic functional groups have characteristic Raman frequency shift ranges just as they have characteristic infrared absorption frequency ranges. The exact frequency of a given functional group Raman peak is a function of the local environment of the group and its position can be used to qualitatively analyze the molecular structure near the functional group of interest. Hibben discusses the effect of substituents on the Raman frequency of common organic functional groups (*90*).

Until the advent of scanning infrared spectrophotometers, Raman spectroscopy was much more important than infrared absorption spectroscopy due to the relative ease with which spectra were obtained.

Compared to infrared spectroscopy, which yields similar data about molecules, Raman spectroscopy has a number of advantages. Water is an excellent Raman solvent and Raman spectroscopy has been used to identify the species present in aqueous solutions of strong acids. Glass sample cells are completely satisfactory. The spectra of noninterfering compounds are equal to the sum of the independent spectra. No mulling for solids is required. Fundamental structure information about a compound can be inferred by comparison with the spectra of compounds with known structures.

Despite the experimental utility of Raman spectroscopy, the conventional technique has disadvantages which have limited its application. Many of these have been due to the exciting source. The most common Raman sources have been the Toronto high and medium pressure mercury arcs, both of which emit an intense line at 4358 Å. Unfortunately, this wavelength excites fluorescence in many compounds. This fluorescence contributes to the background noise and obscures the Raman frequencies. In addition to the fluorescence problem, many compounds absorb the

4358-Å line, thereby precluding the possibility of obtaining a spectrum Finally, the wavelength of the mercury arc is on the fringe of the photochemically active region and initiates decomposition in many compounds, again preventing the procurement of a proper spectrum.

Because of the low intensity of the Raman signal, it is normally necessary to use concentrated solutions or multipass cells. When multipass cells are used, the polarization information obtained is of limited value because of the multiple reflections the exciting light has undergone. The light from the Toronto arc is emitted over a large area and to ensure maximum Raman signal in an experiment, large samples were necessary. In those cases where large samples are not available, for example in studies of natural products, the use of Raman spectroscopy as a tool is prohibited by this requirement.

Finally, the normal configuration in a conventional spectrograph or spectrometer involves collection of the scattered light at 90° from the direction of the exciting light. This arrangement frequently prevented the use of solid samples which were not highly transparent.

In this discussion only continuous lasers are considered as Raman-exciting sources because these are the best for analytical applications. The most common laser used in Raman work has been the helium–neon laser, emitting a red line at 6328 Å. Argon lasers emitting at 4880 Å and 5145 Å and gallium arsenide semiconductor lasers emitting at 8400 Å have also been used (*91*, *92*). While the power available from the argon-ion laser is higher than that from the He–Ne laser (1 W vs. 80 mW), the useful tube life is shorter for the argon laser and less work has been done with it.

In the brief period since the laser has been introduced as a Raman light source, numerous advantages of this exciting source have become apparent. Sampling techniques are much simpler. Beattie reports spectra of samples held in brown reagent bottles *and test tubes* (*10*). Small samples can be used to obtain spectra of excellent quality. Bailey *et al.* (*93*) describe the use of a capillary sample cell which forms an integral part of the spectrometer optical system and permits obtaining high-quality spectra from samples as small as 0.04 ml.

As mentioned above, the wavelength of the most common laser Raman source, the He–Ne laser, is 40% longer than that of the conventional mercury source. This change in wavelength reduces the relative intensity of the Rayleigh scattering which constitutes background noise (*91*). The longer wavelength coupled with the high intensity of the laser permit study of colored compounds which could not have been scanned using a mercury source. A number of papers have reported Raman spectra of such highly colored compounds as molten gallium triiodide, ferri- and ferrocyanides, thiocyanates, and dichromates (*10*, *94*, *95*). The longer wavelength of the He–Ne laser also reduces the likelihood of fluorescence by the sample or the sample cell. A clear demonstration of the superiority of the laser in the latter regard is given by Hawes *et al.*, who present spectra of 9,10-dibromoanthracene excited by the 4358 Å line of the Toronto arc and by the 6328-Å He–Ne line (*96*). In the first case the Raman lines are completely obscured by the fluorescence emission of the sample. The spectrum obtained using the laser source clearly shows the Raman lines (*10*).

At least one of the commercial laser photoelectric Raman spectrophotometers currently available analyzes the light scattered at 180° from the incident light (*96*). This "front surface illumination" allows the use of opaque samples for study. The solids MbO_3 and PCl_5 have been examined briefly in this manner by Beattie (*10*).

The output of the laser is already polarized so that no polarizer is necessary to convert the light from the exciting source from random polarization to one orientation. A half-waveplate is used in commercial units to rotate the plane of polarization when determining depolarization ratios. Depolarization ratios are particularly important in the study of single crystals but are seldom used in quantitative analytical studies.

Finally the half-width of the laser output is smaller than the 4358-Å line of the mercury arc. According to Weber and co-workers (*91*), the half-width of the most common mercury line is approximately 0.2 cm^{-1}. For the argon-ion and helium–neon lasers the half-widths are 0.15 and 0.05 cm^{-1}, respectively. This parameter of the exciting line is important because it determines the maximum resolution possible. The use of the He–Ne laser in high-resolution Raman spectroscopy constitutes a real advance. With this source, the limiting resolution of the Raman technique approaches the resolving limit of infrared spectroscopy (0.025 cm^{-1}) (*97*).

2. Analytical Applications of Raman Spectroscopy

Raman spectroscopy is not a technique commonly used in quantitative analytical chemistry. With the advantages of the method itself and the new advantages of laser-excited spectroscopy it is reasonable to expect that the method will become more important in routine measurements. The qualitative use of laser Raman spectroscopy is very similar to qualitative infrared spectroscopy and the general method will not be discussed in detail. Since quantitative Raman applications may be less familiar, these will be reviewed in some detail, along with the new experimental methods that have been developed in connection with the use of the laser.

3. Quantitative Raman Analytical Methods

Raman analytical work is similar to absorption spectroscopy except that the emitted intensity of a Raman line, obtained from the sample, is used instead of the absorbance. In the literature prior to about 1960, recording of intensities was done photographically. Modern instruments, however, particularly those purchased for primarily analytical use, measure intensity photoelectrically. In general, photoelectric instruments are desirable for methods development and routine use because the measurements are simpler and not subject to the error that can arise in the plate development process. The methods mentioned below will all refer to intensity and are the same whether that quantity is measured photographically or photoelectrically. The accuracy of quantitative Raman spectroscopy is 1–5%.

The most obvious method of determining the concentration of an unknown in a diluent is to establish a series of standard samples with different known quantities of the compound in the diluent, measure the intensity of a line in the knowns, and plot the intensity of the line as a function of the concentration of the compound of interest. The spectrum of the sample of unknown concentration is then obtained and the concentration of the unknown can be read off from the calibration curve.

Two-component systems can be analyzed from their Raman spectra in two ways. The first method involves measurement of the intensity of one line in the spectra of each of the pure components. The intensity of the same two lines is measured in the unknown mixture of the two components. The mole ratio of the two components in the mixture can then be calculated as follows:

$$\frac{\text{Moles of substance 1}}{\text{Moles of substance 2}} = \frac{(I_1/I_2)_{\text{UNK}}}{(I_1/I_2)_{\text{KNOWN}}} \tag{28}$$

Another method of analyzing two-component systems for concentration of both components is to prepare a series of solutions of compounds 1 and 2 of known concentration and subsequently measure the ratio of intensities I_1/I_2 using appropriate lines. The ratios of intensities are then plotted against the ratios of weights for each sample in a calibration curve. The weight ratio of an unknown can then be read off upon determination of the ratio of intensities of the two lines.

For multicomponent systems an empirical averaging technique has been used with some success (*90*). In a series of known mixtures the intensity of the appropriate line for each compound is divided by the intensity of the same line in the pure compound. This figure is then used to divide the known percentage of the compound in the calibrating mixture. The result of this calculation is a correction factor which may be used to obtain the percentage of a component in an unknown mixture. In determining the quantities of the components of interest present in a multicomponent mixture, the line intensities in the unknown are divided by the appropriate line intensities in the pure compounds, and these intensity ratios are then multiplied by their correction factors worked out in the calibration. The more standard-mixture samples used in the calibration procedures, the more accurate the results will be. In setting up the method, the range of component percentages over which the correction factors are approximately constant should be determined, and if possible all unknowns kept within this range. Hibben has demonstrated this method in the analysis of mixtures of *p*-ethyltoluene, pseudocumene, *m*-ethyltoluene, and mesitylene (*98*).

Finally, internal standards can be used in Raman spectroscopy for quantitative analysis. The common internal standards are compounds which have only a few clearly identifiable lines which do not interfere with lines of the substances to be analyzed. Carbon tetrachloride and carbon disulfide have been used as internal standards (*99*). The calibration procedure involves measuring intensity ratios of known mixtures of the substances of interest and a small known weight percent of the standard. The amount of standard to be used is approximately 10%. The results of analysis of the known mixtures are obtained, and for each component the ratio of intensities of substance line to standard line is plotted against weight of substance divided by weight of standard for each component. If the quantity of standard added to the unknown is the same as that added to the knowns, the ratio of intensities determined from the unknown can be used to read off the weight ratio of substance to standard in the unknown from the calibration curve. No routine quantitative use of laser Raman spectroscopy appears to have been reported in the literature.

A number of new applications of the Raman effect have been presented since the first use of the laser as the exciting source. Hendra and coworkers have determined the Raman spectra of species adsorbed on silica gel (*100*, *101*). The Raman spectra beautifully supplement the infrared measurements in an adsorbent–adsorbate system because the adsorbent permits measurement of only high-frequency absorption bands by transmission. The spectrum of acetaldehyde on silica gel resembled that of paraldehyde, probably formed due to the catalytic effect of the silica-gel surface.

Hendra and Willis have also reported on the laser Raman spectra of stretched polymer fibers (*102*, *103*). Vibrational analysis of the polymer molecules are given. The spectrum of isotactic polypropylene obtained from a translucent powder has been published as well (*104*).

Among the more exotic recent uses of laser Raman spectroscopy are studies of molten salts, *F* centers, and solids at high pressures (*105–107*).

In all of these cases, the laser has allowed data to be taken which would otherwise be very difficult to obtain. In the last three references, the measurements are used as a source of structure information. In the case of the *F* centers, the spectra are a means of studying the interaction of the centers with lattice vibrations (*107*).

Other recent laser Raman references indicate that workers interested in molecular structure in the solid and liquid states have fully accepted the utility of this technique (*108–112*).

B. *Absorption Spectroscopy*

As a light source, the laser has properties which make it highly suitable for use in absorption spectroscopy. The characteristics of laser radiation which have shown to be of value in this area are the narrow bandwidth of the emission and the high intensity. The fact that the laser output is relatively fixed in wavelength has been both an advantage and a disadvantage. The lack of ability to broadly tune the wavelength of the laser source has forced the matching of the absorber to the light source. Unfortunately, it is not often possible to do this in analytical chemistry. Gerritsen has presented an excellent discussion of the various means of obtaining more than one wavelength from a laser (*113*).

Houle and Grossaint have compared the helium–neon laser to a tungsten filament lamp as the source in sample absorption spectroscopy of the dye, Fast Acid Green. This dye, the sodium salt of dibenzyldiethyl diamino-O-chlorotriphenylcarbinol disulfonic acid anhydride, is soluble in water and has an absorption maximum in the visible near the 6328-Å emission of the He–Ne laser (*114*).

The absorbance of a series of solutions was measured at two sensitivities and at varying slit widths. When the tungsten lamp was used, changing the slit width of the monochromator in the commercial spectrophotometer used changed the bandwidth of the radiation used to measure the absorbance. Changing the slit width when the laser was the source only varied the light intensity and had no effect on the bandwidth of the radiation. The effects produced by the use of heterochromatic light are a broadening of the absorption band, reduction of the absorbance measured at the absorption maximum, and a falloff in the slope of the Beer's-law plot at high solute concentrations. As the slit width increases, these problems become more pronounced. These authors showed that for high dye concentrations, the plot of absorbance versus concentration using the tungsten lamp was curved on the low side of a proper Beer's-law plot. Reduction of the instrument sensitivity necessitated wider slit widths, and the curvature was worse. Using the laser, however, all points fell on the same straight line in the Beer's-law graph, regardless of slit width or instrument sensitivity. This behavior was also checked in the differential method, where the reference solution was a solution of the dye slightly less concentrated than the most dilute sample. The observations were the same as before. Consequently, when the wavelength–sample-absorbance relationship is right, the laser is the best source for measuring high solute concentrations.

Going the other way, the laser can be used to measure very low absorbances. If the weak absorber is placed *within* the laser cavity, the low absorption is greatly magnified by its effect on the gain of the laser and a large variation in the output power of the laser is observed. In a laser which has a single pass gain of 0.6%, an increase in single pass loss of only 0.01% may change the output power by as much as 5%.

Brandi has described an instrument which works in this fashion (*115*). The device is calibrated by two glass plates within the cavity oriented close to the Brewster angle. The loss of the plates as a function of angle is well known and can be calculated. The instrument was used to measure single pass loss of three-layer dielectric antireflection coatings. In another application of this technique, Powell has shown that the change in output power of a He–Ne laser emitting at 3.391 nm can be correlated to the concentration of alcohol in a vapor stream (*116*). His cell is 2 mm in diameter and 50 cm long and it is placed within the resonant cavity of the laser. This system can detect 50 ppm of alcohol. It is calibrated by observing the effect of gas mixtures of known concentration of the output power of the laser. It has also been shown that the He–Ne laser can be a very sensitive detector of methane by a similar technique (*117*). Powell has also discussed the use of the infrared-emitting He-Ne laser in the determination of ether in gas mixtures (*118*).

If atomic laser transitions in the infrared have only moderate gain, it is possible to displace their emitting frequency by a small amount by subjecting the active medium to a magnetic field (*119*). This amounts to tuning the laser over a small wavelength range. Since the emission bandwidth of the laser is so narrow, this technique may be used for high-resolution spectroscopy of vibrational–rotational lines of polyatomic molecules. It is also quite possible that the same technique can be used for quantitative absorption-spectroscopic analysis of gas mixtures, although this does not appear to have been attempted yet. The laser in the near infrared has a distinct advantage in resolution over conventional infrared spectrometers. Absorption lines separated by less than 0.005 cm^{-1} can be resolved by the laser method. Maximum resolution of the best infrared spectrometers is approximately 0.05 cm^{-1}. The bandwidth of laser emission around 8 μ is 0.001 cm^{-1} (*120*). Brunet has reported the use of a laser filled with either xenon or neon and tuned with an axial magnetic field to measure vibration–rotation bands in methane, ethane, propane, carbon tetrachloride, and sulfur hexafluoride (*120*). The gas was contained in a cell 1 m long external to the resonant cavity. The absorption gas pressure was 5–20 Torr. Jacoby and Long have measured the absorption profile of one vibrational–rotational line in carbon dioxide using a He–Xe laser tuned magnetically (*121*). Their technique has less relevance to analytical chemistry, however, since the absorption cell used was 50 ft long. Sakurai and Shimoda reported the construction details of tunable He–Ne and He–Xe lasers and the spectra of H_2CO, CH_3OH, and HCOOH measured with these lasers (*122*).

Another interesting light-absorption phenomenon involving lasers has been observed and used for quantitative measurements. While this technique has not yet been reported as a routine analytical method it shows great promise of doing so and consequently deserves mention here.

Gordon *et al.* (*123*) discovered that liquids which weakly absorbed the wavelength emitted by a laser were heated by the beam and due to the heating set up a density gradient in the liquid. The density gradient in turn produced a refractive index gradient in such a way that the liquid in the cell became a negative lens. Subsequently, Rieckoff measured the divergence of laser beams passed through a long cell containing various liquids (*124*). The extent of the divergence depended upon the incident power and the liquid. Beam expansions were reported for benzene, water, toluene, cyclohexane, *n*-hexane, acetone, and carbon disulfide; benzene gave the weakest response and carbon disulfide the greatest. Leite *et al.* (*125*) reported that this phenomenon showed utility as a power-limiting device for light beams.

The ability of a light-absorbing liquid to diverge a laser beam is called the "thermal-lens effect." It appears to be very useful in the measurement of compounds with molar absorptivities of the order of 10^{-4} liter/mole-cm at the wavelength of the laser source. It is difficult to measure absorptivities of this magnitude by conventional means due to the difficulty in separating scattering losses from absorption effects. The technique of determining low absorbances by the thermal-lens effect is fairly simple and requires, in addition to a stable laser, only the ability to measure the width of the beam after it has been diverged by the absorber (*126*). This measurement method should be seriously considered by analytical chemists as a tool to be used in the measurement of low concentrations of weak absorbers.

C. *Emission Spectroscopy*

Probably the earliest true analytical use of the laser was in laser-emission spectroscopy. This method provides experimental simplicity with the possibility of relatively high precision in the analysis of small areas without removal from their environment. While the technique in its present state has already proved valuable, particularly to biologists, further refinements that increase its utility may be expected in the near future.

The method of analysis with the laser microprobe is quite simple. The sample is placed at the focus of a "short focal length" lens and the *Q*-switched laser is pulsed. The laser output strikes the sample after passage through the lens and a small quantity of the sample is evaporated. The vapor is partially ionized, and it rises between two pure graphite electrodes which carry a 1–2 kV potential. The plume between the electrodes reduces the resistance between the gap sufficiently to permit a spark to be automatically discharged. In the discharge, the atoms and ions produce atomic or ionic emission lines, very similar to those normally obtained in a spark discharge. The lines produced permit qualitative identification of the species and quantitative estimation of amounts present in the sample. The emission lines are usually recorded photographically with a grating spectrograph in the normal manner.

The method originally used the multiple-spike pulse from a non-*Q*-switched laser without the auxiliary spark excitation. Although this method avoided the emission from impurities in the carbon electrodes, many elements produced no emission spectra when irradiated (*127*). In addition, the emission spectra obtained without supplementary spark excitation are subject to strong self-absorption because of the large concentration of ground-state atoms in the plume (*128*, *129*). Further, the lines obtained without the spark are much broader than those with the spark. Burkhalter showed, theoretically, that the amount of material vaporized with a long pulse would depend on the sample composition (*130*). If this condition prevailed, the amount of material evaporated when the sample was pulsed would vary from sample to sample, and the analysis results with the laser microprobe would be a function of the sample matrix. It was further pointed out that a single-spike, *Q*-switched pulse would evaporate any sample to the same extent regardless of its chemical composition. The high peak powers of the *Q*-switched pulse cause uniform evaporation and cooling of the plume from the sample. Differences in response of the material to the other properties of the laser emission, such as wavelength, are swamped out by the extreme intensity of the light. Rosan *et al.* (*131*) have shown that the crater size formed by a *Q*-switched laser pulse is reasonably constant for metals, glasses, plastics, and gelatins and is unaffected by sample color and melting point. The energy delivered to the sample is much more reproducible if the laser is *Q*-switched. The number of spikes emitted by a non-*Q*-switched laser varies for one voltage input to the flash lamp and in addition is dependent on the voltage (*129*). It is possible that a *Q*-switched laser will emit more than one pulse. This can be prevented by diaphragming the rod so that only its center is exposed to the mirrors (*132*).

The size of the crater formed in the sample varies from 50 to 100 μm in diameter and 5 to 70 μm deep for a 0.1-J pulse. This corresponds to a volatilized sample mass of 0.01 to 0.1 μg and a volume of 1–10 picoliters. Such a tiny weight loss classifies laser microprobe analysis as a nondestructive method. The logarithm of the pit volume is proportional to the logarithm of the energy input to the flash lamp (*132*).

Not all the sample ejected from the crater is vaporized; some drops back onto the sample to form an annular deposit around the hole. This is particularly true of metal samples. In addition, materials sparked in air form stains over an area 4 mm wide and 6 mm long. If not prevented, this staining precludes analysis of adjacent areas. The staining does not occur if the sample is analyzed in an argon atmosphere or if the sample is covered with a perforated file card. The hole in the card should be 1 mm in diameter and should be centered over the area to be examined (*129*).

Precautions are necessary with the laser microprobe to assure good results. The sample must be isolated from ground to prevent arcing from the electrodes to the sample, if the latter is conductive. The electrodes should be close enough to the sample to guarantee adequate use of the vapor from the sample. A gap of 1.25–2 mm between the electrodes and the sample is preferred (*129*). The discharge in the microprobe is not as large as that in a normal dc arc and consequently the optics of the instrument must be designed to fill the slit and utilize as much of the light from the source as possible. The hot-vapor jet ejected from the sample by the laser emission can strike the lens in the objective and corrode it. It is possible to protect the objective by covering it with a 0.75-ml polyester film. The peak power from the *Q*-switched laser is so high that the focusing lenses may be damaged. Removing the glue between the lenses helps, but this problem has not yet been completely overcome (*129*).

The lower detection limit of the technique is reported as 4×10^{-13} mole for magnesium and in powdered samples is comparable to the sensitivity of the dc-arc technique (*131*, *133*). The sensitivity is a function of the exposure levels used in the analysis.

The precision of the laser microprobe is equivalent to that of the dc-arc method. Recently reported results indicate that for 11 replicate analyses of 16 atomic lines in four different matrices, the average coefficient of variation of intensity ratios with internal standard lines was $\pm 10\%$ (*133*). Single-spike operation is more precise than multiple spike. The precision of line intensities depends more on the laser reproducibility than on any of the spark variables (*132*). Sample preparation for laser-spark excitation is quite simple. Frozen microtomed biological samples may be analyzed directly. Inorganic samples are best ground and milled in a mechanical mortar with ethanol. After drying and further grinding in a Wig-L-Bug the powder is then hand-packed in a shallow plastic dish (*133*). It has been suggested that blood serum can be analyzed after spreading the sample on a paper substrate of controlled thickness (*131*). Rosan has suggested an easy technique for sample preparation where the material to be analyzed can be separated from its matrix (*134*). He dissolves or slurries the material and disperses it on unfixed photographic emulsion. The emulsion is readily available, reproducible, takes up solutions rapidly, and dries quickly. Analyzed samples may be stored in the laboratory notebook. The standard application used is 200 μl of solution micropipetted on a 2-cm circle of emulsion which is then oven-dried at 60°C. Drying can also be done under an ir lamp. Contaminations due to the sample support should always be checked. The impurity emission lines obtained from the surface of a chemically clean glass slide have been reported (*134*).

There are a number of advantages in the laser-spark method of analysis. The technique is often compared to the electron microprobe. While the area analyzed by the laser technique is somewhat larger than that scanned by the electron method, the possible range of elements that can be analyzed by the laser is larger. For technical reasons, the electron microprobe cannot be used on elements lower in atomic number than potassium without special equipment. No such restriction exists for the laser-spark emission analysis. All elements in the periodic table can be analyzed by the latter method. In addition, the laser microprobe is a simple and fast technique and no extensive sample preparation is necessary. The time required to strike the spectrum is about 0.01 sec. The analysis is not dependent upon the electrical conductivity of the sample. Finally, and of great interest to biologists and biochemists, the laser microprobe makes possible determination of the elemental composition of localized areas, if necessary, of live specimens without destroying the animal. This ability to investigate the chemical makeup of small areas is also of interest in other professions as will be seen in the discussion of applications.

Chemical applications of sweat analysis, heavy metal poisoning, and dermatitis have been suggested for the laser-spark emission method. The manufacturer of a commercial microprobe illustrates the analysis of trace elements in the epidermis of a live rabbit's ear (*135*). Cation analysis of single red blood cells and single rat spermatozoa has also been reported (*136*). Nonbiological applications of the laser microprobe include analysis of metal discontinuities to detect welds in structurally defective airplane engine components and the determination of elemental composition of defects in glass (*137*). Materials in individual components on small semiconductor devices have been identified. Inclusions in individual particles of a bimetallic alloy on the surface of a ceramic disk and in polyethylene have all been analyzed by laser-emission spectroscopy (*129*).

Ceramicists have used the laser microprobe to study the advancing front of contaminants on a grain-to-grain basis in furnace refractories (*138*). Mineralogists find the microprobe of great utility in the detection of cations in naturally occurring borate phosphate, chromate, tungstate, arsenate vanadate, and silicate minerals (*139*, *140*).

D. *Other Applications of Lasers to Analytical Chemistry*

In addition to the applications already discussed, lasers have been used in a preliminary way in the fields of Rayleigh scattering, mass spectrometry, and fluorescence analysis. In most of the cases to be discussed below, no direct use in analytical chemistry has been reported. The papers cited primarily test the feasibility of the laser in the technique cited. Often the use of the laser is reported to experimentally substantiate some theoretical result predicted earlier that could not be proven with conventional light sources. Several such papers will be referred to as an indication of laser applications which have not yet been reduced to routine use.

1. Rayleigh Scattering

Vand and co-workers (*141*) have reported the use of a 1-mW He–Ne laser as the light source in a study of crystalline imperfections. The laser was used to illuminate impurities and inclusions in lithium fluoride, magnesium oxide, and ruby-laser crystals. A comparison study on monodispersed polystyrene latex suspensions was also carried out using a high-pressure mercury arc as an incoherent source. The light scattered by the particles was photographed at an angle of 90° from incidence. The final objective, which focused light on the scatterers, was moved back and forth so that the entire field of view of the microscope used to collect the light was illuminated. Latex particles as small as 880 Å were visible to the eye. It was found that the high intensity of the laser light allowed shorter photographic exposures than were possible with the incoherent source and reduced the difficulties caused by Brownian motion of the particles. In the work on the crystalline imperfections using long camera exposures and the laser, particles of the order of 300 Å equivalent latex diameter were detected. This is not a lower experimental limit because high-power lasers will permit observation of smaller particles.

The use of a *Q*-switched laser has been suggested by Wyatt as a means of overcoming Brownian motion of bacterial cells when their light-scattering behavior is studied (*142*). He also pointed out that the intensity of scattered light depends on the difference between the refractive indices of the scattering particle and the medium in which it is immersed. When the difference in refractive indices is large, the reduced scattering intensity demands a strong light source for practical observation. A laser is admirably suited as a source for studying scattering under these conditions.

A number of papers have appeared on the light-scattering properties of sols and fogs (*143*). The data reported have little relevance to analytical interests and for this reason are not discussed further here. Workers have also reported interesting scattering phenomena, but again the utility is doubtful (*144*). Borch and Marchessault have studied the scattering of the light from a He–Ne laser by granules (*145*). The scattering patterns obtained permit measurements of the average diameter of the granules in the 10–90-nm region.

It would seem that the laser has been shown very useful in light-scattering work, but this information has not been made known to workers using scattering in a routine manner. Undoubtedly, this situation will be corrected in the near future.

2. Fluorescence Excitation

Lasers have been used to excite the infrared fluorescence of triatomic molecules in studies of energy-loss rates from excited rotational–vibrational states (*146–151*). Infrared-excited fluorescence of polyatomic gases has also been reported (*152*, *153*). The sample is contained in a cell held in or out of the resonant cavity. The sample cell has windows set perpendicular to the direction of excitation, through which the fluorescence is detected. The use of the narrow-bandwidth laser emission in a pulsed mode excites enough molecules to a single vibrational level to permit observation of the radiative decay of this excited state by the fluorescence emitted when the excited molecules revert to lower energy levels. The reciprocal of the time required for the fluorescence intensity to decay to $1/e$ of its original value has been plotted against the pressure of the gas in the sample cell. The resultant plot is usually linear in the 2–50-Torr region. The gases studied so far in this manner are SF_6, CO_2, N_2O, and CH_4. The inverse lifetime–pressure graph could be used in an analytical determination in special cases, although the effects due to the presence of other gases must be determined.

Lasers have been used to excite fluorescence in solids and liquids also. Damen and co-workers used an argon laser to cause fluorescence in rare-earth crystals. They reported that laser-excited fluorescence was 100 to 1000 times brighter than the fluorescence obtained using a high-pressure Hg arc (*154*). Broad-band fluorescence has been observed in clean pure liquids pulsed with a *Q*-switched ruby laser (*155*). Both water and carbon tetrachloride emitted the fluorescence, which suggests that a molecular dissociation process is involved.

The utility of excitation to specific energy levels should be of great value in complicated cases of fluorescence analysis. No application of this feature appears to have been reported in the analytical literature, however.

3. Mass Spectrometry

Mass spectrometry is a field in which the laser has been added to existing instrumentation to obtain information of great scientific interest.

It is well known that high-power pulses from lasers can vaporize most solid materials. The exact nature of the species expelled from the solid is not known, however. The nature of the vapor molecules or ions, when fully described, should permit some description of the detailed process by which the light from the laser creates the vapor. Numerous investigators have used mass spectrometry to determine the *m/e* ratio of products emitted by a variety of laser-heated solids (*155–160*).

Designs of special spectrometers for this application have been reported (*161*). The only apparent true analytical use of this combination of methods of analysis has been the isotopic analysis of primordial rare gases in meteorites by Megrue (*162*). He volatilized the gases from the meteorite samples by pulses from a ruby laser and then separated the gases cryogenically and determined the isotopes with a 60°-sector instrument. The uncertainties in the method were traceable to the uncertainty in the quantity of sample volatilized by the laser. The data obtained agreed within experimental error to isotope abundances obtained by other methods. Mass-spectral analysis, coupled with laser sampling, has not yet been given the opportunity to realize its potential as an analytical technique.

4. Miscellaneous

Additional miscellaneous but routine uses of lasers are as an aid in optical-system alignment and as a means of calibrating spectrometers (*163*).

Finally, in one-of-a-kind applications, the laser has been used for remote opening of containers, of interest in analyses of radioactive reactions producing gases (*164*), for improvement of resolution in time-based spectroscopy (*165*), and as a means of determining the carbon dioxide content of exhaled breath, by using the exhaled breath as the active medium in a carbon dioxide laser (*166*).

Mention of laser chemistry has been deliberately omitted in this survey because the action of coherent light on materials and the ensuing chemical reactions are subjects more suitable for detailed treatment elsewhere rather than cursory review here. The reader is urged to consult the article by Rousseau (*167*) for an excellent introduction to this field.

REFERENCES

1. T. G. Polanyi and W. R. Watson, *J. Appl. Phys.*, **34,** 553 (1963).
2. T. H. Maiman, *Phys. Rev.*, **123,** 1145 (1961).
3. B. A. Lengyel, *Introduction to Laser Physics*, Wiley, New York, 1966, p. 92.

4. H. Statz and G. deMars, in *Quantum Electronics* (B. Lax, ed.), Columbia University Press, New York, 1960, pp. 530–537.
5. F. J. McClung and R. W. Hellworth, *J. Appl. Phys.*, **33,** 828 (1962).
6. G. Bret and G. Gires, *Appl. Phys. Lett.*, **4,** 175 (1964).
7. I. J. D'Haenens and D. A. Buddenhagen, *J. Soc. Motion Picture Television Engrs.*, **71,** 828 (1962).
8. A. L. Schalow, *Sci. Am.*, **209,** 2 (1963).
9. R. C. Hawes, K. P. George, D. C. Nelson, and R. Bickwith, *Anal. Chem.*, **38,** 1842 (1966).
10. I. R. Beattie, *Chemistry in Britain*, 347 (1967).
11. D. A. Buddenhagen, B. Lengyel, F. J. McClung, and G. F. Smith, *IRE Internatl. Conv. Record*, 1961, Pt. 5, p. 285.
12. D. F. Nelson and R. J. Collins, *J. Appl. Phys.*, **32,** 739 (1961).
13. G. Stroke, *An Introduction to Coherent Optics and Holography*, Academic Press, New York, 1966, p. 70.
14. T. H. Maiman, R. H. Haskins, I. J. D'Haenens, C. S. Asawa, and V. Evtuhov, *Phys. Rev.*, **123,** 1151 (1961).
15. P. Rabinowitz, S. Jacobs, and G. Gould, *Appl. Opt.*, **1,** 513 (1962).
16. A. L. Schawlow and C. H. Townes, *Phys. Rev.*, **112,** 1970 (1958).
17. P. Rabinowitz, S. Jacobs, and G. Gould, *Appl. Opt.*, **1,** 513 (1963).
18. D. Roess and G. Zeidler, *Electronics*, **39,** 115 (1966).
19. S. A. Ochs and J. I. Pankove, *Proc. IEEE*, **52,** 713 (1964).
20. J. P. Markiewicz and J. L. Emmett, *IEEE J. Quantum Elec.*, **QE-2,** 707 (1966).
21. G. A. Hardway, *Microwaves*, **3,** 108 (1966).
22. Z. S. Kiss and R. J. Pressly, *Appl. Opt.*, **5,** 1474 (1966).
23. J. E. Jackson and D. M. Yenni, "High Power Laser Operation," Union Carbide Corp., Linde Div., Indianapolis, Ind., Second Interim Tech. Rept., Contract SRCR-66-4, May 1966.
24. R. J. Pressly, J. R. Collard, P. V. Coedertier, F. Sterzer, and W. Zernik, "Solid State Laser Explorations", RCA Laboratories, Princeton, N.J., Tech. Dept. Rept. AFAL-TR-66129, Contract AF 33 (615) 2645, June 1966.
25. C. F. Johnson, J. E. Geusic, and C. G. Van Uitert, *Appl. Phys. Letters*, **7,** 127 (1965).
26. J. P. Lesnick and C. H. Church, *IEEE J. Quantum Elec.*, **QE-2,** 17 (1966).
27. M. Clifton, C. F. Luck, Jr., C. G. Shafer, and H. Statz, *Proc. IRE*, **49,** 960 (1961).
28. C. Bowness, *Appl. Opt.*, **4,** 103 (1965).
29. D. Roess, *ibid.*, **3,** 259 (1964).
30. D. Roess, *IEEE J. Quantum Elec.*, **QE-2,** 208 (1966).
31. C. G. Young, *Appl. Phys. Lett.*, **2,** 151 (1963).
32. A. Heller, *ibid.*, **9,** 106 (1966).
33. A. Lempicki and A. Heller, *ibid.*, **9,** 108 (1966).
34. T. M. Shepherd, *Nature*, **216,** 1200 (1967).
35. P. P. Sorokin and J. R. Lankard, *IBM J. Res. Dev.*, **10,** 162 (1966).
36. P. P. Sorokin and J. R. Lankard, *ibid.*, **11,** 148 (1967).
37. B. H. Soffer and B. B. McFarland, *Appl. Phys. Lett.*, **10,** 266 (1967).
38. Y. Miyazoe and M. Maeda, *ibid.*, **12,** 206 (1968).
39. B. B. Snavely, O. G. Peterson, and R. F. Reithel, *ibid.*, **11,** 275 (1967).
40. O. G. Peterson and B. B. Snavely, *ibid.*, **13,** 238 (1968).

41. A. L. Bloom, *Appl. Opt.*, **5,** 1500 (1966).
42. Ref. *3*, p. 195.
43. A. D. White and E. I. Gordon, *Appl. Phys. Lett.*, **3,** 197 (1963).
44. E. F. Labuda and E. I. Gordon, *J. Appl. Phys.*, **35,** 1647 (1964).
45. Ref. *3*, p. 193.
46. A. D. White, *Appl. Opt.*, **3,** 431 (1964).
47. A. Javan, W. R. Bennett, Jr., and D. R. Herriot, *Phys. Rev. Lett.*, **6,** 106 (1961).
48. W. W. Rigrod, H. Kogelnick, D. J. Brangaccis, and D. R. Herriot, *Appl. Phys. Lett.*, **33,** 743 (1962).
49. Ref. *3*, p. 197.
50. A. G. Fox and R. Li, *Bell System Tech. J.*, **40,** 453 (1961).
51. T. G. Polanyi and W. R. Watson, *J. Appl. Phys.*, **34,** 553 (1963).
52. C. F. Luck, R. A. Paananen, and H. Statz, *Proc. IRE*, **49,** 1954 (1961).
53. C. L. Strong, *Sci. Am.*, **211,** 227 (1964).
54. E. I. Gordon, E. F. Labuda, and W. B. Bridges, *Appl. Phys. Lett.*, **4,** 178 (1964).
55. Ref. *3*, p. 206.
56. E. I. Gordon and E. F. Labuda, *Bell System Tech. J.*, **43,** 1827 (1964).
57. J. B. Goldsborough, E. B. Hodges, and W. E. Bell, *Appl. Phys. Letters*, **8,** 137 (1966).
58. C. K. N. Patel, *ibid.*, **7,** 273 (1965).
59. J. Ehrlich, G. Hutcheson, W. Hales, and T. Roberts, IEEE Conference on Laser Engineering and Applications, Washington, June 1967.
60. C. K. N. Patel, *Sci. Am.*, **219,** 22 (1968).
61. C. K. N. Patel, R. A. McFarlane, and W. L. Faust, *Phys. Rev.*, **113,** A1244 (1964).
62. C. K. N. Patel, *App. Phys. Lett.*, **6,** 12 (1965).
63. T. G. Roberts, J. H. Ehrlich, and C. J. Hutcheson, *J. Alabama Acad. Sci.*, **38,** 71 (1967).
64. F. Horrigan, "High Power Gas Laser Research," **AD 637,023,** U.S. Dept. Comm. Clearinghouse, Washington, D.C., 1966.
65. C. K. N. Patel, W. L. Faust, and R. A. McFarlane, *Bull. Am. Phys. Soc.*, **119,** 500 (1964).
66. C. K. N. Patel, *Phys. Rev. Lett.*, **13,** 617 (1964).
67. C. K. N. Patel, in *Physics of Quantum Electronics* (P. Kelley, B. Lax, and P. Tannenwald, eds.), McGraw-Hill, New York, 1965, p. 643.
68. C. K. N. Patel, *Appl. Phys. Lett.*, **6,** 12 (1965).
69. C. K. N. Patel, P. K. Tien, and J. H. McFee, *ibid.*, **7,** 290 (1965).
70. R. A. Paananen, *Proc. IEEE*, **55,** 2035 (1967).
71. D. Rosenberger, *Phys. Lett.*, **21,** 520 (1966).
72. P. K. Cheo, *Appl. Phys. Lett.*, **11,** 38 (1967).
73. G. B. Jacobs and H. C. Bowers, *J. Appl. Phys.*, **38,** 2692 (1967).
74. T. S. Bridges and C. K. N. Patel, *Appl. Phys. Lett.*, **7,** 244 (1965).
75. W. J. Witteman, *ibid.*, **11,** 337 (1967).
76. R. H. Carbone, *IEEE J. Quantum Elec.*, **QE-3,** 373 (1967).
77. F. M. Taylor, A. Lombardo, and W. C. Eppers, *Appl. Phys. Lett.*, **11,** 180 (1967).
78. F. Harrigan, R. Rudko, and D. Wilson, International Quantum Electronics Conference, Miami, Florida, April 1968.
79. S. Koozenani, J. McCoy, and D. Rensch, *Am. J. Phys.*, **34,** 989 (1966).
80. V. N. Smiley, *Rev. Sci. Instr.*, **34,** 820 (1963).

81. F. H. Fry, unpublished work, 1967.
82. G. J. Desenberg and J. A. Merritt, *Appl. Opt.*, **6,** 1541 (1967).
83. M. A. Kovacs, G. W. Flynn, and A. Javan, *Appl. Phys. Lett.*, **8,** 69 (1966).
84. N. Djeu, T. Kan, C. R. Miller, and G. J. Wolga, *J. Appl. Phys.*, **39,** 2157 (1968).
85. T. J. Bridges, *Appl. Phys. Lett.*, **9,** 174 (1966).
86. O. R. Wood and S. E. Schwarz, *ibid.*, **11,** 88 (1967).
87. P. L. Hanst, J. A. Morreal, and W. J. Henson, *ibid.*, **12,** 58 (1968).
88. A. E. Hill, *ibid.*, **12,** 324 (1968).
89. J. C. D. Brand and J. C. Speakman, *Molecular Structure*, E. Arnold, London, 1960, p. 181.
90. J. H. Hibben, in *Physical Methods of Chemical Analysis* (W. G. Berl, ed.), 2nd edition, Academic Press, New York, 1960, Vol. 1, p. 400.
91. A. Weber, S. P. S. Porto, L. E. Cheesman, and J. H. Barrett, *J. Opt. Soc. Am.*, **57,** 19 (1967).
92. G. W. Chantry, H. A. Geffie, and C. Hilsum, *Nature*, **203,** 1052 (1964).
93. G. F. Bailey, S. Kent, and J. R. Scherer, *Anal. Chem.*, **39,** 1040 (1967).
94. W. Krasser and H. W. Nurnberg, *Naturwissenschaften*, **54,** 135 (1967).
95. D. V. Luu, C. Clement, and R. Lafont, *Compt. Rend.*, **264B,** 1822 (1967).
96. R. C. Hawes, K. P. George, D. C. Nelson, and R. Beckwith, *Anal. Chem.*, **38,** 1842 (1966).
97. E. K. Plyler and E. D. Tidwell, in *Advances in Molecular Spectroscopy*, Pergamon, New York, 1962, pp. 1336–1342.
98. J. H. Hibben, *Ind. Eng. Chem.*, **26,** 646 (1934).
99. Ref. *90*, p. 405.
100. P. J. Hendra and E. J. Loader, *Nature*, **216,** 789 (1967).
101. P. J. Hendra and E. J. Loader, *ibid.*, **217,** 637 (1968).
102. P. J. Hendra and H. A. Willis, *Chem. Commun.*, **1968,** 225.
103. P. J. Hendra and H. A. Willis, *Chem. Ind.*, **1967,** 2146.
104. R. F. Schaufele, *J. Opt. Soc. Am.*, **57,** 105 (1967).
105. J. H. R. Clarke and C. Solomons, 153rd Meeting, American Chemical Society, Miami Beach, Florida, April 1967.
106. Anonymous, *Chem. Eng. News*, **45,** 52 (May 13, 1967).
107. J. M. Worlock and S. P. S. Porto, *Phys. Rev. Lett.*, **15,** 697 (1965).
108. P. J. Hendra and E. R. Lippincott, *Nature*, **212,** 1448 (1966).
109. J. E. Cahill, K. L. Treuil, R. E. Miller, and G. E. Leroi, *J. Chem. Phys.*, **47,** 3678 (1967).
110. T. Barrowcliffe, I. R. Beattie, P. Day, and K. Livingston, *J. Chem. Soc.* (*A*), **1967,** 1810.
111. J. E. Griffiths, *J. Chem. Phys.*, **49,** 642 (1968).
112. J. R. Durig, D. J. Antion, and F. G. Baglin, *ibid.*, **49,** 666 (1968).
113. H. J. Gerritsen, RCA Laser Research and Engineering, Radio Corporation of America, Camden, N.J., 1966, p. 40.
114. M. S. Houle and K. Grossaint, *Anal. Chem.*, **38,** 768 (1966).
115. H. P. Brandi, *Rev. Sci. Instr.*, **39,** 583 (1968).
116. T. Powell, *Proc. Roy. Soc. Med.*, **59,** 904 (1966).
117. C. B. Moore, *Appl. Opt.*, **4,** 252 (1965).
118. T. Powell, *Brit. J. Anesthesia*, **39,** 751 (1967).
119. R. L. Fork and C. K. N. Patel, *Proc. IEEE*, **52,** 208 (1964).

120. H. Brunet, *IEEE J. Quantum Elec.*, **QE-2,** 382 (1966).
121. B. F. Jacoby and R. K. Long, *Appl. Phys. Lett.*, **8,** 202 (1966).
122. K. Sakurai and K. Shimoda, *Japan. J. Appl. Phys.*, **5,** 938 (1966).
123. J. P. Gordon, *Bull. Am. Phys. Soc., Sec. II*, **9,** 501 (1964).
124. K. E. Rieckoff, *Appl. Phys. Lett.*, **9,** 87 (1966).
125. R. C. C. Leite, S. P. S. Porto, and T. C. Damen, *ibid.*, **10,** 100 (1967).
126. R. C. C. Leite, R. S. Moore, and J. R. Whinnery, *ibid.*, **5,** 141 (1964).
127. J. Debras-Guedon and N. Liodec, *Compt. Rend.*, **260B,** 3336 (1963).
128. E. F. Runge, F. R. Bryan, and R. W. Murck, *Can. Spectry.*, **80,** (1964).
129. S. D. Tasberry, B. F. Scribner, and M. Margoshes, *Appl. Opt.*, **6,** 81 (1967).
130. J. H. Burkhalter, *Fed. Proc.*, **24,** Suppl. 14: S-311 (1965).
131. R. C. Rosan, F. Brech, and D. Glick, *ibid.*, Suppl. 14: S-126 (1965).
132. S. D. Rasberry, B. F. Scribner, and M. Margoshes, *Appl. Opt.*, **6,** 87 (1967).
133. A. B. Whitehead and H. H. Heady, *Appl. Spectry.*, **22,** 7 (1968).
134. R. C. Rosan, in *Spectroscopic Tricks* (L. May, ed.), Plenum, New York, 1967, p. 17.
135. Jarrell Ash Co., Waltham, Mass., Bulletin 1-A, 1968.
136. E. S. Beatrice, I. Harding-Barlow, and D. Glick, 18th Meeting, The Histochemical Society, Chicago, Illinois, April 1967.
137. J. B. Ryan, C. B. Clark, and E. Ruh, 67th Meeting, American Ceramic Society, Phila., Pa., May 1956.
138. T. H. Malim, *Iron Age*, Feb. 4, 1965, p. 59.
139. H. Moenke and L. Moenke-Blankenburg, *Jena Rev.*, **11,** 166 (1966).
140. J. A. Maxwell, *Can. Mineralog.*, **7,** 727 (1963).
141. V. Vand, K. Vedam, and R. Stein, *J. Appl. Phys.*, **37,** 2551 (1966).
142. P. J. Wyatt, *Appl. Opt.*, **7,** 1879 (1968).
143. G. C. Sherman, F. S. Harris, Jr., and F. L. Mose, Jr., *ibid.*, **7,** 421 (1968), and included references.
144. C. A. Arguello, D. L. Rousseau, S. P. S. Porto, L. E. Cheesman, and J. F. Scott, *ibid.*, **7,** 1913 (1968).
145. J. Borch and R. H. Marchessault, *J. Colloid Interface Sci.*, **27,** 355 (1968).
146. L. D. Hocker, M. A. Kovacs, C. K. Rhodes, G. W. Flynn, and A. Javan, *Phys. Rev. Lett.*, **17,** 233 (1966).
147. C. B. Moore, R. E. Wood, B. L. Hu, and J. T. Yardley, *J. Chem. Phys.*, **46,** 4222 (1967).
148. J. T. Yardley and C. B. Moore, *ibid.*, **45,** 1066 (1966).
149. J. T. Yardley and C. B. Moore, *ibid.*, **46,** 4491 (1966).
150. H. Burnet, *Compt. Rend.*, **264B,** 1721 (1967).
151. R. D. Yates, Jr., G. W. Flynn, and A. M. Ronn, *J. Chem. Phys.*, **49,** 1432 (1968).
152. A. M. Ronn, *ibid.*, **48,** 511 (1968).
153. C. Borde, A. Henry, and L. Henry, *Compt. Rend.*, **262B,** 1389 (1966).
154. T. C. Damen, S. P. S. Porto, and F. Varsanyi, American Physical Society Meeting, Chicago, Illinois, March 1963.
155. P. A. Barnes and K. E. Rieckhoff, *ibid.*, Stanford, Calif., Dec. 1966.
156. J. Berkowitz and W. A. Chupka, *J. Chem. Phys.*, **48,** 5726 (1968).
157. N. C. Fenner and N. R. Daley, *Rev. Sci. Instr.*, **37,** 1068 (1966).
158. F. J. Vastola and A. J. Pirone, 151st Meeting, American Chemical Society, Pittsburgh, Pa., March 1966.

159. W. K. Jay, W. R. Ladner, and E. Pritchard, *Nature*, **217,** 640 (1968).
160. W. H. Giles and W. R. Brandkamp, *General Electric Technical Information Series*, Document No. 67SD 315, 1967.
161. E. Bernal, G. L. P. Levie, and J. F. Ready, *Rev. Sci. Instr.*, **37,** 398 (1966).
162. G. H. Megrue, *Science*, **157,** 1555 (1967).
163. H. G. Lipson and D. M. Clark, *Appl. Opt.*, **7,** 558 (1968).
164. D. L. Horrocks and M. H. Studier, *Anal. Chem.*, **13,** 1782 (1965).
165. J. R. Lalanne and J. J. Piand, *Compt. Rend.*, **264B,** 939 (1967).
166. D. J. Hunt, *J. Sci. Instr.*, **44,** 408 (1967).
167. D. L. Rousseau, *J. Chem. Educ.*, **43,** 566 (1966).

CHAPTER 3

ANALYTICAL METHODS AND TECHNIQUES FOR ACTINOMETRY

Hugh A. Taylor

CHEMSTRAND RESEARCH CENTER
DURHAM, NORTH CAROLINA

I. Introduction

One of the principal experimental problems facing the photochemist is that of measuring the intensity of a light source used in a photochemical reaction. Unless the investigation is of an extremely crude nature, the determination of this quantity is an unavoidable and somewhat bothersome task. One may recite numerous situations in which data of this type are essential. For example, investigations into the kinetics and mechanisms of photochemical reactions are based on such data. The determination of luminescence quantum yields would not be possible without these measurements. In addition to such situations requiring absolute measurements, relative intensity values are required for most photochemical work. Since light sources may fluctuate during operation and certainly deteriorate with age, a device is usually required to measure the relative if not the absolute amount of light emitted. Many investigations are dependent on a knowledge of the exact spectral distribution of a light source or light source/monochromator combination. Again the need for a device capable of light measurement is obvious. Because of the many applications for the measurement of light in photochemical investigations, it was thought essential to include a chapter in this text discussing the various methods of light measurement with particular emphasis on chemical actinometry.

A. *Devices for Measuring the Intensity of Light*

Instruments for the measurement of ultraviolet and visible radiation fall into three general categories: radiometric devices, photoelectric devices, and chemical systems. Each of these has its own advantages and disadvantages and may find usefulness in different applications. Let us now consider the nature of the devices that fall into each of these categories.

1. Radiometric Devices

Radiometric detectors are devices which convert light energy into heat and are capable of responding in some manner to the resultant temperature change. Included in this category are bolometers, thermistors, and thermopiles. Bolometers and thermistors are devices which show a change in resistance with a change of temperature. Thermopiles are basically a series of thermocouples situated behind a blackened receiver and operate by generating a potential corresponding to the heat produced by the absorbed radiation. Only this latter type has found widespread use in visible and ultraviolet photochemistry. Radiometric detectors are virtually the only devices available that possess a response which is independent of wavelength from the ultraviolet through the infrared. Thermopiles are by far the most popular type. For this reason they are considered to be the standards to which all other measuring devices are compared. Their operation is quite complex and will be discussed only briefly here.

Thermopiles are constructed such that each metal–metal junction behind the blackened receiver is matched by another junction which is shielded from the radiation. The potential thus generated by the "hot" junction is compared against the "cold" junction and fluctuations in ambient temperature are automatically compensated. The geometry of the blackened receiving surface may vary with the desired end use. Line thermopiles are used for radiation from monochromators while circular receivers are available for the measurement of larger radiation surfaces. In measuring the intensity of a beam of light, it is necessary to use a receiver that is larger in all dimensions than the spot of light to be measured. If this is not possible, an integrating procedure may be used wherein the thermopile is uniformly moved across the beam and the values measured at each position totaled. If the beam is diverging, gross errors may result with incorrect positioning of the receiver. In the event that the beam is very small as compared with the area of the receiver, errors may be introduced if the sensitivity of the receiver is not uniform across its surface. Although the sensitivity of a thermopile is ideally uniform at all wavelengths, selectivity has been reported. It is thus not recommended that thermopiles be used for sources having widely different spectral distributions than those used for calibration. Another possible source of error is in the system used to measure the potential. Since these potentials are quite small, extremely sensitive readout equipment is required. If measurements of the light entering a reaction cell are to be made, the absorbances and reflectances of all intervening surfaces must be accurately known.

Calibration of thermopiles is accomplished through the use of standard lamps. These may be obtained from the National Bureau of Standards or from the Eppley Laboratory in Newport, Rhode Island. The researcher need not perform his own calibrations since a wide variety of calibrated instruments is available from the Eppley Laboratory.

In spite of the many sources of error mentioned above, with painstaking care the thermopile is capable of extremely accurate measurements due largely to its uniform wavelength response. The interested reader is referred to Leighton and Leighton (*1*) or Calvert and Pitts (*2*) for a complete description of the use of thermopiles.

2. Photoelectric Devices

Photoelectric devices are basically of two types: phototubes and photovoltaic cells. A phototube makes use of the well-known photoelectric effect; a photovoltaic, or barrier-layer, cell consists of a strip of iron on which is deposited a semiconductor such as selenium, on top of which is placed a collector electrode. Light causes a current to flow from the semiconductor to the collector. This type of device is used in the photographic light meter. Neither of these photoelectric devices lends itself to quantitative measurements in photochemistry due to their extreme wavelength dependence. The particular response curve of a phototube depends on the cathode material. A variety of materials is available to extend the response from the near infrared through much of the vacuum ultraviolet. One of the principal advantages of the phototube lies in its sensitivity. Through the incorporation of dynodes to facilitate secondary emission of electrons, amplification factors of up to 10^6 are possible. These modified phototubes are, of course, the familiar photomultiplier tubes. Another inherent advantage of the phototube is its linear response to light intensity. For these latter two reasons phototubes have found widespread use in spectrophotometers. Their use in photochemistry has been very limited primarily because of their wavelength selectivity. The reader is referred to Koller (*3*) for a more complete description of photoelectric devices.

3. Chemical Systems

The third system which the photochemical investigator may choose for the measurement of light is the chemical actinometer. Such a system is simply a photochemical reaction possessing a known and accurately reproducible quantum yield. Ideally this quantum yield should be fairly constant over the useful wavelength range. Systems have been developed for wavelengths ranging from the visible spectrum through much of the vacuum ultraviolet. Most of the acceptable liquid systems, however, are only useful from 2000 to 5000 Å. The supreme advantage of this device over either the radiometric or photoelectric system is that of simplicity. Actinometers are usually prepared from easily available reagents and thus the only equipment required is that for analyzing the exposed solutions. This hardly ever amounts to more than conventional glassware or a spectrophotometer which may be found in most laboratories. Another distinct advantage is the freedom from many of the sources of error which might occur in thermopile operation. Yet another advantage is that measurement with an actinometer may be made in the same photolysis cell as that used for the reaction being studied. This completely eliminates the problems of reflectances, uneven light distribution, or divergence of the beam, which can play havoc with thermopile or photoelectric cell measurements. For these reasons it was thought to be desirable to devote the remainder of this chapter to the theory and use of actinometers in general and of several systems in particular. Emphasis will be placed on the measurement of light in the ultraviolet and violet portions of the visible spectrum. Systems for the vacuum ultraviolet will be mentioned only briefly. Before considering further the actual use of actinometers, a brief history of the early development of actinometry will be presented in order to place the subject in its proper perspective.

II. History of the Early Development of Chemical Actinometry

That light is an essential factor in certain naturally occurring processes such as the growth of plants and the bleaching of dyes was well known by the ancient Greeks. The nature of such reactions was open to much qualitative and ineffective speculation since the nature of the activating force was not understood. For many centuries inquiries into the mechanisms of photochemical reactions were based on the several theories of light which had been proposed by the ancient philosophers, such as the normative theory of Plato, the corpuscular theory of Pythagoras, and the undulatory theory of Aristotle.

The scientific investigations of the alchemists contributed little to the studies of photochemical reactions. Their writings contain references to the action of light but the connotations are certainly far too vague to be of use. The first empirical study took place at the end of the 17th century with the observations by Ray, the botanist, on the action of light in plant growth.

The fundamental basis for the quantitative understanding of photochemistry was laid in 1817 when Grotthus enunciated the fundamental law of photochemistry (*4*). As stated by him: "The substance which responds most markedly to a given colored light shows in its natural state the complementary color to this light." In other words, chemical change must be initiated by absorbed light. This law was rediscovered some 30 years later by Draper (*5–7*), and it is his name that is usually associated with it. Although the Grotthus–Draper law is purely qualitative in character, it represents the first stage in the development of photochemical theory and the understanding of light-initiated reactions.

During the next half-century the rapid development of photography tended to delay the scientific development of photochemistry. Early in the present century, however, the introduction of the quantum theory by Planck and its application, in 1912, by Einstein to photochemistry (*8*, *9*) created a renewed interest in the theoretical aspect of this science. Einstein's Law of Photochemical Equivalence has provided the fundamental basis for the mechanistic interpretation of photochemical reactions during the past 50 years.

The problem with which the photochemist is faced in the confirmation or application of the equivalence law is a determination of the absolute number of quanta being emitted by a light source during the course of a photochemical reaction. The nature of the task involves experimental difficulties which have led to many erroneous conclusions in photochemical studies. William Crookes (*10*) in 1868 expressed this problem well in his statement: "The measurement of the intensity of a ray of light is a problem the solution of which has been repeatedly attempted, but with less satisfactory results than the endeavors to measure the other radiant forces."

The first chemical actinometer was developed by the Swiss physicist Horace Bénédict de Saussure (*11*) in 1790 and entailed a measurement of the amount of oxygen liberated from a solution of chlorine water by the action of the sun. He noted that the rate of development of the gas was proportional to the intensity of the light and suggested its use as a photometer. The term actinometer originated some 40 years later with F. W. Herschel (*12*). The definition given by Herschel, however, referred to a system which would measure the heating effect of the sun's rays. In 1839 Becquerel developed a photochemical silver halide actinometer and spoke of it as a device for measuring the chemically active intensity of the sun's rays (*13*). Becquerel noted in his studies that the various portions of the sun's spectrum produced different effects on his actinometer.

The first gas-reaction actinometer was developed by J. W. Draper (*14*) in 1843 and consisted of a mixture of equal portions of hydrogen and chlorine gases. The hydrogen chloride evolved was absorbed and measured. He referred to this device as a tithonometer. This crude apparatus of Draper's was greatly modified about 20 years later by Bunsen and Roscoe (*15*). They determined the quantum efficiency against standard flames and defined the first set of photometric units on the basis of the change produced in their actinometer by these flames. Their studies were principally directed at measurement of the sun's intensity at various times of day, in different months of the year, and at different geographical locations.

The origin of solution actinometry may also be attributed to J. W. Draper (*16*), who, in 1857, suggested the use of an aqueous solution of peroxalate of iron which decomposed to carbon dioxide and a precipitate of the protoxalate of iron (i.e., ferrous oxalate) when exposed to light. He states that the ray which chiefly affects it is indigo. Several methods are suggested for the analysis of the actinometer, the method employed by Draper being the measurement of the gas evolved. A modification of this actinometer by J. C. Draper (*17*) in 1859 suggested the addition of gold chloride to the iron peroxalate solution. The precipitate of gold formed under the action of light was weighed. Other suggestions for the use of this actinometer were made by T. Woods (*18*) in 1860 and C. Lea (*19*) in 1867.

The investigation of the photochemical reactions of ferrioxalate solutions was not continued until 1895 when Lemoine (*20*) studied the photochemistry of mixtures of oxalic acid and ferric chloride. In 1928 Kornfeld (*21*) continued the study of this system, but later works by Allmand and Webb (*60*) in 1929 and Parker (*61*) in 1953 defined its quantitative usefulness. The results of these investigations will be discussed in some detail later.

Another system containing mercuric chloride and ammonium oxalate in aqueous solution (Eder's solution) appears in the early literature but was never very popular. A reference to its use can be found as recently as 1961, however (*22*). Much work in the early part of this century was concerned with uranyl-sensitized photodecomposition of oxalate. It has subsequently become one of the most popular systems for actinometry. It will also be discussed in some detail later.

To supplement this brief history, the reader is referred to Wightman's paper (*23*) which presents a very interesting history of actinometry.

III. General Considerations and Techniques for the Use of Actinometry

A. *Theoretical Considerations*

In the process of either developing a new actinometer system or choosing one of the available systems for photochemical work, a number of factors must be taken into consideration both for convenience and accuracy. For the sake of discussion these factors may be divided into two categories: those pertaining to the basic physics and chemistry of the system, and those relating to the physical manipulation or analysis of the system. Let us consider some of the factors in each group and their effects on the operation of the system.

The primary prerequisite for an actinometer is that it provide accurate results. Among the factors which will influence this accuracy are its range of absorption, its quantum efficiency, the presence or absence of an inner-filter effect, the temperature coefficient, adherence to the reciprocity law, and the amount of dark reaction. More specifically, an acceptable actinometer must possess an absorption and a constant quantum efficiency over a wide range of wavelengths. A system which has a narrow absorption range will not be sufficiently flexible for general use. If the quantum efficiency is not constant over the useful wavelength range, undue complications will arise in the measurement of polychromatic radiation. This again is the same drawback that was previously discussed with regard to photoelectric detectors. The inner-filter effect is a problem which has precluded the widespread use of many actinometers. It is the absorption of incident light by photo-products which in effect, causes a steady decrease in the quantum efficiency. Systems possessing such a deficiency should be avoided. The temperature coefficient and dark reaction are two factors which can be compensated for experimentally but merely inject unnecessary complications into an otherwise simple system. Actinometers are available which possess negligible dark reactions and temperature coefficients that are essentially unity within reasonable limits. The adherence of a system to the reciprocity law, which states that the product of time and intensity remain constant, is essentially the same as the requirement that the quantum efficiency be constant with intensity. An actinometer in which the product of time and intensity did not remain constant would be severely limited.

In addition to the basic properties which are directly related to the physics and chemistry of an actinometer system, the actual manipulation and operation must be considered. For the sake of convenience, it is to the operator's advantage to choose a system which makes use of easily available materials and which can be easily and precisely analyzed. Since systems are available which can be prepared from commercial reagents, there is no particular advantage in selecting one that requires extensive purification of reagents. On the contrary, if extensive purification is necessary, the possible effects of impurities on the accuracy of the system are too great and the system should be avoided. The sensitivity of an actinometer depends both on its quantum efficiency and the method of analysis. The latter factor is by far the most important since orders-of-magnitude differences in sensitivity can be achieved by proper selection of analysis procedures. Again, however, convenience must be considered. For example, titrimetric methods have simplicity and may possess the sensitivity of other methods. Their innate simplicity may be offset, however, by the difficulty of detecting a colored end point in a darkroom, where all actinometer analyses must be carried out. Extremely sensitive gas-chromatographic procedures have been developed, but maintaining such an instrument in a state of readiness may be inconvenient for some laboratories. Many spectrophotometric procedures have been developed. In general they possess both simplicity and sensitivity and require the use of instruments which are found in most laboratories.

Actinometers have been developed for all three phases of matter. Liquid systems are probably the easiest to use and are to be recommended for most work. However, the most convenient choice of a system is one which is in the same state as the reaction in conjunction with which it is being used. This facilitates the use of the same photolysis cells and avoids many of the problems inherent in the use of other measuring devices. In the choice of a system for actinometry, therefore, the investigator must weigh the various pros and cons of the many available systems and choose one which best fits his needs.

B. *Experimental Techniques*

Although the experimental techniques involved in the use of chemical actinometry are very simple when compared with radiometric or photoelectric devices, a number of precautions must be observed to ensure accurate results. No particular experimental setup is required since actinometers, when used as light monitors for a photochemical reaction, are exposed in the reaction vessel. For other purposes the size and shape of the vessel are dictated by the particular application. Details regarding photolysis vessels are presented in other chapters; several points are worthy of mention here, however.

It has not infrequently occurred that photochemical experiments have been carried out in a cell on one side of a light source while an actinometer is being exposed in another cell on the other side of the light. This is an extremely dangerous procedure since the output of a source is not always symmetrical about its axis. Multiple exposures with a line source can only be carried out by use of a rotating drum on which the samples or reaction cells are mounted. This is the procedure used in carbon and xenon arc fadeometers. Several procedures can be used to exactly duplicate the light falling on the reaction of interest. If short-term variations in the source are small, the actinometer may be exposed both before and after the other reaction. The intensity is then assumed to be linear between these two readings. If simultaneous monitoring is necessary, exposure may be carried out in a divided reaction cell, the reaction of interest on one side and the actinometer on the other. This procedure is poor if the light beam is not perfectly uniform. A better procedure is to make use of a beam splitter and two photolysis cells. In this case the relative amounts of light entering the two cells are determined by simultaneously exposing the actinometer in both cells.

One of the principal advantages of actinometers for monitoring is that the reflections and absorptions of intervening surfaces need not be determined. However, if the absolute intensity of a source must be known, then these factors must be measured and the intensities suitably corrected. Reflections from an optically perfect surface may be calculated from Fresnel's law. Values for some commonly occurring interfaces are given by Noyes and Leighton (*24*).

In using an actinometer for any reason, it is extremely important to determine the amount of light absorbed by the system. For simplicity it is advantageous to adjust the path length of the actinometer so that all of the light is absorbed. If this is not possible, corrections must be made for the transmitted light and the light reflected back from the back of the cell. It is also very important to adjust the concentration of the actinometer so that all of the light is not absorbed in a very short path length. If this occurs, there may be local depletion of the photoactive species at the front face of the cell and problems from the inner-filter effect may be encountered. If the absorption of a system is too high at a particular wavelength, dilution is recommended. To help avoid any problems from the inner-filter effect, stirring of the actinometer during exposure is always recommended wherever possible. This may be accomplished by bubbling a stream of nitrogen through the cell or by placing a magnetic stirring bar in the bottom of the cell and turning it with a magnetic stirring motor placed under the cell.

Most actinometers should be kept in the dark following exposure since many are sensitive to room light. Analysis may be carried out in a darkroom with either red or yellow safelight. The speed with which the analysis should be carried out depends entirely on the system. Some actinometers are stable for a considerable time following exposure while others suffer reversal in the dark or possess dark reactions which could lead to erroneous results.

Some of the many actinometer systems which have been reported in the literature are discussed in the following section. Several of the most common systems are presented in sufficient detail to facilitate their use by the interested investigator.

IV. Theory and Operation of Some Actinometer Systems

Of the many actinometer systems available for the measurement of light intensities, several liquid systems have been widely used. Five such systems are described in this section together with a discussion of the advantages and disadvantages of each.

A. *Malachite Green and Crystal Violet Leucocyanides*

A chemical actinometer making use of the coloration produced on irradiation of an alcoholic solution of crystal violet leucocyanide was first proposed by Weyde and Frankenburger (*25*) in 1930. The quantum efficiency was determined to be unity for wavelengths of 2537 and 3130 Å A paper in 1949 (*26*) described the use of this system for the measurement of ultraviolet light. Harris *et al.* (*27*, *28*) in 1935 suggested the use of malachite green leucocyanide as an alternative. Their work and that of Calvert and Rechen (*29*) demonstrated the advantages of the malachite green system. It is to be preferred over the former system for a number of reasons.

Irradiation of an ethanolic solution of malachite green leucocyanide causes the formation of a green color having a broad absorption in the visible region with a maximum around 6200 Å. One of the disadvantages of this system lies in the fact that the photo-product absorbs some of the ultraviolet light, which in very accurate work, therefore, necessitates a correction for this inner-filter effect. This correction is larger for the crystal violet than for the malachite green system. The region of applicability is limited to the region from 2480 to 3340 Å. Quantum efficiencies have been reported for several wavelengths in this range (*29*) but are essentially unity. Another advantage of the malachite green over the crystal violet system is the fact that the former is 30 times more soluble than the latter, thus permitting shorter cell lengths. The quantum efficiency has been demonstrated to be constant over a 10-fold range of concentrations and a 25-fold range of intensities. The temperature coefficient is unity at temperatures below 27°C. A rather serious disadvantage of this system lies in the fading of the color of the photo-compound in the dark. This is presumably due to hydrolysis and may be retarded by addition of hydrochloric acid.

The exact experimental details for the use of this actinometer are fairly straightforward. The preparation of a suitably pure sample of malachite green leucocyanide is described by Calvert and Rechen (*29*). Solutions of this material may be prepared for use in either 95% or absolute ethanol at a concentration of about 10^{-4} *M*. To each 3.8 ml of actinometer solution 0.02 ml of 0.3 *M* HCl should be added prior to irradiation. Following irradiation the absorption of the exposed solution is measured at 6200 Å using an unexposed solution as blank. The molar absorptivity of the photo-product is 9.49×10^4. Thus one of the principal advantages of this system is its sensitivity. Exposure time should be regulated such that the concentration of the photo-product is about 10^{-6} *M*. Although stirring may not be absolutely essential, it is recommended. Low quantum efficiencies may be obtained if the light intensity is too high. This is due to an increase in the concentration of the photo-product at the cell face giving rise to the inner-filter effect. When using this system the operator must always keep this problem in mind and take all precautions to avoid erroneous results which could ensue. It is the principal disadvantage of this system.

A variation of this system, making use of filter papers impregnated with the cyanides of malachite green or pararosaniline, has been developed by Chalkley (*30*). The dried paper is stable for over three years and produces a color when irradiated with light of wavelengths shorter than 3132–3342 Å.

These papers are sensitive down to 2144 Å. The reciprocity relation has been shown to hold. These papers were developed mainly for use in measuring sunlight at wavelengths below those which window glass transmits and are thus similar to the solution actinometers in their limited wavelength range.

The principal advantages of this system are its sensitivity and nondestructive analysis. The disadvantages of the inner-filter effect and limited wavelength range preclude its use for many applications where more versatile systems are available.

B. *The Monochloroacetic Acid Actinometer*

This system, whose quantum efficiency at 2537 Å was first incorrectly determined by Rudberg (*31*) in 1924, makes use of the photochemical hydrolysis of monochloroacetic acid by light of wavelengths lower than 3000 Å according to Eq. (1).

$$ClCH_2COOH + H_2O \rightarrow HOCH_2COOH + H^+ + Cl^- \qquad (1)$$

The value of unity obtained by Rudberg was in good agreement with the value of 0.9 as determined by Farkas (*32*). This latter value, however, was obtained in the 1850–2000-Å region by comparison with an incorrect quantum efficiency of 0.25 for the photolysis of ammonia. For some years, therefore, Rudberg's value was used by photochemists much to the detriment of the profession. A redetermination of the quantum efficiency by Harris and Kaminsky (*33*) in 1935, by comparison with the malachite green system, also agreed with earlier work, yielding a value of 1.07. It was not until the work of Smith, Leighton, and Leighton (*34*, *35*) in 1938–1939 and Thomas (*36*) in 1940 that the correct quantum efficiency was determined. They report values in the vicinity of 0.32 at 25°C. There remain inconsistencies in the literature, however, for Küchler and Pick (*37*) reported a value of 0.62 in 1940.

A full understanding of the operation of this actinometer and its associated problems is essential prior to its use. The work of Smith, Leighton, and Leighton (*34*, *35*) provides this information. Among the disadvantages of the monochloroacetic acid actinometer are its lack of versatility since it is only operative below 2537 Å, the associated dark reaction, and the high temperature coefficient. The dark reaction was shown to be 1% at 25°C and 57% at 69°C. The temperature coefficient causes a variation in the quantum efficiency from 0.31 at 25°C to 0.69 at 69°C.

In actual operation the monochloroacetic acid should be distilled so that a freshly prepared solution does not contain any free chloride ion. Solution concentrations of 0.5 to 2.0 *M* have been shown to be operable, and irradiation intensities of 7.6 to 97.6 ergs/mm^2 · sec have been used. The chloride produced may be analyzed by any conventional method. Smith, Leighton, and Leighton (*34*, *35*) titrated the chloride potentiometrically using 0.0075 *M* silver nitrate in a cell consisting of a calomel electrode connected to a AgCl–Ag electrode by a saturated potassium nitrate bridge (Cl^- free). Kemula and Grabowska (*38*) suggested that the polarographic determination of chloride would yield increased sensitivities and thus afford shorter exposure times. The anodic wave at 0.270 V versus the calomel electrode was used. Fitzgerald and Smith (*39*) have used a turbidimetric method for the determination of Cl^- in the presence of Ag^+. This latter method should provide a significant increase in sensitivity.

In view of the problems involved with this actinometer, it is not recommended for general use. Its low quantum efficiency, high temperature coefficient, dark reaction, and limited wavelength range are unnecessary complications which may easily be avoided by the choice of another system.

C. *Uranyl Oxalate Actinometer*

The photochemical oxidation of oxalate solutions containing uranyl ions as sensitizers was for years the standard actinometer in most photochemical studies. It maintained this position virtually unchallenged until Parker (*61–64*) developed the potassium ferrioxalate system. Since that time its popularity has declined somewhat, although it is still widely used. Early studies of the uranyl oxalate system by Boll (*40*), Büchi (*41*), Bowen and Watts (*42*), and Anderson and Robinson (*43*) yielded quantum efficiencies varying from 0.04 to 50. Quantitatively reproducible results were not available until 1930 with the classic work of Leighton and Forbes (*44*). They determined the following quantum efficiencies at 25°C for solutions 0.01 *M* in uranyl sulfate and 0.05 *M* in oxalic acid: 0.602 at 2537 Å; 0.582 at 2650 Å; 0.570 at 3000 Å; 0.561 at 3130 Å; 0.492 at 3660 Å; 0.563 at 4060 Å; and 0.584 at 4350 Å. At higher wavelengths the quantum efficiency drops sharply. A recent redetermination (*45*) obtained mean values of 0.492 and 0.573 at 3655 and 4358 Å, respectively. Brackett and Forbes (*46*) furthered the investigation using a zinc spark source and determined the following values: 0.48 at 2080 Å; 0.63 at 2530 Å; and 0.59 at 2780 Å. The adherence of this system to the reciprocity principle was demonstrated in these studies. Forbes and Heidt (*47*) showed that reagent purification was greatly simplified if uranyl oxalate was substituted for the sulfate and that the quantum efficiency remained constant for a tenfold dilution of the actinometer.

This reaction is not greatly affected by temperature. The 10° temperature coefficient from 10 to 25°C at 3660 and 3130 Å was shown by Leighton and Forbes (*44*) to be 1.03 ± 0.025 and by Norton (*48*) to be 1.03 ± 0.015 at 3130 Å between 5 and 58°C. The 10° temperature coefficient at 2537 Å was shown by Norton (*49*) to be 1.02 from 25 to 85°C for solutions 0.001 *M* in uranyl sulfate and 0.005 *M* in oxalic acid and decreasing to unity and below at lower concentrations.

The principal difficulty with this actinometer is that of sensitivity. Since the conventional analysis procedure calls for titration of the oxalate with standard permanganate, determination of the amount of oxalate decomposed on exposure to low-intensity sources is limited to a small difference in two large titers unless an inordinately long exposure is carried out. Several methods for overcoming this difficulty have been suggested. Pitts *et al.* (*50*) proposed the use of ceric sulfate instead of permanganate as standard oxidant. The former has a molar absorptivity of 5.41×10^3 at 3200 Å and may thus be determined spectrophotometrically with great sensitivity. Rogers (*51*) has suggested a modification of the analysis involving addition of a measured excess of permanganate followed by addition of KI and starch. The iodine liberated by the excess permanganate is thus titrated with sodium thiosulfate using the starch–iodine end point. Due to the intensity of this latter end point, much more dilute solutions can be used. Zill (*52*) has proposed the manometric determination of the CO and CO_2 produced from the decomposed oxalate as being more sensitive than the permanganate titration. Application of the uranyl oxalate actinometer to microchemistry was suggested by Porter and Volman (*53*) and Volman and Seed (*54*). They developed a gas-chromatographic technique using flame-ionization detection for analysis of the CO produced after hydrogenation to CH_4. Since the stoichiometry of the uranyl oxalate system does not exactly follow Eq. (2), but varies with acidity,

$$H_2C_2O_4 \rightarrow H_2O + CO_2 + CO \tag{2}$$

concentration, etc., it was necessary for them to determine the quantum efficiencies for CO production under the conditions of their method. They reported the following values at 30°C for solutions 0.01 *M* in uranyl oxalate and 0.05 *M* in oxalic acid: 0.33 at 2537 Å; 0.26 at 3660 Å; and

0.31 at 4350 Å. These values are independent of intensity, wavelength, and temperature, but are dependent on concentration and acidity. Their procedure utilizes 10-μl samples of actinometer and they thus claim a sensitivity 50–1000 times greater than that of other colorimetric or titrimetric procedures. The disadvantages in the equipment involved may be a drawback for some laboratories.

The applicability of this system to a wide variation in light intensities has been demonstrated by Christie and Porter (*55*). Their work with flash discharges has shown the quantum efficiency to be linear over a thousandfold variation in light intensities. The widespread popularity that the uranyl oxalate actinometer has gained over the years is certainly not without justification. For high-intensity sources such as the General Electric A-H6 high-pressure mercury lamp or a high-intensity xenon source, it may be quite useful as originally developed by Leighton and Forbes. For sources of lower intensity its usefulness in an unmodified form is seriously limited by the length of exposure required. For low-intensity work, either a modified version may be used or the potassium ferrioxalate system of Parker may be chosen. For those who may wish to use the uranyl oxalate system in its original form, the following detailed directions are presented.

Although the original investigations were carried out using uranyl sulfate and oxalic acid, the advantages of substituting uranyl oxalate for the uranyl sulfate have been demonstrated. Uranyl oxalate may be prepared by cooling a hot solution of either uranyl nitrate or uranyl sulfate and oxalic acid. The resulting crystals of $UO_2C_2O_4 \cdot H_2O$ are washed and dried in air at 110°C. The actinometer solutions to be used may be made up at various concentrations depending on the wavelength of light to be measured. The relative concentrations must always be adjusted so as to maintain a 5:1 ratio of oxalic acid to uranyl oxalate. The concentrations should be chosen so that all of the incident light is absorbed, but must be sufficiently dilute to avoid total absorption in a very thin layer of solution. The actinometer should be stirred during exposure to avoid problems from the inner-filter effect. Much of the past work has been carried out using solutions 0.01 *M* in uranyl oxalate and 0.05 *M* in oxalic acid. Such a solution is prepared by dissolving 3.761 g of uranyl oxalate and 6.302 g of oxalic acid dihydrate in sufficient water to make 1 liter. At wavelengths of 3130 Å and below, a tenfold dilution of this concentration is recommended. It is possible using uranyl oxalate to decompose up to 75% of the total oxalic acid, but a lower extent of decomposition is recommended.

The actinometer solutions, both before and after exposure, are titrated in sulfuric acid solution with standardized potassium permanganate. The procedure may be found in any analytical chemistry text. Of course the actinometer solutions must be kept in the dark at all times. Titrations should also be carried out in the dark as far as is practicable with a safelight being used for all but the final end-point detection.

D. *Vanadium(V), Iron(III) Oxalate Actinometer*

One of the newest and least-studied systems for chemical actinometry is that using a mixture of the oxalates of iron(III) and vanadium(V). Its operation is based on the photochemical reduction of ferric oxalate to produce ferrous ions which in turn reduce the vanadium from the 5+ to the 4+ state and are reoxidized to the ferric state. Since this latter reaction is essentially instantaneous, the concentration of the photoactive species remains constant until almost all of the vanadium(V) is consumed. The presence of the vanadium(V) is merely as an internal indicator. The vanadium(IV) produced is blue and can be analyzed spectrophotometrically. This system, therefore, possesses two advantages over many others. Its ease of analysis can be matched only by malachite green leucocyanide, and the constancy of the photoactive species is without equal in other systems.

Unfortunately the two tremendous advantages offered by this system are somewhat offset by its inherent disadvantages. The most notable of these is the dark reaction. Vanadium(V) reacts in the dark with oxalate to produce vanadium(IV). This reaction has been studied in some detail (*56*). It is not of sufficient magnitude to present problems at room temperature but elevated temperatures should be avoided. In any case compensation for the dark reaction must be made. This dark reaction also precludes the possibility of storing the prepared actinometer for more than a day or two. However, the actinometer is easily prepared from stock solutions and need not be kept for any great length of time. Another disadvantage in some situations is its lack of sensitivity. Since the molar absorptivity of the vanadium(IV) in this medium is only 22.3, the overall sensitivity is of the same order of magnitude as that of the unmodified uranyl oxalate system and is considerably less than that of the malachite green or potassium ferrioxalate system.

Early studies of the photochemistry of this system were carried out by Bricker and Schonberg (*57*). Quantum efficiencies were measured by Van Tassel (*58*) and Taylor (*59*) using comparison techniques against uranyl oxalate and potassium ferrioxalate as standard actinometers. Values were determined at the major emission lines of mercury using filters and filter solutions for isolation of the desired wavelengths. Satisfactory values were obtained at 5460, 4350, 4040, and 3130 Å; they are 0.11, 1.01, 1.00, and 0.91, respectively. At 3660 Å, however, a discrepancy in the values was reported depending on which standard actinometer was used for comparison. The two values obtained were 1.21 and 1.02. Although no explanation for this discrepancy is available, an average value of 1.12 would suffice for semiquantitative work. The studies at 2537 Å are inadequate and are unworthy of further discussion. This work was carried out at 25 and 5°C. No temperature effect was noted at any wavelength but 5460 Å. At this wavelength the temperature coefficient was found to be 4.5% per degree. For this reason and due to the low quantum efficiency and low molar absorptivity, this actinometer is not recommended for use at 5460 Å.

In summary, therefore, the principal advantages of this system are its ease of analysis, the nondestructive nature of the analysis, a quantum efficiency that is almost linear with wavelength, the high molar absorptivity throughout the ultraviolet and into the visible, the constancy of the photoactive species, and the negligible temperature coefficient at all wavelengths but 5460 Å. Its disadvantages are the dark reaction and, for some applications, the low sensitivity.

The reagent solutions required are the following: 0.14 *N* solution of sodium vanadate, 0.5 *N* solution of ferric ammonium sulfate, 0.074 *M* solution of sodium oxalate, and 6 *N* sulfuric acid. The reader is referred to Bricker and Schonberg (*57*) or Taylor (*59*) for the exact procedures for preparing these reagents. The actinometer solution is prepared by addition of 75 ml of the sodium oxalate solution, 10 ml of the sodium vanadate solution, 5 ml of the iron(III) solution, and 8 ml of 6 *N* sulfuric acid to sufficient water to make 100 ml.

Determination of vanadium(IV) in the exposed actinometer is carried out spectrophotometrically at 7500 Å. A sample of solution is removed from the photolysis cell with a clean, dry pipette and placed in a clean, dry spectrophotometer cuvette. The absorbance is measured versus an unexposed sample which is stored in the dark at the same temperature as the exposed solution. Thus any dark reaction will be automatically compensated. Following analysis the exposed solution may be returned to the photolysis cell for further exposure. With careful technique the volume of solution lost by this procedure is less than 0.5 ml. If the volume of the photolysis cell is sufficiently large, this loss will cause a minimal error. If small-scale work is being done, exposure directly in the photometric cuvette is possible.

E. *Potassium Ferrioxalate Actinometer*

Of the many liquid actinometers available for light measurement, one of the most studied systems involves the photochemical reduction of iron(III) in acidic oxalate solutions. In spite of the many investigations of this system dating back to the 19th century, no quantitative data were reported until the work of Allmand and Webb (*60*) in 1929. Prior to that time the system was plagued by the inevitable precipitation of the ferrous oxalate produced and by poor methods of analysis of the exposed actinometer. Allmand and Webb overcame the former of these problems by the use of excess oxalate, and the latter by titration of the iron(III) with a hydrochloric acid solution of titanous chloride using thiocyanate as indicator. However, the results were still poor, the reported quantum efficiencies varying with changes in the conditions of exposure. The position of prominence which the system occupies in present-day photochemistry is almost entirely due to the investigations of Parker and co-workers (*61–64*) in the period 1953 to 1959. The success of their studies is due largely to their use of orthophenanthroline as an analytical reagent for measurement of the ferrous produced. Due to the large molar absorptivity of the ferrous orthophenanthroline complex at 5100 Å (i.e., 1.11×10^4), this actinometer is extremely sensitive and is admirably suited to the measurement of low light intensities. Its use is not limited to low intensities, however, for the quantum efficiencies remain constant from 5×10^{-11} to 2×10^{-4} Einsteins $cm^{-2}\ sec^{-1}$. The operator may experience some difficulty at high intensities in exposing the actinometer for a sufficiently short period. Other advantages to be noted are the low temperature coefficient and the fairly uniform quantum efficiency over a wide range of wavelengths extending throughout the ultraviolet and well into the visible. It is also notable for the lack of any inner-filter effect or dark reaction.

The recommended quantum efficiencies for this system at 22°C are presented in Table 1. The temperature coefficients vary from -2 to $+5\%$ per 10°C at all wavelengths but 5090 Å where its value is $+16\%$ per 10°C. Therefore the effect of temperature is minimal and accurate control is unnecessary for all but the most critical work. The exact values of the temperature coefficient may be found in the original paper by Hatchard and Parker (*63*). Due to the widespread use of this system, the experimental procedure will be presented in detail.

The potassium ferrioxalate solutions required for this actinometer may be prepared either from the pure solid as recommended by Parker or from ferric alum and potassium oxalate. The validity of this latter technique was confirmed by the work of Baxendale and Bridge (*65*). There do not appear to be any advantages to the use of either system and the choice is left up to the discretion of the individual. The former procedure will be

TABLE 1
QUANTUM EFFICIENCIES OF THE POTASSIUM FERRIOXALATE ACTINOMETER AT 22 C[a]

Wavelength, Å	Concentration, *M*	Accepted value of Φ
5770/5790	0.15	0.013
5460	0.15	0.15
5090	0.15	0.86
4800	0.15	0.94
4680	0.15	0.93
4360	0.15	1.01
4360	0.006	1.11
4050	0.006	1.14
3650/3660	0.15	1.18
3650/3660	0.006	1.22
3340	0.006	1.23
3130	0.006	1.24
2970/3020	0.006	1.24
2540	0.006	1.25

[a] From Parker (*66*).

described here. Pure potassium ferrioxalate is prepared by mixing 3 vol of 1.5 *M* potassium oxalate and 1 vol of 1.5 *M* ferric chloride with vigorous stirring. The precipitated $K_3Fe(C_2O_4)_3 \cdot 3H_2O$ is then recrystallized three times from warm water and dried with a stream of air at 45°C. This procedure should be carried out in a darkroom illuminated only with a safelight. The reagent must be stored in the dark at all times since exposure to light will cause decomposition. The 0.006 *M* actinometer is prepared by adding 2.947 g of the crystals and 100 ml of 1.0 *N* sulfuric acid to sufficient water to make 1 liter. This concentration should be adequate for most work at wavelengths below 4000 Å. At higher wavelengths a 0.15 *M* solution is recommended. This is prepared in a similar manner using 73.68 g of the pure crystals. In any case it is recommended that a sufficient depth of solution be used to absorb all of the incident light. The reagents required for analysis of this actinometer are a 0.1% solution of *o*-phenanthroline monohydrate in water and a buffer solution prepared by mixing 600 ml of 1 *N* sodium acetate and 360 ml of 1 *N* sulfuric acid with sufficient water to make 1 liter. For the calibration, a standardized 0.1 *N* solution of ferrous sulfate in 0.1 *N* sulfuric acid is required. Alternatively, a 0.1 *N* solution of ferrous ethylenediamine sulfate in 0.1 *N* sulfuric acid may be used. (The author recommends this latter reagent which can be obtained in pure form from G. F. Smith Chemical Co., Columbus, Ohio. It has the advantages of purity, ready solubility, and a high molecular weight which facilitates weighing.) The 0.1 *N* solution is diluted with 0.1 *N* sulfuric acid to give a solution containing 0.4×10^{-6} mole/ml of ferrous ion.

Calibration is accomplished by adding the following to a series of 25-ml volumetric flasks: 0.0 to 5.0 ml of standard ferrous solution, sufficient 0.1 *N* acid to make the total volume 10.0 ml, 2 ml of the *o*-phenanthroline solution, and 5 ml of buffer. The flasks are then diluted to volume, mixed, and allowed to stand for at least 1 hr. The absorbance is then measured in 1-cm cells at 5100 Å. A plot of the absorbance versus concentration of ferrous ion should give a straight line and the slope should provide a molar absorptivity of about 1.11×10^4 liters/mole · cm.

The actinometer is analyzed in a similar manner. Exposure should be regulated such that the concentration of ferrous ion produced is about 0.2×10^{-6} mole/ml. A 5-ml aliquot of the exposed actinometer may then be combined with the other reagents as above in a 25-ml volumetric flask and analyzed spectrophotometrically after 1 hr in the dark. A similarly prepared unirradiated solution is used as a blank. The length of the development period may be decreased to 30 min according to Baxendale and Bridge (*65*) if 1 ml of 2 *M* ammonium fluoride is added. The developed solution is stable in the dark for several hours.

Stirring the solution is not absolutely essential for all work but is always recommended where possible. If it is to be avoided, the cell path length must be sufficiently long to afford a complete absorption of all the incident light.

F. *Other Chemical Actinometers*

In addition to the five actinometer systems already discussed, many others have been proposed and are frequently used in photochemical studies. The actinometers presented in this section will be divided into liquid, solid, and gaseous systems. None of these will be covered in any detail since they are generally less versatile than those already discussed.

The purpose of this section is to acquaint the reader with some different types of photochemical light-measuring devices which might suit a particular need better than those previously mentioned.

A liquid actinometer that has been used in large part for the measurement of the ultraviolet component of sunlight is the acetone–methylene-blue system (*67–69*). Light of wavelengths shorter than 3200 Å causes a bleaching of the color of a solution of methylene blue in acetone. Unfortunately this change is reversed in the dark and it is absolutely essential to exclude air in order to obtain reproducible results. Another system has been reported in which the fluorescence created by irradiation of aqueous benzoic acid solutions with ultraviolet light below 3000 Å is measured (*70*). A number of compounds may be added to enhance the fluorescence response and thus the sensitivity. Few actinometers are available for use primarily with visible light. One such system is a manometric actinometer for the region 4200–6700 Å (*71*) employing a pyridine solution of thiourea and ethylchlorophyllide for red, yellow, or blue light and a pyridine solution of protoporphyrin for green or blue light. Quantum yields of the order of 0.98 based on oxygen uptake have been measured. Another chemical actinometer making use of manganese-(III) oxalate has been suggested (*72*). Since 1 mole of CO_2 is evolved for each Mn^{3+} destroyed, it is possible either to titrate the Mn^{3+} iodimetrically or to measure the CO_2 given off. A radiometric technique has been developed employing the photochemical exchange between methyl iodide and iodine labeled with ^{131}I at 2537 Å (*73*). The quantum yield of the iodine exchange was shown to be unity.

The field of solid-state actinometry has not been exploited fully in the past except for a few isolated cases. Dyed papers using malachite green or pararosaniline have already been mentioned and the use of blue-dyed papers and cloths have been used for some time to determine light fluxes in carbon arc fadeometers. A rather novel solid actinometer has recently been proposed (*74*) making use of a KBr disc of *o*-nitrobenzaldehyde. The extent of reaction was measured by following the disappearance of the nitro group by infrared spectroscopy. The procedure was checked by preparing a KBr disc of potassium ferrioxalate and following the growth of an infrared band at 2350 cm^{-1} due to the formation of CO_2 in situ. A more recent paper (*75*) suggests the use of *o*-nitrobenzaldehyde in poly(methyl methacrylate) as an actinometer for the region 2800–4100 Å. The potential for further work in this field would seem to be virtually unlimited. The need for such actinometers especially in the field of polymer degradation is very great indeed.

Gaseous actinometers have been available practically from the inception of actinometric investigations. In general these systems lack the wavelength versatility of many of the better liquid actinometers. Most gaseous actinometers are useful only in the far and vacuum ultraviolet. Since gaseous systems are most convenient to use in conjunction with gas-phase photochemical studies, some of the available systems will be briefly mentioned. The interested reader is referred to Calvert and Pitts (*76*) for a more complete review of some of these systems.

Two of the more common gaseous actinometers for use in the near ultraviolet are those using the photolysis of acetone and diethyl ketone. They are both applicable to the 2500–3200-Å region and their quantum efficiencies with respect to CO formation are near unity. Both systems require high degrees of purification. The photolysis of hydrogen bromide has been suggested for the region 1800–2500 Å. Some of the systems that have been found to be useful in the vacuum ultraviolet are gaseous oxygen (1300–1900 Å), nitrous oxide (1470–1849 Å), and carbon dioxide (below 1700 Å). Except in the case of studies involving gas-phase photochemistry, the use of gaseous actinometers is not recommended. There are certainly no analytical advantages for this type of system and, in general, the manipulative problems can be avoided by the use of a suitable liquid or solid system.

REFERENCES

1. P. A. Leighton and W. G. Leighton, *J. Phys. Chem.*, **36**, 1882 (1932).
2. J. G. Calvert and J. N. Pitts, Jr., *Photochemistry*, Wiley, New York, 1967, pp. 771–778.
3. L. R. Koller, *Ultraviolet Radiation*, Wiley, New York, 1965, pp. 272–299.
4. Abhandlung über Electrizität und Licht, Ostwald's Klassiker, No. 152, p. 101.
5. *Phil. Mag.*, **19**(3), 195 (1841).
6. *Phil. Mag.*, **23**, 401 (1843).
7. *Phil. Mag.*, **27**, 327 (1845).
8. *Ann. Phys.*, **37**(4), 832 (1912).
9. *Ann. Phys.*, **38**, 881 (1912).
10. W. Crookes, *Brit. J. Photogr.*, **15**, 402 (1868).
11. H. B. de Saussure, *Mémoires de l-Academie de Turin* **IV**, 44 (1790).
12. J. F. W. Herschel, *Report of the Third Meeting of the British Association for the Advancement of Science*, *Cambridge*, John Murray, London, 1834, pp. 379–381.
13. E. Becquerel, *Compt. Rend.*, **9**, 145 (1839).
14. J. W. Draper, *Phil. Mag.*, **23**(3), 401 (1843).
15. R. W. Bunsen and H. E. Roscoe, *Phil. Trans. Roy. Soc. London*, **147**, 355 (1852).
16. J. W. Draper, *Phil. Mag.* (4), **14**, 161 (1857).
17. J. C. Draper, *Photogr. J.*, **6**, 24 (1859).

18. T. Woods, *ibid.*, **6,** 135 (1860).
19. M. C. Lea, *Brit. J. Photogr.*, **14,** 464 (1867).
20. K. Lemoine, *Ann. Chim.-Phys.*, **6,** 433 (1895).
21. G. Kornfeld, *Z. Elektrochem.*, **34,** 598 (1928).
22. M. Frank, *Arch. Physik. Therapie*, **13,** 119 (1961).
23. E. P. Wightman, *Photogr. Sci. Eng.*, **3,** 64 (1959).
24. W. A. Noyes, Jr., and P. A. Leighton, *The Photochemistry of Gases*, Reinhold, New York, 1941, p. 84.
25. E. Weyde and W. Frankenburger, *Trans. Faraday Soc.*, **27,** 561 (1931).
26. Y. Miyake, *Bull. Chem. Soc. Japan*, **22,** 105 (1949).
27. L. Harris, J. Kaminsky, and R. G. Simard, *J. Am. Chem. Soc.*, **57,** 1151 (1935).
28. L. Harris and J. Kaminsky, *ibid.*, **57,** 1154 (1935).
29. J. G. Calvert and J. L. Rechen, *ibid.*, **74,** 2101 (1952).
30. L. Chalkley, *J. Opt. Soc. Am.*, **42,** 387 (1952).
31. E. Rudberg, *Z. Physik.*, **24,** 247 (1924).
32. L. Farkas, *Z. Physik. Chem.*, **B23,** 89 (1923).
33. L. Harris and J. Kaminsky, *J. Am. Chem. Soc.*, **57,** 1154 (1935); quoted elsewhere.
34. W. G. Leighton, R. N. Smith, and P. A. Leighton, *ibid.*, **60,** 2566 (1938).
35. R. N. Smith, P. A. Leighton, and W. G. Leighton, *ibid.*, **61,** 2299 (1939).
36. L. B. Thomas, *ibid.*, **62,** 1879 (1940).
37. L. Küchler and H. Pick, *Z. Physik. Chem.*, **B45,** 116 (1940).
38. W. Kemula and A. Grabowska, *Rocz. Chem.*, **29,** 834 (1955).
39. J. M. Fitzgerald and E. J. Smith, Abstracts, 154th National Meeting of American Chemical Society, Chicago, Illinois, Sept. 1967, No. B46.
40. M. Boll, *Compt. Rend.*, **156,** 1891 (1913).
41. P. F. Büchi, *Z. Physik. Chem.*, **111,** 269 (1924).
42. E. J. Bowen and H. G. Watts, *J. Chem. Soc.*, 1607 (1926).
43. W. T. Anderson and F. W. Robinson, *J. Am. Chem. Soc.*, **47,** 718 (1925).
44. W. G. Leighton and G. S. Forbes, *ibid.*, **52,** 3139 (1930).
45. C. A. Discher, P. F. Smith, I. Lippman, and R. Turse, *J. Phys. Chem.*, **67,** 2501 (1963).
46. F. P. Brackett, Jr., and G. S. Forbes, *J. Am. Chem. Soc.*, **55,** 4459 (1933).
47. G. S. Forbes and L. J. Heidt, *ibid.*, **56,** 2363 (1934).
48. B. M. Norton, *ibid.*, **56,** 2294 (1936).
49. B. M. Norton, *Ohio J. Sci.*, **66,** 421 (1966).
50. J. N. Pitts, Jr., J. D. Margerum, R. P. Taylor, and W. Brim, *J. Am. Chem. Soc.*, **77,** 5499 (1955).
51. G. T. Rogers, *Chem. Ind.*, 572 (1956).
52. L. P. Zill, *Arch. Biochem. Biophys.*, **59,** 473 (1955).
53. K. Porter and D. H. Volman, *J. Am. Chem. Soc.*, **84,** 2011 (1962).
54. D. H. Volman and J. R. Seed, *ibid.*, **86,** 5095 (1964).
55. M. I. Christie and G. Porter, *Proc. Roy. Soc.* (*London*), **A212,** 390 (1952).
56. J. R. Jones and W. A. Waters, *J. Chem. Soc.*, 4757 (1961).
57. C. E. Bricker and S. S. Schonberg, *Anal. Chem.*, **30,** 922 (1958).
58. J. H. Van Tassel, *Princeton University Technical Report*, 1961.
59. H. A. Taylor, Doctoral Dissertation, Princeton Univ., Princeton, N.J., 1963.
60. A. J. Allmand and W. W. Webb, *J. Chem. Soc.*, 1518 (1929).
61. C. A. Parker, *Proc. Roy. Soc.* (*London*), **A220,** 104 (1953).

62. C. A. Parker, *Trans. Faraday Soc.*, **50,** 1213 (1954).
63. C. G. Hatchard and C. A. Parker, *Proc. Roy. Soc.* (*London*), **A235,** 518 (1956).
64. C. A. Parker and C. G. Hatchard, *J. Phys. Chem.*, **63,** 22 (1959).
65. J. H. Baxendale and N. K. Bridge, *ibid.*, **59,** 783 (1955).
66. C. A. Parker, *Photoluminescence of Solutions*, American Elsevier, New York, 1968, p. 210.
67. T. A. Webster, L. Hill, and A. Eidinow, *Brit. J. Radiol.*, B.I.R. Section **29,** 335 (1924).
68. T. A. Webster, L. Hill, and A. Eidinow, *Lancet*, **I,** 745 (1924).
69. T. W. B. Osborn and A. D. Stammers, *Brit. J. Actinotherapy*, **5,** 68 (1930).
70. H. Moroson and A. Gregoriades, *Nature*, **204,** 676 (1964).
71. O. Warburg and V. Schocken, *Arch. Biochem.*, **21,** 363 (1949).
72. B. Rackow, *Z. Chem.*, **5,** 397 (1965).
73. G. M. Harris and J. E. Willard, *J. Am. Chem. Soc.*, **76,** 4678 (1954).
74. J. N. Pitts, Jr., J. K. S. Wan, and E. A. Schuck, *ibid.*, **86,** 3606 (1964).
75. G. A. Cowell and J. N. Pitts, Jr., *ibid.*, **90,** 1106 (1968).
76. J. G. Calvert and J. N. Pitts, Jr., *Photochemistry*, Wiley, New York, 1967, pp. 782–783.

Part II

USE OF PHOTOCHEMISTRY IN ANALYTICAL DETERMINATIONS

CHAPTER 4

ANALYTICAL METHODS EMPLOYING PHOTOCHEMICAL PRETREATMENT

William M. Riggs

E. I. DU PONT DE NEMOURS AND COMPANY
CENTRAL RESEARCH DEPARTMENT
WILMINGTON, DELAWARE

I. Introduction

Many of our traditional methods of chemical analysis depend upon a chemical reaction which takes place before the performance of the actual analytical measurement. For example, gravimetric determinations depend upon precipitation reactions, oxidimetric titrations often depend upon prior reduction reactions, and colorimetric determinations usually depend upon a color-forming reaction of some kind. This chapter is devoted to the description and illustration of the ways in which photochemical reactions may be involved in that key pretreatment step which allows a precise and useful analytical measurement to be made.

II. General Principles

A. *Types of Photochemical Pretreatment*

In many cases it is possible to utilize a photochemical reaction of a chemical species to convert it to another species which can be easily measured. An example is the photodecomposition of ferrioxalate ions in aqueous solution which has been used for the determination of both iron (*1*) and oxalate (*2*, *3*). With this reaction iron(III) is photoreduced to iron(II), which is easily determined either titrimetrically or colorimetrically. Since the reaction is stoichiometric it can be used for the analysis of either iron(III) or oxalate.

In other cases a photochemical reaction may be used to generate a reagent in situ. For example, thorium has been determined gravimetrically as the iodate when it was quantitatively precipitated from solution by photochemical conversion of periodate to iodate (*4*). Closely related to in situ reagent generation is the idea of photochemically removing a masking agent from a precipitant, allowing it to precipitate the ion of interest. The advantage of this approach over the simple addition of the reagent is that the slow release of the reagent in solution may promote the formation of large, pure, and easily filtered crystals of precipitate. This method has been used for the photochemical precipitation of tantalum selenite (*5*).

Another interesting approach to analysis utilizing photochemical reactions is the so-called photokinetic method (*6*). This method takes advantage of the fact that the photobleaching of some dyes is catalyzed by the presence of certain metal ions. The mechanism involves the transfer of an electron from a photoexcited dye molecule to the metal ion which in turn reduces oxygen or some other convenient oxidant. The net effect is the oxidation and bleaching of the dye. The rate of the reaction is proportional to the metal-ion concentration if the experimental conditions are properly controlled. Thus, measurement of the rate of bleaching, or equivalently, the *extent* of bleaching for a fixed irradiation time and intensity, can be used to determine the metal-ion concentration.

Applications of photochemical reactions to instrumental techniques constitute another important aspect of photochemical pretreatment. Photoionization with high-energy ultraviolet radiation has been used to ionize molecules for mass-spectrometric analysis. Photoionization has also been applied as the basis for detectors in gas chromatography.

Molecular-structure characterization is another area of application of photochemical pretreatment. Organic compounds with similar structural features give similar reaction products when they undergo mercury-sensitized photolysis. By combining photolysis with subsequent gas-chromatographic analysis of the reaction products, organic molecular structures may be characterized.

Detailed examples of the various types of photochemical pretreatment, including those briefly mentioned above, are discussed later along with other applications.

B. *Types of Photochemical Reaction*

Photochemical reactions may be conveniently classified in four categories: photooxidation–reduction, photoisomerization, photosubstitution, and photoaddition reactions. Photooxidation–reduction reactions are by far the most important in photoanalytical applications, because oxidation or reduction of a compound frequently brings about a dramatic change in its chemical and physical properties. This change provides the "handle" which the analytical chemist needs to make his measurement. Photoisomerization reactions may also provide the needed "handle" if the product of the reaction has spectral features or chemical reactivity differences from the reactant which allow a measurement to be made. Likewise, photosubstitution reactions may bring about changes which will be useful for an analytical measurement. For example, the spectral changes brought about by photoaquation of some complex metal ions in aqueous solution (*7*) might be measured and related to the amount of the original complex present. Alternatively, the species *released* in the substitution may be readily measurable. Photoaddition reactions may also result in physical and chemical changes in the parent molecule which will be of potential analytical use, or it may be possible to measure the amount of the added species which is taken up. Thus it is clear that the analytical chemist has a number of possible ways to approach the design of an analysis utilizing a photochemical reaction. The key requirement for such a reaction is that some change be brought about, the results of which can be readily measured.

C. *Absorption Spectra*

Since light must be absorbed before a photochemical reaction can occur, knowledge of the absorption characteristics of a compound is important. Thus, an absorption spectrum should be obtained as the first step in a photochemical study. Examination of the spectrum shows (a) what wavelengths of light will be efficiently absorbed, (b) how efficiently any given wavelength will be absorbed, and (c) to some extent, what type of photochemistry one may expect from excitation at a given wavelength. The first point is important since, obviously, light will be absorbed only in spectral regions where the compound exhibits a finite absorbance. The molar absorptivity, ϵ, at any point in the spectrum can be used to calculate, using the Beer–Lambert absorption law, the depth of penetration of light of a given wavelength into a solution of known concentration (*8*, *9*). This is important in the design of reaction vessels and the selection of concentration ranges in which to work.

Of the kinds of information we wish to extract from an absorption spectrum, predictions about the expected photochemistry of the compound are the most difficult to obtain, and the most subject to error. Because of this, analytical chemists will generally utilize well-known photochemical reactions in the design of photoanalytical methods. Nonetheless, it is desirable to have a rudimentary knowledge of spectroscopic and photochemical principles so that the number of surprises encountered may be minimized. The reader is referred to standard textbooks on photochemistry (*8*) and to reviews (*10*) on the subject for this sort of background information.

D. *Characteristics of Photoanalytical Reactions*

A major requirement for a photochemical reaction which is to be used in an analytical procedure is that it be fairly efficient, i.e., its quantum yield should be high. If this is not true, it will take an excessively long time to bring the required photochemical change to completion. Secondly, it is highly desirable that the reaction be stoichiometric, as with all reactions used in analysis. If the reaction is not stoichiometric, empirical calibration is required, and conditions must be controlled carefully to assure that the degree of nonstoichiometry remains constant. Third, it is best if the detailed mechanism of the photoreaction is known. This understanding allows the chemist to control experimental conditions to the best advantage. Knowledge of the wavelength dependence of the reaction is also important. The quantum yield of the reaction nearly always depends on the wavelength, and proper selection of the wavelength will allow maximum efficiency to be achieved and possibly avoid interfering reactions. The temperature dependence should also be known. Fortunately, many photochemical reactions have a negligible temperature dependence near room temperature, a desirable situation for analytical work since this makes it unnecessary to provide for temperature control.

E. *Measurement of Reaction Product*

The analytical chemist who wishes to utilize photochemical reactions has at his disposal all the techniques of analytical chemistry for the measurement of reaction products. Among the techniques more commonly used in photoanalytical procedures are redox titrations and spectrophotometric determinations since these are easy to use and widely applicable. Other methods which have been less commonly used include measurement of the volume of an evolved gas and determination of the weight of a precipitated solid. The ultimate range of methods available for this phase of a photochemical analysis is limited only by the imagination and ingenuity applied to the problem.

III. Methods and Applications

A. *Experimental Techniques*

1. Lamps

There is a variety of types of lamps available for photochemical use (*8*). Generally the most useful of these are the mercury vapor lamps, of which there are three kinds: low, medium, and high pressure. These lamps are discussed in Chapter 1 of this book.

Other types of lamps such as resonance lamps utilizing metals other than mercury, and inert-gas resonance lamps are available for photochemical studies. Fortunately, the low- and medium-pressure mercury vapor lamps are readily available, easy to use, and they satisfy the great majority of the needs of the photoanalytical chemist. These are generally the light sources of choice for analytical applications. The high-pressure lamps are generally too unstable and cumbersome for convenient analytical use.

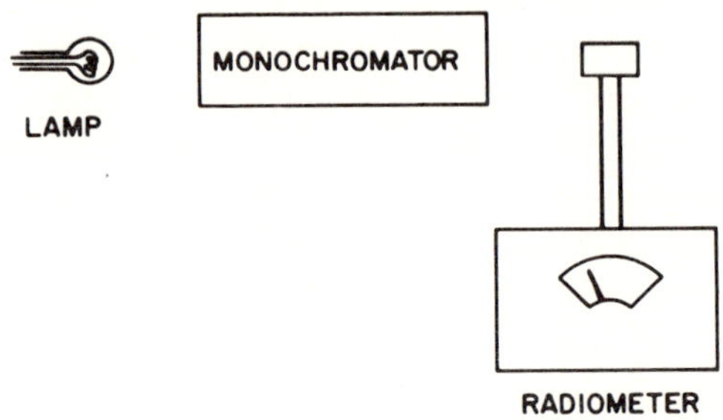

FIG. 4-1. Apparatus for measurement of emission spectra of photochemical light sources.

Mention should be made of one other light source which has played an important role in the development of photoanalytical methods: the sun. The sun has the advantages of being inexpensive and available to everyone. However the available intensity is highly variable and not extremely dependable. Furthermore, the solar spectrum, as observed at ground level, is not rich in the ultraviolet which is so useful in many applications. Thus, the sun will probably play an ever smaller role in laboratory applications.

Many lamp manufacturers do not supply emission spectra with the lamps they provide. Thus it is sometimes necessary for the user to determine the spectrum himself. The following two methods of measuring lamp spectra can be applied when necessary with fair ease.

The spectrum of a lamp may be obtained with the simple apparatus block diagrammed in Fig. 4-1. The lamp is placed in a fixed position at the entrance slit of a Beckman DU monochromator. At the exit slit is fixed the probe of a YSI-Kettering radiometer, or its equivalent. This instrument is sensitive to wavelengths from about 200 nm to 3 μ and the output is linear with energy incident on the probe. The probe is best mounted on an adjustable platform such as a lab jack so that it can be easily and sensitively adjusted for maximum beam interception. The lamp is turned on and the intensity of light incident upon the probe is measured at intervals from 2 to 50 nm over the wavelength range 250–800 nm. The intensity readings are then plotted versus wavelength after correction for the dispersion change of the prism over the range investigated. The dispersion data can be obtained from information published by the Beckman Instrument Co. (*11*). A typical spectrum for a medium-pressure mercury vapor lamp is shown in Fig. 4-2 with intensity plotted in

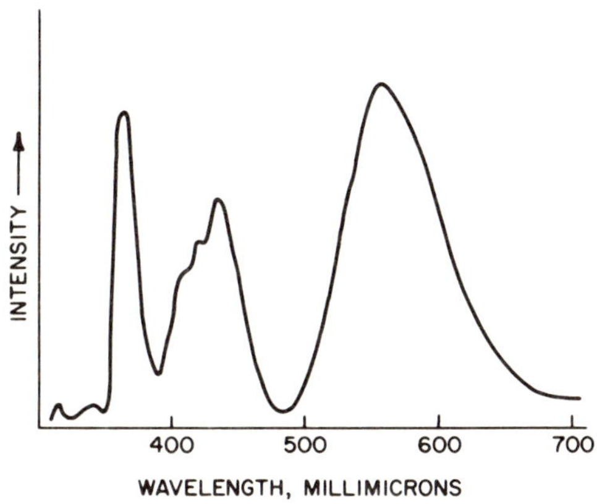

FIG. 4-2. Emission spectrum of medium-pressure mercury vapor lamp obtained with YSI-Kettering radiometer and monochromator.

arbitrary units which are proportional to the number of photons per square centimeter-second incident on the radiometer probe.

To complement the information obtained by the above procedure, the lamp spectrum can be measured by a second method which gives better resolution but which entails some uncertainty in the relative intensities found. The major uncertainty arises from the wavelength dependence of the photomultiplier output. This measurement may be accomplished by placing the mercury lamp in the position normally occupied by the source lamp of a suitable recording spectrophotometer. The lamp is then turned on and the spectrum is recorded by scanning the monochromator with the apparatus in the emission mode. This can be done at different full-scale sensitivities in order to obtain a detailed picture of the output of the lamp. The author has found the Bausch and Lomb Spectronic 505 particularly

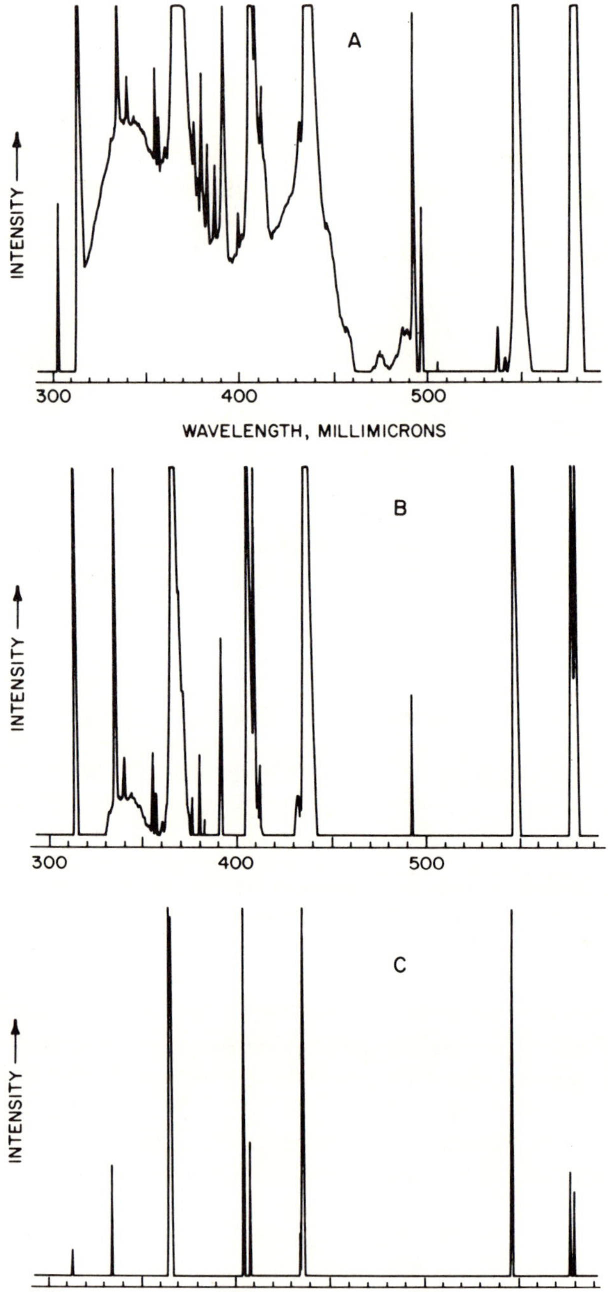
A
INTENSITY
300
400
500
WAVELENGTH, MILLIMICRONS
B
INTENSITY
300
400
500
C
INTENSITY
300
400
500

useful for these measurements since the source housing is easily removed from the instrument.

Typical resulting spectra for the same medium-pressure mercury lamp are shown in Fig. 4-3. A comparison of Figs. 4-2 and 4-3 provides a good qualitative picture of the wavelength distribution of the output of the lamp and of the energy available at each wavelength.

2. Wavelength Isolation

There are two possible solutions to the problem of isolation of a single wavelength from the spectrum of a lamp. One solution is the use of filters; the other is the use of a monochromator. Filters have the advantages over monochromators of relatively low cost and of higher transmission intensity. There are numerous kinds available including interference filters, colored glasses, and colored solutions. An excellent discussion of filters, including directions for preparation of a set of them for the isolation of prominent mercury resonance lines, is available (*8*). Monochromators have the great advantage of convenience in that one can easily adjust them for the desired wavelength and bandpass, usually by turning a few knobs. However, they are not widely used in photoanalytical work because of the large intensity losses involved with their use. Further details of wavelength isolation are given in Chapter 1.

3. Temperature Control

Fortunately, temperature control is rarely a problem in photochemical reactions of analytical usefulness. If it does become necessary, conventional liquid-bath thermostats are available in most laboratories, and may be used over a moderate temperature range above and below room temperature. The reaction vessel may be wholly or partially immersed in the bath and it may be irradiated either from above or through the side of the bath. In the event that higher temperatures are needed, specially designed ovens for photochemical studies at elevated temperatures may be used (*8*).

FIG. 4-3. Emission spectrum of medium-pressure mercury vapor lamp obtained with Bausch and Lomb 505 spectrophotometer. (A) High sensitivity—note large number of weak lines; (B) medium sensitivity; (C) low sensitivity—note that only most intense lines are seen at this setting.

4. Reaction Vessels

Reaction vessels for photoanalytical work can be of virtually any shape and size which is convenient. The primary requirement, which although obvious must always be kept in mind, is that some part of the container must be transparent to the radiation needed for the reaction. Whenever possible, it is a good practice to check the transmission spectrum of every vessel intended for use in a photoanalytical system. An example illustrating the value of this practice can be drawn from the development of a photochemical determination of oxalate (*3*). Different degrees of interference were observed from organic acids when two similar 50-ml Erlenmeyer flasks were used as the reaction vessel. Examination revealed that the flasks were of two different brands, both of borosilicate glass, but with slightly different transmission spectra. One of them transmitted the 303-nm mercury line, while the other filtered it out. This was sufficient to create a significant interference from photolysis of the interfering acid due to the 303-nm line. This type of problem can be prevented or diagnosed by placing the flask in the sample cell of a visible-uv recording spectrophotometer and scanning versus air reference. Taking this precaution costs only a few minutes, and can be quite important. For this purpose, a spectrophotometer with a large sample compartment is particularly useful.

B. *Applications*

1. Quantitative Inorganic Analysis

a. Determination of Iron(III). Ferric iron may be photochemically reduced in aqueous solution by a variety of organic anions. The application of many of these reactions to the determination of iron(III) has been investigated (*1*, *12*, *13*). Although citrate, mandelate, lactate, oxalate, and other anions of organic acids may be utilized as the reductant, oxalate is the most favorable due to the high efficiency of the photoreaction of iron(III) oxalate, and to the insensitivity of the reaction to solution acidity. The procedure is as follows: An excess of oxalate is added to a sulfuric acid (0.1–1.0 *N*) solution of the iron(III) sample to be determined. Irradiation of the sample with either bright sunlight or the full spectrum of a medium- or high-pressure mercury vapor lamp brings about the reaction:

$$2Fe^{3+} + C_2O_4^{2-} \xrightarrow{h\nu} 2Fe^{2+} + 2CO_2 \qquad (1)$$

The reaction mechanism (*14–16*) involves iron(III) oxalate complexes which absorb light in the charge transfer band, ultimately leading to the observed redox reaction through a series of rapid steps. The iron(II) which is formed is then titrated with standard sodium vanadate solution. Stronger oxidants such as dichromate and ceric cannot be used because they will oxidize the unphotolyzed oxalate. Syrupy phosphoric acid is added before the titration to lower the oxidation potential of the iron(III)/iron(II) couple. Diphenylamine, diphenylbenzidine, or diphenylamine sulfonate may be used as indicators for this titration (*12*).

A photochemical method for the determination of submicrogram amounts of iron(III) has been reported (*17*). This method takes advantage of the catalytic effect of iron(III) on the photooxidation and attendant bleaching of methyl orange by atmospheric oxygen. The iron(III) ion acts as a photochemical electron acceptor, and the reduced iron is subsequently reoxidized by dissolved oxygen. The rate of bleaching is proportional to the amount of iron(III) present in the 0–1.0 μg/ml range if the pH is properly controlled.

This method is an interesting one, but unfortunately it is somewhat cumbersome and imprecise ($\pm 10\%$ reported precision), and is therefore not likely to enjoy widespread use.

b. DETERMINATION OF MERCURY(II). Determination of mercury(II) chloride in aqueous solution may be carried out through utilization of the photochemical reduction of Hg(II) to Hg(I) by oxalate. The photochemical reaction is represented by Eq. (2).

$$2HgCl_2 + Na_2C_2O_4 \xrightarrow{h\nu} Hg_2Cl_2 + 2NaCl + 2CO_2 \qquad (2)$$

Both uranyl ion (*18*) and ferric ion (*19*) are effective sensitizers for this reaction, and can be used to speed the analytical procedure. The photochemical reaction may be carried out with sunlight, a mercury vapor lamp, or if iron(III) is used as the sensitizer, a tungsten filament lamp. When the reaction is complete, the precipitated Hg_2Cl_2 is dissolved with a measured excess of iodide–iodine solution according to Eq. (3).

$$Hg_2Cl_2 + 6KI + I_2 \rightarrow 2K_2HgI_4 + 2KCl \qquad (3)$$

The residual iodine is then titrated with standard thiosulfate solution with starch as the indicator. The amount of Hg(II) originally present can then be calculated.

c. DETERMINATION OF URANIUM(VI). Several photochemical methods have been proposed for the determination of U(VI) (*20–27*). One general approach is to utilize the U(VI)-sensitized photodecomposition of oxalic acid according to Eq. (4).

$$H_2C_2O_4 \rightarrow H_2O + CO_2 + CO \tag{4}$$

Under controlled conditions of pH, oxalic acid concentration, and irradiation time, the volume of gas given off can be measured (*22*) and related to the amount of uranium present. Alternatively, the amount of oxalic acid decomposed can be determined by permanganate titration before and after irradiation (*23*) or the measurement can be made photometrically by the following procedure: The oxalate-containing solution is added to a standard ferric thiocyanate solution. Oxalate will preferentially complex ferric ion, causing partial decolorization of the thiocyanate solution. The extent of decolorization is measured both before and after irradiation, and the difference can be related to UO_2^{2+} concentration in the original solution.

The gas-volume method allows measurement of as little as 15 μg of UO_2^{2+}, but the technique is very difficult and time-consuming and the apparatus is expensive. The method is therefore impractical for any kind of routine analytical application. The second method can also measure as little as 15 μg of UO_2^{2+}, but the accuracy is poor ($\pm 10\%$ average error). However, this method seems more promising in terms of applicability of the technique in analytical laboratories.

A generally more satisfactory approach has been to utilize the photochemical reaction of uranyl ion with compounds such as lactic acid (*21*) and various alcohols, especially ethanol (*20*, *24*, *25*), which will bring about the reduction of U(VI) to U(IV). The reduced uranium can then be determined by an oxidimetric titration with vanadate (*20*, *21*, *24*, *25*). Alternatively, the oxidized organic reductant can be determined photometrically (*27*) or the intensely colored arsenazo III complex of U(IV) can be formed and determined photometrically (*26*).

The method involving photoreduction of U(VI) with lactate (*21*) seems to work satisfactorily, but the exposure time required with a 200-W mercury vapor lamp is fairly long (1 hr) and even then the results are very slightly low. The use of ethanol as the reductant (*20*, *24*, *25*) is faster and does not give low results. It may be used in either sulfate or phosphate media. In fact, it appears that sulfate and phosphate ions play an important part in the reaction by complexing and stabilizing U(IV) sufficiently to prevent it from being rapidly reoxidized by dissolved oxygen from the air. Other metal ions such as Fe(III), Mo(VI), and W(VI) which are photochemically reduced and can be subsequently titrated with vanadate will interfere. Some metals such as Cu(II), Ti(IV), Fe(III), and Mo(VI) are known to promote the oxidation of U(IV) back to the U(VI) state in sunlight in the presence of oxygen. Thus the presence of these metals may interfere in this way if oxygen is not excluded from the solution. Arsenate, perchlorate, citrate, Mn(II), nitrate, chloride, Ni(II), Ca(II), Sr(II), Al(III), NH_4^+, Na^+, and K^+ do not interfere appreciably.

The photometric methods for analysis of traces of UO_2^{2+} have the advantage of great sensitivity, being capable of detection of less than 1 μg of UO_2^{2+} (*26*, *27*). As is the case with many trace determinations, the available techniques require careful manipulation and close control of the experimental conditions.

The two methods available both utilize photochemical reduction of U(VI) with ethanol. The first of these utilizes the formation and photometric measurement of the intensely colored arsenazo III complex of U(IV). The difficulty with the method is that the photoreduction must be carried out in phosphate or sulfate media while the colored complex must be formed and measured in HCl solution (*26*). The presence of sulfate or phosphate at the color-developing stage causes incomplete development of the color and thus diminishes the sensitivity. This problem is solved in practice with a compromise: the photochemical reaction is run in 0.5 *N* H_2SO_4 solution, and the attendant 5% loss in sensitivity is tolerated when the color is subsequently developed in HCl solution. The sulfuric acid concentration must be accurately controlled since the extent of color formation depends on this variable. A second minor disadvantage is that the absorbance of the arsenazo III–U(IV) complex must be measured in a 10-cm cell if the maximum sensitivity is to be achieved.

The second method utilizes the photometric determination of the acetaldehyde formed when U(VI) is photochemically reduced by ethanol (*27*). A 70% aqueous ethanol solution which also contains sulfuric acid and the U(VI) to be determined is irradiated for a measured time period. The solution is purposely exposed to air so that the U(IV) can be reoxidized to U(VI) and be available to oxidize more ethanol. In this way a multiplication of the sensitivity is obtained: longer irradiation times give higher sensitivity. A 10-min irradiation allows the determination of 2–40 μg of U(VI), while irradiation for 20 min reduces the lower limit to below 1 μg. After irradiation an aliquot of the solution is treated with fuchsin–H_2SO_3 reagent, developing a red-violet color, the absorbance of which is measured at 530 nm. The relative error of the determination is about $\pm 7\%$ in the 2–10-μg range and $\pm 3\%$ in the 20–40-μg range.

d. DETERMINATION OF THORIUM. Precipitation of thorium iodate from solutions containing periodate and the thorium to be determined can be brought about through the photochemical reduction of periodate to iodate utilizing 2537-Å radiation from a mercury vapor lamp (*4*). In this way, the precipitating reagent is generated in situ and a dense and easily handled precipitate is formed. Subsequent chemical treatment ultimately gives thorium oxalate which is ignited to thorium dioxide and weighed. The accuracy achieved is in the region of $\pm 0.2\%$. Unfortunately, the treatment required after the initial photochemical precipitation is fairly time-consuming, somewhat decreasing the attractiveness of the method.

It has been pointed out (*28*) that this type of procedure cannot, strictly speaking, be regarded as a precipitation from homogeneous solution since the amount of light absorbed, and thus the amount of reagent generated, decreases exponentially with distance from the irradiated surface of the solution. The result, however, is quite similar to the result of precipitation from homogeneous solution.

e. DETERMINATION OF TANTALUM. A photochemical method for the precipitation of tantalum selenite from solution has been reported (*5*). It does not give a stoichiometric precipitate, and is therefore of marginal analytical usefulness. It is, however, noteworthy as an example of the photochemical removal of a masking agent. A solution containing the stable oxalate complex of tantalum, selenous acid, bromine, and manganous ion as a catalyst, is exposed to a 40-W "daylight" (tungsten filament) lamp. The oxalate ligand is photochemically oxidized by bromine under these conditions, slowly releasing tantalum into the solution and bringing about the precipitation of the tantalum selenite. This precipitate is compact, granular, and may be easily filtered, dried, and weighed. A study of the factors affecting the nonstoichiometric nature of this precipitate ($Ta_2O_5:SeO_2 = 1:1.3$) would be quite interesting.

f. DETERMINATION OF GASEOUS OXYGEN. An ingenious colorimetric method for the determination of oxygen in gas streams has been reported (*29*). It utilizes photochemical generation of a colored reagent which is subsequently bleached by the oxygen sample. The difference in the absorbance of the reagent solution before and after reaction with the oxygen in a sample is related to the quantity of oxygen present. The sequence of chemical steps in this determination is represented by Eq. (5).

$$\begin{matrix}\text{Colorless} & & \text{Colored} & & \text{Partially} \\ \text{reagent} & \xrightarrow{h\nu} & \text{reagent} & \xrightarrow{O_2} & \text{decolorized} \\ \text{solution} & & \text{solution} & & \text{solution}\end{matrix} \tag{5}$$

The reagent solution consists of a mixture of methyl viologen (1,1′-dimethyl-4,4′-bipyridinium dichloride), proflavine (3,6-diaminoacridine), and EDTA (ethylenediaminetetraacetic acid). The colored compound is the deep-blue stable radical cation of reduced methylviologen. The

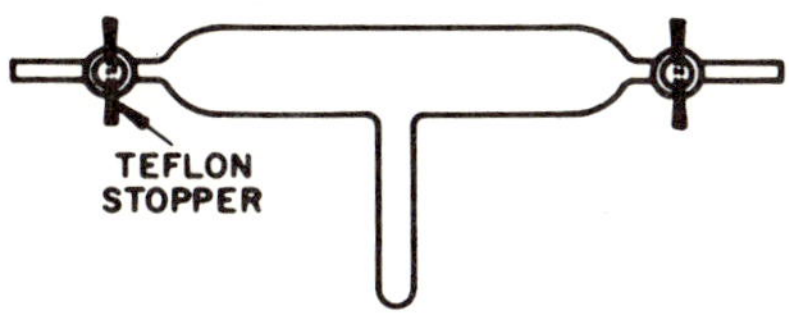

FIG. 4-4. Sampling tube for determination of oxygen in gases. Reprinted from Sweetser (*29*, p. 980) by courtesy of the American Chemical Society.

mechanism of photochemical formation of this compound is apparently as follows: The proflavine absorbs a photon of incident light and in its excited state extracts an electron from an EDTA molecule. The electron is then transferred to the methylviologen dication, thus forming the colored cation radical.

The analytical determination is carried out in a gas-sampling tube which has been modified by the addition of a test-tube-like arm which contains the reagent solution and also acts as the sample cell for the colorimetric measurement (Fig. 4-4). The reagent solution and the gas sample are added to the reaction vessel. The reagent is then exposed to light until the absorbance of the solution at 580 nm is 0.7–0.9 absorbance units. Then the solution is shaken vigorously in order to absorb and react all of the oxygen in the cell. At the completion of this procedure the absorbance is again measured. The change in absorbance is related to the amount of oxygen present in the sample by reference to a calibration curve.

With a 500-ml gas-sampling tube, this method is sensitive to less than 1 ppm O_2. The reagent solution can be used repeatedly until the large excess of EDTA is consumed since neither the proflavine nor the methylviologen is destroyed in the reaction sequence. Strongly oxidizing gases such as bromine, chlorine, and fluorine will interfere, and must be removed before sampling.

2. Quantitative Organic Analysis

a. Determination of Oxalate. Oxalate may be determined conveniently by photochemical oxidation with iron(III), using sunlight (*2*) or a medium-pressure mercury vapor lamp (*3*) as the light source. The well-known photodecomposition of iron(III) oxalate (*14–16*) proceeds stoichiometrically to give two iron(II) ions for each oxalate ion consumed. The reaction is carried out in 0.1–1.0 N H_2SO_4 solution. The iron(II) which is formed may be titrated with either vanadate (*2*) or dichromate (*3*) if the sample size is in the milligram range. For traces of oxalate (as little as 5 μg $Na_2C_2O_4$) the determination of the iron(II) formed is carried out colorimetrically (*3*), using 1,10-phenanthroline to develop the intense orange color of ferrous 1,10-phenanthroline. The absorbance is measured at 508 nm and the iron(II) concentration is found from a calibration curve prepared with primary standard ferrous ethylenediammonium sulfate tetrahydrate. The original oxalate concentration can then be calculated, or if desired, the calibration curve can be converted to read directly in oxalate concentration.

These methods are unique among analytical determinations of oxalate in terms of ease of use, applicability to trace amounts, and the ability to tolerate large excesses of species such as formate, acetate, propionate, malonate, and succinate, as well as many other organic compounds which interfere strongly with the classical oxidimetric methods for oxalate.

b. Determination of Organic Contaminants in Water. One of the best ways to obtain ultrahigh-purity water, free of all organic residue, is by exhaustive radiolysis or photolysis (*31*). Under this treatment the organic material is converted to gaseous products, mainly CO_2, which escape from the liquid. Advantage may be taken of this fact to perform an analytical measurement of the amount of carbonaceous material originally present (*30*). A stoichiometric conversion of organic matter to CO_2 and a sensitive method for measurement of the CO_2 released are necessary for a successful analytical procedure.

The first of these requirements is met by exhaustive photolysis using the

2537-Å mercury resonance line with iron(III) present in the solution as a sensitizer. Under these conditions the carbonaceous material present is converted entirely to CO_2. Mass-spectrometric measurement of the CO_2 formed by the photolysis satisfies the requirement for a sensitive method for quantitative detection of the reaction product.

The analysis is carried out by making the water under test about 0.02 *M* in iron(III), as the sulfate. The sample is placed in a quartz tube which is then connected to the photolysis apparatus. The apparatus is purged with nitrogen, and then the irradiation is carried out. After 3 hr of irradiation with a medium-pressure mercury vapor lamp, the CO_2 is collected in a trap at liquid-N_2 temperature. The trap is then connected to the inlet system of a mass spectrometer for the subsequent gas analysis.

This method is a general one, capable of analyses of less than 1 μmole of CO_2. It represents the only known approach to some trace organic analysis problems. The limitations of the technique are rather severe, however. The accuracy is poor ($\pm 50\%$) although at such low levels this will normally be acceptable. The requirement of a mass spectrometer calibrated for analysis of submicromolar amounts of CO_2 will place this technique beyond the capability of many laboratories.

c. DETERMINATION OF IODOFORM AND METHYL KETONES. A quantitative analysis for iodoform can be performed by utilizing the photochemical decomposition of iodoform to iodine in bright sunlight (or other near-ultraviolet source) in ether–benzene solution (*32*). The iodoform sample is dissolved in the ether–benzene solvent. This solution is placed in a flask with a fourfold excess of standard aqueous 0.01 *N* sodium thiosulfate solution. The mixture is thermostated at 20°C and exposed to the light source. Iodine is rapidly liberated by photooxidation of iodoform by dissolved oxygen. As the reaction proceeds, the iodine colors the ether–benzene portion of the mixture red, thus slowing the reaction by filtering out the actinic light. To counteract this effect the reaction mixture is frequently shaken, causing the iodine to be reduced by the thiosulfate solution, thereby preventing the autofiltration from occurring. At the end of the reaction the mixture is removed from the light and the excess thiosulfate is titrated with standard iodine solution, after the addition of solid sodium iodide and starch indicator solution. A blank determination is necessary, and is carried out in the same way except for the addition of the iodoform sample. This method is capable of accuracy consistently better than $\pm 0.5\%$.

Acetone may also be determined in the presence of some other ketones using this same procedure, by first converting the acetone to iodoform. The accuracy achieved in this case averages about ±1.5%. Obviously, other methyl ketones interfere.

The speed, accuracy, and simplicity of this method are obvious advantages. The most serious disadvantage is interference from other iodine-substituted organic compounds, most of which will liberate some iodine under the experimental conditions of this procedure.

d. DETERMINATION OF TRYPTOPHAN AND ERGOTOXINE. The analytical determination of tryptophan in proteins can be carried out by condensation of tryptophan with *p*-dimethylaminobenzaldehyde followed by oxidation of the condensation product to develop a blue color (*33*). The blue color is utilized for a colorimetric determination at 600 nm. Normally the oxidation step is performed using sodium nitrite as the oxidant (*34*, *35*). It is possible, however, to induce the oxidation reaction by irradiation of the test solution (*36*). The oxidant in this case is presumably dissolved oxygen from the air. A blank for the colorimetric reaction is prepared using the same procedure as for the sample, but with omission of the tryptophan. The precision of the determination is ±1.0% compared to ±0.9% for the sodium nitrite method. This method is recommended in any situation in which the use of sodium nitrite is not feasible.

A closely similar method is possible for the determination of ergotoxine (*13*). The ergotoxine may be extracted, reacted with *p*-dimethylaminobenzaldehyde, and the product then irradiated to form the blue color which is used for a colorimetric determination.

e. DETERMINATION OF ASCORBIC ACID. Methylene blue is bleached by reaction with ascorbic acid, even in the dark. The photochemical bleaching reaction is much more rapid than the dark reaction, however, and it can be used for the quantitative determination of ascorbic acid (*37*). Both qualitative (*38*) and quantitative (*39*, *40*) measurements of ascorbic acid in blood serum have been made using this reaction. After precipitation of the protein matter in the blood sample, ascorbic acid is titrated with a standard methylene blue solution. The titration vessel is strongly illuminated with an incandescent lamp during the determination. The titration is carried out slowly with 30-sec waiting periods between titrant additions to allow photolysis to occur. When the blue color persists beyond this time, titration is complete.

It would seem that the full potential of this reaction has not yet been utilized for analytical purposes. It should be possible to determine small traces of ascorbic acid by colorimetric measurement of the extent of bleaching of a standard methylene blue solution. Another possibility is the use of photometric end-point detection for the titrimetric determination described above. The technique for such a photochemical–photometric method has been described (*16*) and is relatively simple to use. The apparatus is based on the titration cell–centrifugal pump recommended by Higuchi (*42*) and Siggia (*43*). The cell, which is diagrammed in Fig. 4-5,

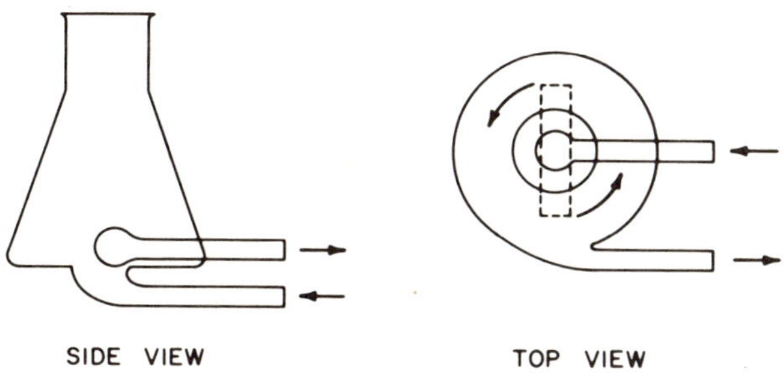

FIG. 4-5. Pumping vessel for photometric titrations.

consists of a 50-ml Erlenmeyer flask with an outlet tube sealed tangentially to its side near the bottom. In the center of the bottom is sealed an inlet tube. These tubes are bent to extend outward from the flask parallel to one another. A magnetic stirring bar turning in the bottom of the flask forces the solution out the side arm and allows it to return through the center of the bottom. This apparatus can be used to pump solution through the sample cell of a spectrophotometer such as the Bausch and Lomb Spectronic 20. In this way the absorbance of the solution can be continuously monitored. Titrant is added into the Erlenmeyer, and irradiation of the solution can be performed through the side of the flask. Plots of titrant volume versus absorbance give curves typical of photometric titrations.

3. QUALITATIVE ANALYSIS AND SPECIAL APPLICATIONS

a. QUALITATIVE TESTS. Several photochemical reactions have been utilized for qualitative indication of the presence of various substances. Most of these tests are simple color-development reactions. Among the substances for which such tests have been developed are silver (*44*), lead tetraalkyl, nickel carbonyl (*45*), organic nitroso compounds (*46*), strychnine, codeine, and morphine (*47*).

b. Photolysis Gas Chromatography. The techniques of reaction gas chromatography, especially pyrolysis gas chromatography, are well established as useful analytical tools for qualitative identification of organic compounds and for organic structure determination (*48*, *49*). One of the major disadvantages of pyrolysis gas chromatography is the great difficulty commonly experienced in obtaining results which may be reproduced from laboratory to laboratory. In the effort to overcome this disadvantage, Juvet and co-workers (*50*, *51*) have introduced the technique of photolysis gas chromatography. Using mercury-sensitized photolysis in place of pyrolysis gives simpler fragmentation patterns and a high degree of reproducibility. In addition to these advantages, functional groups may be identified by characteristic constants related to the retention of certain peaks in the photolysis chromatogram. Using these constants, structural information can be obtained without identification of the photolysis products.

In order to study a compound by the photolysis gas-chromatographic technique, the pure sample, normally a liquid,* is first saturated with elemental mercury by shaking liquid mercury and the sample together. Mercury is soluble in most organic liquids to the extent of about 1×10^{-6} moles per liter (*53*, *54*). The mercury-saturated sample is placed in a 1-mm o.d. quartz capillary and is deaerated on a vacuum line. The capillary is sealed and transferred to the irradiation apparatus. Irradiation is carried out with 2537-Å light from a low-pressure mercury vapor lamp. This radiation is resonantly absorbed by the elemental mercury in the sample and the energy is subsequently transferred by collisional processes to the organic sample, bringing about sensitized photolysis. The photolysis times are kept short (1 sec to 15 min) so that the product yields are below 2%. This decreases the likelihood of secondary reactions, thereby maintaining good reproducibility.

After photolysis the capillary containing the sample is transferred to the solid sampler attachment of a gas chromatograph where the capillary is broken and the contents are swept onto the chromatographic column. The chromatograph is operated in the linear temperature program mode and the resulting retentions of the various photolysis products are expressed as linear temperature program retention indices (*55*). The use of these

* Applications with vapors and solids are also under development (*52*).

retention indices permits direct comparison of data obtained in different laboratories using different carrier-gas flow rates and temperature program rates.

The photolysis chromatograms exhibit three kinds of peaks which can be used for identification purposes and to obtain structural information. These peaks are called *homologous*, *common*, and *distinctive*. *Homologous* peaks exhibit retention indices which are different from that of the parent peak by an amount which is characteristic of the dominant functional group of the parent compound. All members of a homologous series will exhibit peaks differing in retention from the parent peaks by a characteristic, constant amount. Thus, homologous peaks may be used as an indication of what functional group is present in a test compound. *Common* peaks are peaks due to compounds which are always produced by the photolysis of a member of a given homologous series. For example, acetone is always produced in the photolysis of methyl ketones. Therefore, in the photolysis chromatogram of a methyl ketone, there will always be a peak at the retention index value corresponding to acetone. Such common peaks are very valuable in characterization of the sample compound. In addition to homologous and common peaks, each photolysis chromatogram shows peaks which do not appear to follow any simple series relationship. These are called *distinctive* peaks. They must be individually tabulated and are useful when a catalog of photolysis chromatograms is available for comparison with the data from an unknown compound.

Photolysis gas chromatography is a promising technique, complementary to pyrolysis methods. Photolysis requires larger samples than pyrolysis methods, and its application to solids is still in such an early state of development as to be of unproven value. However, the high reproducibility and simple fragmentation patterns are advantages of considerable importance.

c. PHOTOIONIZATION DETECTORS FOR GAS CHROMATOGRAPHY. If a molecule absorbs light of energy above its ionization potential, the result will be immediate ionization of the molecule. This principle is the basis for the recent development of photoionization detectors for gas chromatography (*56–62*). In these detectors, the effluent stream from the gas chromatograph is directed through a cell in which an arc discharge is maintained with oxygen, nitrogen, hydrogen, or one of the noble gases. The ultraviolet emission from the gas in the discharge photoionizes compounds in the effluent stream. A pair of sensing electrodes collects the ion current, which is recorded as a function of time, yielding a gas chromatogram. Figure 4-6 illustrates the components of a typical photoionization detector.

Recent work (*62*) has shown that these detectors are capable of outstanding sensitivity, comparing favorably with the best of the flame-ionization detectors currently used in gas chromatography. It is also possible, at least in principle, to achieve some degree of selectivity with photoionization detectors (*60*). This may be accomplished by utilizing ultraviolet radiation with energy below the ionization potential of some

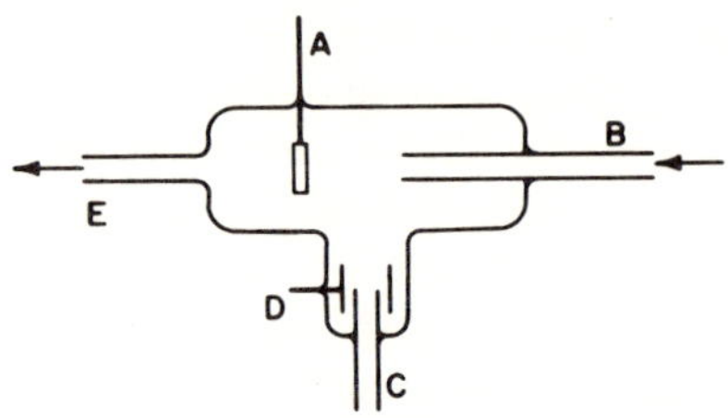

FIG. 4-6. Schematic version of a typical photoionization detector. *A*, discharge anode; *B*, discharge cathode and inlet tube for the discharge gas; *C*, ion-collector anode and inlet for column effluent; *D*, ion-collector cathode; *E*, outlet. (See also Refs. *73* and *74*.)

compounds but above that of others. In such cases, only the latter compounds will respond to the detector.

Gas chromatography with photoionization detectors has been successfully applied to problems requiring great sensitivity, such as air-pollution analysis (*60*) and analysis of hydrocarbons found in meteorites (*62*).

d. PHOTOIONIZATION MASS SPECTROMETRY. The photoionization principle using high-energy ultraviolet radiation has also been applied to mass spectrometry. In this application, an intense source of high-energy ultraviolet radiation such as a microwave plasma (*63*, *64*) or an arc or spark discharge (*65*) replaces the conventional electron-impact source. If a monochromator is employed with the photoionization source, it is possible to select very precisely the photon energy available in the ionization chamber. This precise control can be utilized to accurately measure ionization potentials of molecules (*66*, *67*) by observing the photon energy at which ions are first detected in the mass spectrometer. This is probably the most significant current use of photoionization in mass spectrometry.

Photoionization sources have also been utilized for mass-spectrometric gas analysis (*65*). The advantage is that the spectrum is usually somewhat simpler than obtained with electron-impact ionization, and the molecular ion typically is more abundant. The major drawback is that the sensitivity is lower with photoionization than with electron impact, since it is more difficult to produce high photon fluxes than it is to produce copious electrons.

e. Flash Sampling in Mass Spectrometry. The use of intense light flashes to vaporize and ionize solid samples has been introduced for use with time-of-flight mass spectrometers. A laser (*68–71*) or other intense flash source, such as a xenon flash lamp (*72*), is used as the light source. The irradiated sample is vaporized and in some cases ionized by the light. The time-of-flight mass spectrometer then performs mass analysis of the species in the vapor. This technique has been applied to the analysis of solid materials (*70*, *72*, *73*), to the study of surface interactions (*69*), and to the study of laser-induced breakdown of organic vapors (*68*). A discussion of laser operation may be found in Chapter 2, "Lasers in Analytical Chemistry."

f. Laser Sampling for Emission Spectroscopy. Vaporization of a sample by a laser pulse has also been utilized in emission spectroscopy (*74*). The laser pulse produces a plume of intensely heated sample vapor. The light emitted from the ions and atoms in the plume can be dispersed and recorded with an emission spectrograph providing qualitative and quantitative analysis. Further excitation of the material in the plume using a spark source improves the sensitivity and diminishes matrix effects. Chapter 2 should be consulted for more details of this technique.

IV. Summary

The application of photochemical reactions to the problems of analytical chemistry is a new and promising area of chemical research. The use of such reactions in the pretreatment step before an analytical measurement is an important part of this new field. Photochemical reactions frequently impart unusual and analytically useful attributes to a system. For example, the ability to make a species such as oxalate behave as a strong reductant toward another species such as iron(III) when under irradiation, and yet not interfere in the subsequent oxidimetric titration of iron(II), is the unique contribution of photochemistry and can be of great importance in analytical chemistry. Because of the utility of many such photochemical reactions, their use by analytical chemists should continue to increase at a rapid pace.

REFERENCES

1. G. G. Rao and V. Madhusudhana Rao, *Current Sci.* (*India*), **13,** 317 (1944); *Proc. Nat. Inst. Sci.* (*India*), **12,** 217 (1946).
2. G. G. Rao and G. Aravamudan, *Anal. Chim. Acta*, **13,** 415 (1955).
3. W. M. Riggs and C. E. Bricker, *Anal. Chem.*, **38,** 897 (1966).
4. M. Das, A. H. A. Heyn, and M. Z. Hoffman, *Talanta*, **14,** 439 (1967).
5. Y. Jen-yin and Y. Wei, *Sci. Sinica* (*Peking*), **13,** 343 (1964).
6. I. P. Kharlamov, E. I. Dodin, and A. D. Mantsevich, *Zh. Anal. Khim.*, **22,** 371 (1967).
7. E. E. Wegner and A. W. Adamson, *J. Am. Chem. Soc.*, **88,** 394 (1966).
8. J. G. Calvert and J. N. Pitts, *Photochemistry*, Wiley, New York, 1967.
9. Galen W. Ewing, *Instrumental Methods of Chemical Analysis*, McGraw-Hill, New York, 1960, Chap. 3.
10. E. L. Wehry, *Quart. Rev.*, **22,** 213 (1968).
11. Beckman Instruments, Inc., Fullerton, Calif., Operating Manual, Beckman Model DU Spectrophotometer, pp. 22–3, 1954.
12. G. G. Rao, G. Aravamudan, and N. C. Venkatamma, *Z. Anal. Chem.*, **146,** 161 (1955).
13. G. G. Rao and G. Aravamudan, *Anal. Chim. Acta*, **13,** 328 (1955).
14. C. A. Parker and C. G. Hatchard, *J. Phys. Chem.*, **63,** 22 (1959).
15. J. B. Holden, Jr., Doctoral dissertation, Princeton Univ., Princeton, N.J., 1961.
16. W. M. Riggs, Doctoral dissertation, University of Kansas, Lawrence, Kansas, 1967.
17. I. P. Kharlamov, E. I. Dodin, and A. D. Mantsevich, *Zh. Anal. Khim.*, **22,** 371 (1967), English transl., p. 323.
18. G. G. Rao and P. T. Ramacharlu, *Proc. Nat. Inst. Sci.* (*India*), **8,** 383 (1942).
19. P. T. Ramacharlu and G. G. Rao, *ibid.*, **9,** 67 (1943).
20. G. G. Rao, V. P. Rao, and N. C. Venkatamma, *Z. Anal. Chem.*, **150,** 178 (1956).
21. G. G. Rao, V. P. Rao, and M. V. R. Rao, *Anal. Chim. Acta*, **15,** 97 (1956).
22. A. E. Taylor, G. E. Heckler, and D. R. Percival, *Talanta*, **7,** 232 (1961).
23. A. K. Babko and L. M. Ginzburg, *Zh. Anal. Khim.*, **21,** 1070 (1966), English transl., p. 951
24. A. A. Nemodruk and E. V. Bezrogova, *ibid.*, **21,** 1210 (1966), English transl., p. 1074.
25. A. A. Nemodruk and E. V. Bezrogova, *ibid.*, **22,** 366 (1967), English transl., p. 319.
26. A. A. Nemodruk and E. V. Bezrogova, *ibid.*, **22,** 881 (1967), English transl., p. 744.
27. A. A. Nemodruk and E. V. Bezrogova, *ibid.*, **23,** 388 (1968); C. A. **69,** 15917C (1968).
28. J. M. Fitzgerald, R. J. Lukasiewicz, and H. D. Drew, *Anal. Lett.*, **1,** 173 (1967).
29. P. B. Sweetser, *Anal. Chem.*, **39,** 979 (1967).
30. J. Beattie, C. E. Bricker, and D. Garvin, *ibid.*, **33,** 1890 (1961).
31. H. Fricke and E. J. Hart, *J. Chem. Phys.*, **4,** 418 (1936).

32. S. Bose, *Anal. Chem.*, **30,** 1137 (1958).
33. J. R. Spies and D. C. Chambers, *ibid.*, **22,** 1209 (1950).
34. J. R. Spies, H. S. Bernton, and H. Stevens, *J. Am. Chem. Soc.*, **63,** 2163 (1941).
35. J. R. Spies and D. C. Chambers, *Anal. Chem.*, **20,** 30 (1948).
36. J. R. Spies and D. C. Chambers, *J. Am. Chem. Soc.*, **70,** 1682 (1948).
37. E. Martini and A. Bonsignore, *Boll. Soc. Ital. Liol. Sper.*, **9,** 388 (1934).
38. H. Lund and H. Lieck, *Nature*, **137,** 784 (1936).
39. H. Lund and H. Lieck, *Klin. Wochschr.*, **16,** 555 (1937).
40. H. Lund and E. Trier, *ibid.*, **18,** 80 (1939).
41. E. Weiland, *Biochem. Z.*, **317,** 168 (1944).
42. C. Rehm, J. I. Bodkin, K. A. Conners, and T. Higuchi, *Anal. Chem.*, **31,** 483 (1959).
43. S. Siggia, *Quantitative Organic Analysis*, Wiley, New York, 1963, p. 161.
44. A. J. Velculescu, *Z. Anal. Chem.*, **90,** 111 (1932).
45. R. E. Steiger, *Mikrochem. Ver. Mikrochim. Acta.*, **22,** 22 (1937).
46. F. Feigl, *Spot Tests*, Elsevier, Amsterdam, 1946, p. 132.
47. J. Aloy, *Bull. Soc. Chim. France*, **39,** 792 (1926).
48. M. Beroza and R. A. Coad, in *The Practice of Gas Chromatography* (L. S. Ettre and A. Zlatkis, eds.), Wiley-Interscience, New York, 1967, pp. 461–510.
49. R. S. Juvet, Jr., and S. Dal Nogare, *Anal. Chem.*, **40,** 33R (1968).
50. R. S. Juvet, Jr., and L. P. Turner, *ibid.*, **37,** 1464 (1965).
51. R. S. Juvet, Jr., R. L. Tanner, and J. C. Y. Tsao, *J. Gas Chromatogr.*, **5,** 15 (1967).
52. R. S. Juvet, Jr., private communication, 1968.
53. R. R. Kuntz and G. J. Mains, *J. Am. Chem. Soc.*, **85,** 2219 (1963).
54. J. N. Spencer and A. F. Voigt, *J. Phys. Chem.*, **72,** 464, 471, 1913 (1968).
55. H. Van Den Dool and P. Dec. Kratz, *J. Chromatogr.*, **11,** 463 (1963).
56. C. F. Robinson and W. M. Brubaker, U.S. Patent 2,959,677, Nov. 8, 1960, filed May 2, 1957.
57. J. E. Lovelock, *Nature*, **188,** 401 (1960).
58. M. Yamane, *J. Chromatogr.*, **9,** 162 (1962); **11,** 158 (1963); **14,** 355 (1964).
59. J. F. Roesler, *Anal. Chem.*, **36,** 1900 (1964).
60. D. C. Locke and C. E. Meloan, *ibid.*, **37,** 389 (1965).
61. B. J. Gudzinowicz, in *The Practice of Gas Chromatography* (L. S. Ettre and A. Zlatkis, eds.), Wiley-Interscience, New York, 1967, p. 245.
62. J. G. W. Price, D. C. Fenimore, P. G. Simmonds, and A. Zlatkis, *Anal. Chem.*, **40,** 541 (1968).
63. C. E. Brion, *ibid.*, **37,** 1706 (1965).
64. C. E. Brion, *ibid.*, **38,** 1941 (1966).
65. W. Poschenreider and P. Warneck, *J. Appl. Phys.*, **37,** 2812 (1966).
66. J. Mottle, T. Nakayama, and K. Watanabe, *Final Report on Ionization Potentials of Molecules*, University of Hawaii, Honolulu, Hawaii, 1959.
67. V. H. Dibeler and S. K. Liston, *J. Chem. Phys.*, **49,** 482 (1968).
68. A. H. Adelman, *ibid.*, **45,** 3152 (1966).
69. G. E. Bernal, L. P. Levine, and J. F. Ready, *Rev. Sci. Instr.*, **37,** 938 (1966).
70. N. C. Fenner and N. R. Daly, *ibid.*, **37,** 1068 (1966).
71. D. J. Fabian, *Metals Mater.*, **1,** 27 (1967).
72. K. A. Lincoln, *Anal. Chem.*, **37,** 541 (1965).
73. J. F. Eloy and J. I. Dumas, *Method Phys. Anal.*, **1966,** 251.
74. M. Margoshes and B. F. Scribner, *Anal. Chem.*, **40,** 223R (1968).

Note added in proof: References 75–80 are to literature which has appeared since manuscript preparation and are not cited in the chapter.

75. A. A. Nemodruk and E. V. Bezrogova, *Zh. Anal. Khim.*, **23,** 884 (1968), English transl., p. 768.
76. A. A. Nemodruk and E. V. Bezrogova, *ibid.*, **24,** 404 (1969), English transl., p. 292.
77. A. A. Nemodruk and E. V. Bezrogova, *ibid.*, **24,** 859 (1969), English transl., p. 686.
78. A. A. Nemodruk and E. V. Bezrogova, *ibid.*, **24,** 1044 (1969), English transl., p. 834.
79. A. A. Nemodruk and E. V. Bezrogova, *ibid.*, **24,** 1534 (1969), English transl., p. 1239.
80. A. A. Nemodruk and E. V. Bezrogova, *ibid.*, **24,** 1704 (1969), English transl., p. 1381.

CHAPTER 5

PHOTOCHEMICAL TITRATIONS

J. M. Fitzgerald

DEPARTMENT OF CHEMISTRY
UNIVERSITY OF HOUSTON
HOUSTON, TEXAS

I. Introduction to the Method

Photochemical titrations may be thought of as an extension and refinement of the use of photochemical pretreatment for analytical purposes. Historically and conceptually, the method of photochemical titrant generation is based upon earlier photo-pretreatment methods which had established that photochemical reactions could be made to proceed stoichiometrically and reproducibly enough to be used for analytical determinations. Therefore, it will be assumed that the reader is generally familiar with the problems and applications of photochemical pretreatment as discussed in the preceding chapter.

The various stoichiometric and spectroscopic limitations which must be considered before the application of a photochemical reaction for pretreatment can be made must also be considered in developing a photochemical titration. But in addition, and equally important, the rate of the photochemical reaction must proceed at a reproducible rate from one day to the next. Bricker and Schonberg (*1*) first introduced this method and proved that titrations reproducible to better than 2% could be obtained with relatively simple and inexpensive equipment. This fact remains one of the central attractions of the method, for if elaborate techniques were required, then practical application would be severely limited. The simplicity is a fact, however, and the major limitation at the present time seems to be commitment of time and effort to the development of titrations. The theory and experimental approach are firmly established.

Photochemical titrations have many of the advantages of photochemical pretreatment. The two most important practical advantages are the ability to add "reagent" to a sample without dilution, and the ability to add that "reagent," i.e., photons, from some remote source, across a finite distance, to the sample. Although no published accounts regarding titration of highly radioactive or other hazardous materials have been reported, it is anticipated that such applications of photochemical titrations may well be developed in the future. It should also be noted that most photochemical rate constants have very small temperature coefficients so that elaborate temperature regulation is usually not required.

An additional practical advantage is that the equipment required is neither extensive nor expensive, at least by modern standards. A source of ultraviolet radiation, some form of voltage or current regulation to stabilize the intensity, a shutter and timing system, and a suitable chemical or instrumental indicator are the major requirements once a chemical system with a reproducible photochemical reaction rate has been selected.

As one of several forms of titrimetry, photochemical generation must be compared with alternative methods such as volumetric and coulometric titrations. All titration methods stem historically and conceptually from the early volumetric work. Photochemical titration is no exception and work to date has been concentrated on inorganic redox reactions. The determination of chromium(VI) and vanadium(V) with photogenerated iron(II) was used for the first demonstration of the method (*1*).

In both the first work of Bricker and Schonberg (*1*) and the following study of Kuwana (*2*), the strong analogy between photochemical and electrochemical titrant generation was emphasized. In fact, Bricker and Schonberg used the word *photonometric* titration to describe the method in order to emphasize the similarity with coulometry. Unfortunately the words photonometric and photometric are easily confused. Since it is quite common to carry out photochemical titrations with photometric end points, speaking of a photonometric–photometric titration almost completely confuses the issue.

The word *photonometric* is strictly incorrect as well as potentially confusing since it is implied that the quantity of photons added to the sample is somehow measured, just as the electrons are counted in a coulometric titration. There are three basic problems which render the counting of photons added to the sample extremely difficult, if not impossible.

First, the sample solution must absorb all of the incident radiation. According to Beer's law either an infinite depth or an infinite concentration of absorber is required to reduce the transmitted intensity (I_T) to zero. That is, the absorbance of a solution is never infinite. As we shall see shortly, this theoretical problem is easily overcome in practice.

Second, and of great practical importance, is the fact that the movement of electrons can be confined easily to a solid conductor, which facilitates measurement of the current passed through the sample with a coulometer or ammeter connected in series. However, photons cannot be confined in a fixed column from any radiation source other than a laser. The photons from a conventional ultraviolet source radiate in a 4π solid angle from that source and the intensity at any distance is given by the inverse-square law. Thus the measurement of the instantaneous photon flux can only be carried out with some device connected "in parallel" with the sample. No other point in space is exposed to exactly the same instantaneous flux as is the sample. Some efforts have been made to monitor the intensity of the source with a phototube or actinometer situated near the sample (*2*, *3*). However if there is any "wander" in the arc, then it follows that the monitor in parallel with the sample may not be exposed to the same change of intensity. Therefore, one is quickly reduced to measuring the intensity of the source by photolyzing a suitable actinometer in the sample container and repeating this calibration from time to time as required. This method has been successfully used by Kuwana (*2*). However it requires that the source intensity remain stable with time. This is, of course, a primary requirement for a reproducible titrant generation rate.

A third problem still frustrates the experimentalist even if the flux of photons on the sample container were accurately known; this is the fact that only a few solution photochemical reactions have overall quantum yields of unity. This means that the quantum efficiency, which corresponds to the current efficiency in coulometry, is seldom exactly 100%. However, just as is the case in coulometry (*4*), it is possible to develop useful methods based on "working" or calibration curves, but a knowledge of the exact value of photon flux is small help in the construction of such a working curve. Of course, an order-of-magnitude value of the intensity incident on the sample container is useful in predicting the concentration range which can be conveniently titrated. The intensity usually turns out to be on the order of microeinsteins per minute of photolysis.

Therefore, since the originally suggested name can be confusing and is somewhat inaccurate, the present author feels it would be better to call the method *photochemical titration*. As a matter of fact, as will be shown, photochemical titration can be considered as a special application of photochemical kinetics, and the entire area of research might perhaps be better termed *photochemical kinetic analysis*.

Finally, the very strong dependence of photochemical titration on the field of chemical actinometry should be mentioned. Since the development of a new titration method requires a generating reaction which is reproducible, independent of intensity, and relatively unaffected by such solution variables as ionic strength and pH, it follows that only well-characterized reactions can be widely applied for titrant formation.

The interested reader should consult Chapter 3 on actinometry in order to appreciate the efforts involved in developing an actinometer, which accounts in part for the limited number of truly successful actinometers. Conversely, any photochemical titrant-generating reaction which is really well characterized is essentially a chemical actinometer.

II. Theory of Photochemical Titrant Generation

Photochemical titrant generation can be considered as a specialized application of photochemical kinetics. Consider a photochemically active species A which is converted to the inactive (and nonabsorbing) species B by the energy contained in a photon, $h\nu$.

$$A + h\nu \rightarrow B \tag{1}$$

The basic rate equation for this reaction is given by:

$$\frac{-dC_A}{dt} = \frac{+dC_B}{dt} = \Phi I_{\text{abs}} \tag{2}$$

where Φ is the quantum yield in units of molecules of A converted to B per photon absorbed, and I_{abs} is the intensity of radiation absorbed in units of number of photons per unit time. By dimensional analysis then, Eq. (2)

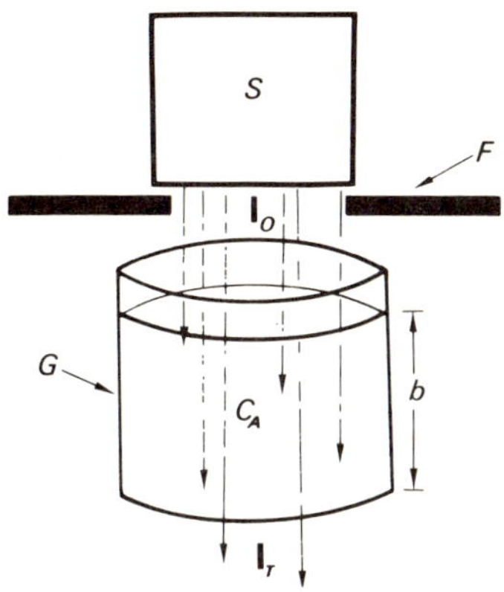

FIG. 5-1. Geometry of in situ photochemical conversion of A to product B. S, photolysis source; F, shutter mechanism, includes filters and beam-defining baffles if used; G, generation cell, containing solution of A; b, depth of solution of concentration C_A exposed to beam (note that b is not single valued if the beam is not collimated onto cell G); I_0, incident intensity of photolytically active wavelengths; I_T, transmitted intensity of photolytically active wavelengths (note $I_T = I_0 - I_{\text{abs}}$).

has units of molecules of B formed per unit time. This is exactly the quantity we wish to know. The general picture of a photochemical experiment in which A is converted to B is shown in Fig. 5-1.

The intensity of absorbed radiation can be obtained by application of Beer's law. For the case illustrated in Fig. 5-1, the absorbed intensity is simply

$$I_{\text{abs}} = I_0 - I_t \tag{3}$$

However, the transmitted intensity is related to concentration of A and to the incident intensity by Beer's law:

$$I_t = I_0 \exp(-\text{Abs}/0.434) \tag{4}$$

where Abs represents the absorbance of the solution in question. Now if the conditions are such that A is the only absorber in the solution (which means product B and all other solution constituents are transparent in the ultraviolet) and if the depth of the solution is in centimeters, then

$$I_t = I_0 \exp(-a_A b C_A/0.434) \tag{5}$$

Combining Eqs. (5) and (3) and substituting into (2), we obtain

$$\frac{+dC_B}{dt} = \Phi I_0[1 - \exp(-a_A b C_A/0.434)] \tag{6}$$

For the case where polychromatic light is used for the photolysis, the expression must be either summed for a line source or integrated for a continuous source over the range of photochemically effective wavelengths. This means that the correct values of the wavelength-dependent terms in Eq. (6), namely the quantum yield, Φ, incident intensity, I_0, and molar absorptivity of the generator, a_A, must be used for the computation.

There are three very important consequences of Eq. (6). First, for values of absorbance greater than 2.00 the exponential term becomes smaller than 0.01; that is, the transmitted intensity is less than 1% of the total incident. For this condition Eq. (6) reduces to

$$\frac{dC_B}{dt} \approx \Phi I_0 \tag{7}$$

This means that for sufficiently large concentrations of photogenerator A the rate of formation of B becomes independent of the concentration of generator. This is usually the desired situation for photochemical titrations. However one should avoid the temptation to make the concentration of A too large, for a second consequence, the depth effect, is seen when Eq. (6) is differentiated with respect to depth of solution, b.

$$\frac{d}{db}\left(\frac{dC_B}{dt}\right) = -\frac{(a_A C_A)}{0.434} \Phi I_0 \exp\left(\frac{-a_A b C_A}{0.434}\right) \tag{8}$$

This means that the rate of titrant generation will decrease with increasing depth of solution. The extent of this decrease depends on both the molar absorptivity and the concentration of the generator A. It is impossible to have homogeneous generation of B, and the situation becomes increasingly inhomogeneous as the concentration of A increases (*5*). This problem is a basic one in photochemistry and has complicated the situation in many cases (*6*, *7*). The solution can, of course, be rendered nearly homogeneous. Bricker and Schonberg, in their initial publication in the field, found that stirring was required, but the exact rate was not critical. On the other hand, Yokoyama and Ikeda (*8*) later found that stirring was not necessary when small volumes (5 ml) and short path lengths (10 mm) were used for titrations. Under such conditions, diffusion of reactants and products is apparently enough to approximate homogeneity.

Finally, if one considers the form of Eq. (6) for values of absorbance less than 0.05, a simplified form is obtained by a Taylor-series expansion of the exponential term; the result is identical to the equation for fluorescence intensity as a function of concentration at low absorbances (*9*, *10*).

$$\frac{dC_B}{dt} = \Phi I_0 a_A b C_A \tag{9}$$

This means that the rate of formation of B at any time is a function of the instantaneous concentration of A. This is the basis for photochemical kinetic analysis (*9*). Since this case applies only for very small concentrations of A, it is the most sensitive photochemical method for the determination of A. Also, just as in fluorescence (*10*), the sensitivity of the determination can be increased by increasing the intensity of photolytic radiation.

Returning to the case of importance for photochemical titrations given by Eq. (7), we see that the concentration of B at time t, $C_{B,t}$, can be obtained by integration of Eq. (7) over the time interval Δt,

$$C_{B,t} = \Phi I_0 \, \Delta t \tag{10}$$

Since the generation is carried out with no dilution, the amount of B formed is directly proportional to the time of photolysis, Δt. This is the fundamental basis for the linear relation found between the photolysis time to the end point and the amount of sample originally taken for photochemical titration. The way in which Eq. (10) is implemented to establish a calibration line is considered in the following two sections.

III. Principles of in Situ Titrant Generation

To date, most photochemical titrations have used in situ generation, in which the sample to be titrated and a titrant generator are irradiated together. The major advantage of the in situ method is that the sample is not diluted during titration. The major disadvantage is that the absorption spectrum of each constituent must be considered; substances present which absorb ultraviolet and thus decrease the intensity available for titrant generation will yield high results for the substance sought due to the prolonged titration time.

The three possibilities for using photochemical titrant generation for in situ analysis are called *direct*, *indirect*, and *cyclic* titration. The distinction is made on the basis of the interrelation of the generation and titration steps.

A. *Direct*

In a direct titration, the sample *A* is titrated directly with photons to product *B*. The concentration of either the sample or one of the photolysis products can be followed by some technique which is a linear function of concentration. The titration reaction is represented by Eq. (1); if the concentration of *A* is large enough, then Eqs. (7) and (10) apply, and the titration rate depends only on the incident intensity. At some point the concentration of *A* decreases to a point where the rate expression becomes more complex (*5*, *6*, *9*). However, an extrapolation of data points from early in the titration allows one to obtain an end point where the titration should have been complete had light absorption remained constant. Such

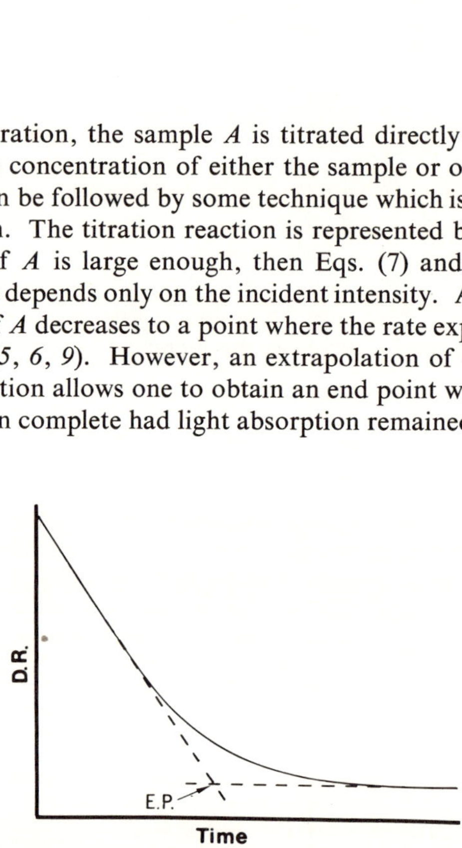

FIG. 5-2. Determination of end point for direct photochemical titration. D.R., detector response, in relative units, from the device used to follow concentration of sample *A*. Linear response with change in C_A assumed; Time, time of exposure to photolysis radiation, in minutes; ——, trace of value of C_A experimentally observed (as in recorder trace of D.R. vs. time); – – –, extrapolation of initial and "infinite time" portions of experimental trace; E.P., extrapolated end point. Titration time which would have been required if light absorption had remained near 100%.

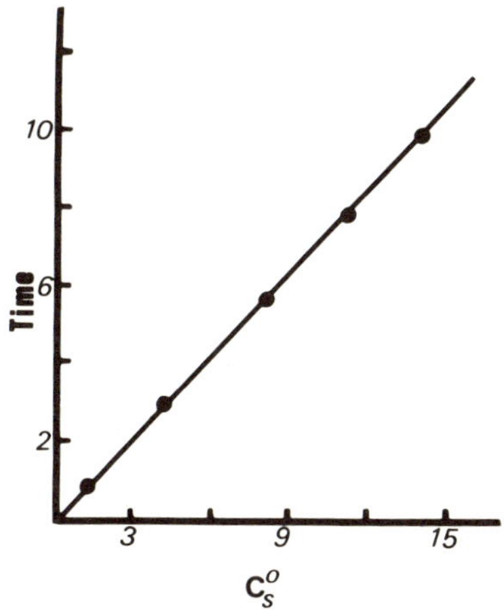

FIG. 5-3. Typical calibration plot for photochemical titrations. C_s^0, original concentration of sample taken, in micromoles or microequivalents; Time, time to extrapolated end point (shown in Fig. 5-2), in minutes.

an extrapolation is illustrated in Fig. 5-2. An example is the direct titration of quinine (*11*). Fluorescence intensity was used to follow the titration.

The final step in the method is the construction of a calibration plot. The extrapolation technique results in a "forced fit" of the titration to the linear time–concentration relation given by Eq. (10). Therefore a plot of initial concentration of A versus time to the extrapolated end point yields a straight line as shown in Fig. 5-3. For increased precision, a mathematical least-squares fit can be used. If this is done, one is not limited by the readability of the graph paper.

B. *Indirect*

The next most complicated situation is an indirect titration in which the titrant generation step is given by Eq. (1). This first step is followed by:

$$B + C \rightarrow D + E \tag{11}$$

in which the titrant is now B; this titrant is photochemically generated and reacts with the sample C to yield product(s). As a general principle, the concentration of A should be much greater than any other constituent (sample or products or other "inert" absorber) in the solution. This assures that: (1) the concentration of A will be such that the generation rate remains constant; and (2) the photolytic radiation is essentially absorbed only by substance A, which protects the rest of the sample via the "inner-filter effect" (*12*).

With indirect photochemical generation, the rate of formation of B can be described by Eq. (7) and the concentration of B at any time is calculated from Eq. (10). This means that sharper titration curves are obtained as shown in Fig. 5-4. The only cause of rounding in the indirect case lies

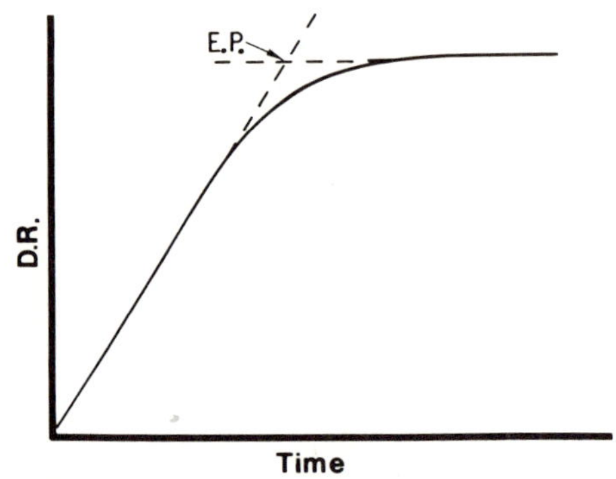

FIG. 5-4. Determination of end point for indirect or cyclic photochemical titrations. D.R., Detector response, as defined in Fig. 5-2 (shown here for detector which follows product of titration reaction); Time, photolysis time, as defined in Fig. 5-2; ——, trace of concentration of products (deviation from linearity near E.P. due to incomplete reaction at equivalence); – – –, extrapolation to locate end point; E.P., end point.

with the equilibrium constant for the titration [Eq. (11)], which determines the completeness of the reaction at the equivalence point.

Calibration plots of photolysis time to the end point versus concentration of sample C are constructed in the same manner as for the direct titration case (Fig. 5-3).

Application of indirect generation to date has been largely for precipitation titrations (*13*, *14*) where the major problem turns out to be the light-scattering properties of the precipitate itself. Intensity losses due to scattering can cause deviations in the titrant generation rate (*15*). In this case, nonlinear calibration curves are obtained. The explanation is simple: As the amount of sample taken increases, the amount of precipitate formed is larger and thus light scattering increases. This phenomenon determines the upper concentration limit which can be titrated. The lower limit will be set by the K_{sp} of the precipitate involved (*14*) if the instrumental limit is not reached first (*13*).

C. *Cyclic*

Probably the most desirable titration condition is one in which the generator is itself regenerated in the titration step. This is termed cyclic in situ generation, and is common in redox titrations (*1*, *2*, *8*). The reaction following the titrant generation [Eq. (1)] is given by

$$B + C \rightarrow D + A \tag{12}$$

Since the titrant generator A is a product of the titration reaction, the concentration of generator remains constant until all of sample C has been consumed. The first photochemical titration (*1*) used such a system, in which iron(III) oxalate was photoreduced to iron(II), which in turn reacted in situ with the oxidants which were to be determined so that iron(III) was regenerated. An excess of oxalate was required since it was irreversibly oxidized in the generation reaction.

It can be seen that the titrant generation rate, given by Eq. (6), will remain constant for any concentration of A so long as the value is a constant. Therefore, the exact value of C_A should be unimportant if a cyclic system is used. However, for reproducibility from sample to sample, it is best to maintain the absorbance of the generator greater than 2 so that the rate is concentration independent [Eq. (7)]. As a general rule, cyclic titrations are preferred since any amount of sample can be titrated, given enough irradiation time. Of course, both the sample and all the titration products must be only weak ultraviolet absorbers if their concentrations are comparable to that of the generator.

The titration curves obtained with cyclic generation and the calibration plots prepared from such titrations are identical to those which result from indirect titrant generation (Figs. 5-3 and 5-4).

The important requirements for a good in situ titrant generator are relatively few, but rigorous. It is essential that the quantum yield Φ be quite constant and insensitive to a variety of solution conditions such as pH and ionic strength. If there is a wavelength used for generation where the quantum yield of the generator is of questionable reproducibility (*16*) or where the molar absorptivity is quite small, it is best to remove that particular wavelength region from the generating beam with filters (*15*).

A large molar absorptivity is in fact not an essential requirement if the concentration of the generator can be large. For example, mono- and trichloroacetic acids, which have molar absorptivities less than 10 at 254 nm, can be used to generate chloride ion (*14*). The solid reagent is added just prior to titration, and the amount is not critical so long as the minimum weight to yield an absorbance of 2 is exceeded.

The final factor for successful in situ generation is the requirement of a constant incident intensity, I_0. The stabilizing of ultraviolet sources has been considered in detail elsewhere (*3*, *17*) and is also discussed in Chapter 1. In essence, photochemical titration systems which have been used to date have established that intensity can be stabilized to better than 1% (short term) by control of either voltage (*1*, *8*, *11*) or lamp current (*2*). All workers using medium-pressure mercury sources have reported a slow deterioration of the lamp over several months (*1*, *2*, *11*). This effect should be investigated further. Presently, one can only say that calibration curves constructed on a given day will serve for several weeks thereafter. Low-pressure mercury arcs should probably be investigated further due to their long lifetimes and extreme stability (*18*, *28*) although they operate at lower intensity. Of course, the solid angle subtended by the titration container must be reproduced fairly exactly so that the incident intensity is the same for all samples. This fact was recognized in the first photochemical titrations (*1*) and all subsequent workers have used some clamping (*2*, *28*) or mounting system (*11*, *14*, *21*).

IV. Principles of External Titrant Generation

It should be obvious by now that the major difficulty for in situ generation is that *any* extraneous substance which competes with the titrant generator in the absorption of the photolytically effective radiation makes the method unsuitable for routine analytical work. One solution to this problem is to physically separate the location of the generation and titration reactions. The technique was originally developed for coulometric titrations; the titrant is generated in a flowing stream and then added to the sample (*19*). It has been demonstrated that the same idea can be used in photolytic titrant generation (*18*). In this way, the titrant generator can be converted completely to active titrant, and competing absorbers in the sample have no effect. The price that one pays for this convenience is additional complication of the apparatus and end-point methods. Of primary importance is the fact that the sample is diluted.

In addition, apparatus for providing a constant flow rate, and thereby a constant rate of titrant addition, is required. At 100% photoconversion, the intensity of the photolysis source is no longer important and a low-intensity source can be used due to the longer exposure times possible.

Only one example of external generation has been reported (*18*). The most important finding seems to be that photolysis in a flowing stream can be carried out with precision better than 1%. The use of photochemical analytical methods for flowing streams is a possibility. Whether or not external generation for photochemical titrations is applied to other systems remains to be seen at this time. Certainly in situ methods will prevail in the case where the phototitrant is an extremely unstable and transient species. Oxygen-sensitive titrants are also best handled with enclosed in situ systems (*2*, *20*, *28*).

Apparatus for carrying out photolytic reactions in flowing streams is discussed in Chapter 1. The major experimental advantage of the photolysis is that the system can be completely closed and the generating radiation introduced via a quartz window (*18*) or other suitably transparent material, depending on the generating wavelengths used (*20*).

Since dilution of the sample occurs, logarithmic end-point detection such as potentiometry (*18*) is preferable to linear detection such a photometry (*1*). If a constant flow rate is used, then the time of titrant addition required to reach the end-point is proportional to the amount of sample taken. A calibration plot identical to one obtained with in situ generation (Fig. 5-3) can be constructed (*18*).

V. Review of Established Photochemical Titration Systems

With the general principles described above, we can now consider the photochemical titrations which have been reported to date. A summary of these is presented in Table 1. There are three generalizations about existing methods which can be drawn from this summary.

First, one sees that the usual sample concentration determined in photochemical titration is usually between one and a few hundred micromoles. The limits of concentration are determined by the quantum yield for the particular photochemical reaction employed for titrant generation. The output from lamps with suitable stability is usually a few microeinsteins of photons per minute. With the realistic limitation that a titration should take no more than 10 to 20 min, the sample size is therefore

TABLE 1. SUMMARY OF PHOTOCHEMICAL TITRATIONS

Titrant	Photogenerator	Class[a]	Substances titrated	Amount determined[b]
Fe(II)	Fe(III) in oxalate–H_2SO_4	Cyclic	Cr(VI), V(V)	150–1200 μequiv 25–1300 μequiv
"Reduced anthraquinone"	Anthraquinone in basic methanol	Cyclic	O_2, Cu(II)	0.4–2 μmoles 0.7–5 μmoles
Co(II)	Co(III)(oxalate)$_3$	Indirect	$Fe(CN)_6^{3-}$	20–500 μmoles
Cl^-	Mono- or tri-chloroacetic acid	Indirect	Ag(I)	0.5–100 μmoles
Photons	None	Direct	Quinine sulfate	0.1–4 ppm
W(V)	W(VI)–glucose/HCl	Cyclic	Cr(VI), V(V), $Fe(CN)_6^{3-}$	3–30 μequiv
Fe(II)	Fe(III) in oxalate–H_2SO_4	External	V(V)	10–200 μequiv
U(IV)	U(VI) in lactic acid	Cyclic	Tl(III)	10–100 μmoles
Co(II)	Co(III)(oxalate)$_3$	Indirect	Mg(II), Zn(II), Cd(II), Cu(II)	1–15 μmoles
Fe(II)	Fe(III) in citrate buffer, pH 4	Indirect	1,10-phenanthroline, 2,4,6-tripyridyl-*s*-triazine	1–40 μmoles
"Reduced anthraquinone"	Anthraquinone in NH_3-i-propanol	Cyclic	O_2, Cu(II)	0.6–30 μmoles

[a] Refers to method of titrant generation. See Secs. III and IV.
[b] As reported in paper or estimated from calibration plots.

limited to the low-millimolar or micromolar region. On the other hand, with such low sample concentration, it is seen that only a few millimoles of generator are necessary in order to absorb virtually all of the incident radiation, and also to assure that more than enough generator is available in the case of indirect titrant generation.

Second, it can be seen in Table 1 that the most popular end-point

Reported precision,[c] in %	End point[d]	Comments	Ref.
3 1.5	Spectrophotometric	Established feasibility. Reproducible location of sample container under source important variable. Induced reactions with oxalate found.	*1*
5	Photometric or amperometric	"Absolute" method. Used actinometry and known Φ for anthraquinone reduction. Nonaqueous solvent.	*2*
2	Turbidimetric	Precipitation titration. Titrant generator "substitution inert" complex.	*13*
3	Turbidimetric or nephelometric	Hg arc used as both a generation and a photometric source.	*14*
1.6	Fluorimetric	Hg arc used for both titration and excitation source.	*11*
3	Photometric	Detected excess titrant. O_2-sensitive titrant so closed system used for titration.	*20*
1	Potentiometric	Demonstrated external titrant generation with both 100% and incomplete photoconversion.	*18*
5	Photometric	Demonstrated that stirring not required in small volumes. O_2 interference "pretitrated."	*8*
1.5	Fluorimetric	Ligand displacement (fluorescent to nonfluorescent complex) titration.	*21*
1	Photometric	Complexation titration. Buffer part of photogenerator, also facilitates complex formation.	*27*
2	Electron spin resonance	ESR follows reduction of paramagnetic Cu(II) to diamagnetic Cu(I).	*28*

[c] RSD or other estimate of reproducibility. As reported in paper, or estimated from data given.
[d] Instrumental method used for measurement of photolysis time to end point. (Some are incremental, and some continuous measurements.)

detection system is some form of spectrophotometry. This is understandable because a detailed knowledge of the absorption spectra of the various constituents is needed before a titration system can be set up. So it is tempting to use the spectral information already available to develop an end-point detector. On the other hand, photometric detectors cannot be directly exposed to intense radiation of the photolysis source. Various methods have been used to prevent exposure of the detector to the generation beam; most of these methods are discussed in Chapter 1. In general, a double-shutter system (*1*), selective filters (*2*, *11*, *14*, *21*), and appropriate alignment of the photolytic beam at right angles to the analytical radiation (*8*, *15*, *20*) have all been used. Electroanalytical end-point detection is preferred since electrode processes are only slightly affected by ultraviolet radiation (*7*). Amperometry (*2*) and potentiometry (*18*) have been used already, and related techniques should probably be investigated further. Electron spin resonance has recently been reported as an end-point technique (*28*). As will be discussed shortly, methods which can be used to follow the titration continuously are both more convenient and more precise.

Third, as summarized in Table 1, the precision for all of the titrations to date is in the 1–5% range. The exact sources of the errors are not known in most cases. A combination of variables such as minor lamp-intensity fluctuations, small changes in sample placement in the photolytic beam, and reading errors in end-point estimation usually all contribute to the standard deviation. With one exception the relationship between the photolysis time to the end point and the concentration of sample taken has been established by use of calibration with standards. Kuwana (*2*) used the ferric oxalate actinometer (see Chapter 3) to establish the photon flux in the sample cell, and then replaced the actinometer with the copper solution to be titrated and measured the photolysis time to produce the first excess of phototitrant. The total number of photons incident on the sample can thus be calculated. The quantum yield of the anthraquinone photolysis is then used to calculate the amount of copper originally present. This method is absolute provided that (1) the quantum yield of generation does not change during the titration, and (2) the lamp intensity and geometry of sample orientation do not change between the actinometry and titration. Since the precision obtained by this rather tedious method is not better than that obtained by other workers with standard solutions to set up a calibration plot, the choice of which standardization method to use should be made strictly on the basis of convenience.

A fourth major consideration in the application of a photochemical titration method is not listed in Table 1. This is the choice between carrying on continuous monitoring of the sample during photolysis (*2*, *11*, *14*, *15*, *18*, *20*, *21*) and using an incremental method; in the latter, photolysis is carried out for some time interval, then a shutter between sample and source is closed and a measurement is made (*1*, *8*, *13*). There is no fundamental reason why the various titrations in Table 1 could not be carried out with either monitoring method. However, as with all analytical methods which depend on the accurate measurement of a time interval, some consideration must be given to those factors which cause error in the measurement. Incremental methods must include the sum of the start–stop errors in the clock used (*4*). However, if there is a sluggish reaction between titrant and sample [Eq. (12)], interruption of the titration at intervals can be useful in driving the reaction to equilibrium at all points (*8*, *15*, *27*).

Continuous monitoring of the progress of the titration with a strip-chart recorder provides some signal averaging, and an extrapolation such as that shown in Fig. 5-2 can be made more acurately. The time to the end point is conveniently read directly from the chart paper. However, the precision is thereby limited by the readability and reproducibility of the chart paper. If an appropriate end-point signal can be obtained just at the time of the equivalance point, digital time-measurement techniques (*22*) can be used; and it is anticipated that increased precision should result. No application of digital measurement in photochemical titrations has been reported as yet.

The careful reader will note that Table 1 contains no information regarding interferences in the various titrations. The interferences in such titration systems have not been investigated extensively. However, in the case of redox and precipitation titrations, most of the interferences in the titration reaction are already well known, and so only the interferences for the generation step need to be evaluated. Interferences in generation rate fall into two groups. First are those substances which affect the quantum yield of the generation reaction. If the photochemistry of the generation reaction has been well studied, then one can predict interferences. The second group of interfering substances are those which are good uv absorbers. Substances such as nitrate can seriously decrease the intensity available for titrant formation (*23*). This is the most serious potential type of interference since it is the most difficult to detect a priori.

Photochemical titrations with determination of two or more constituents in a single sample are possible. However, all such titrations to date utilize selective end points rather than any differentiation between constituents by the titrant due to solution conditions (such as masking). For example, Kuwana (*2*) was able to distinguish between oxygen and copper in a mixture, both of which oxidized the titrant, by using amperometric detection of the reductions of Cu(II) to Cu(I) and Cu(0). Similar selectivity for the same titration system has been obtained by monitoring the ESR signal of the ammonia complex of Cu(II) (*28*). However,

Yokoyama and Ikeda (*8*) were unable to distinguish between oxygen and thallic ion since they followed the absorbance due to excess phototitrant [U(IV)] and obtained no signal until after the sample was entirely consumed. In the latter case, the oxygen in the sample caused a sizable intercept in the calibration plot for determination of Tl(III); thermostating the sample was used to maintain the oxygen concentration essentially constant from sample to sample. In a compleximetric photochemical titration, Beck *et al.* (*21*) found that they could not differentiate mixtures of Zn(II), Cd(II), and Mg(II) since the 8-hydroxyquinoline-5-sulfonic acid complexes of all three had similar fluorescence emission spectra; pH masking was not tried, but the formation constants for the three complexes are so similar that it is doubtful whether such a procedure would be successful. Bricker and Schonberg (*1*) differentiated Cr(VI) and V(V) in mixtures by following the absorbance due to V(IV) which does not form until after all the Cr(VI) has reacted. Unfortunately, the oxalate medium used for titrant generation also caused induced reactions with both oxidants (*18*) so that the precision for mixtures was poorer than if the two substances were titrated separately.

To summarize the present status of photochemical titrations, it is clear that the general range of applicability of photochemical generation has been adequately demonstrated; only acid–base titrations have not been carried out photochemically. The method is moderately precise and sensitive. Realistically, the best application should be for routine titration of a large number of samples. Certainly the absence of sample dilution and elimination of large volumes of standard titrant are attractive features. Whether or not such applications develop remains to be seen at the present time.

Some speculation about the future development of new photochemical titrations is in order; these observations are based on the review of the literature summarized in Table 1, the author's own research program, and some present general trends in photochemistry. One can expect advances in both the chemical systems used and the instrumentation and methodology employed.

Certainly the area of energy transfer (*24*, *25*) should be investigated for potential photochemical titration systems. For energy transfer to occur an absorber is excited by irradiation and the excited-state donor then transfers its energy to an acceptor molecule in the near vicinity. The second substance can undergo some sort of photochemical reaction, while the original absorber returns to the ground state. The reaction sequence is

$$D + h\nu \rightarrow D^* \tag{13}$$

$$D^* + A \rightarrow A^* + D \tag{14}$$

$$A^* \rightarrow \text{Products} \tag{15}$$

in which A is the acceptor and D is the donor (*25*). Comparison of Eqs. (1) and (12) with (13)–(15) shows that this system could be used for a cyclic photochemical titration of acceptor A. Since many donors are known and their energy properties are tabulated (*24*) there is reason to believe that titrations of various organic compounds in a variety of nonaqueous solvents could be developed. The details of the internal conversion of singlet donor to triplet and the requirements of spin conservation for energy transfer do pose some complications, however (*24*, *25*).

Titrations in sealed systems and in nonaqueous solvents where use can be made of the "remote addition" feature of photons will probably be reported in the next few years. Low-dielectric solvents which limit the use of electrochemistry should present no problem for photochemical generation.

In the area of instrumentation, one can expect that feedback stabilization of the photolysis source (*26*) and digital time measurement (*22*) will be applied to improve the precision of the method. In addition, sources other than mercury arcs should be explored. There are now an array of commercially available lamps with rich ultraviolet outputs. The intensity requirement can be reduced somewhat by using smaller samples. In this regard, the recent work of Yokoyama and Ikeda (*8*) is significant since they used only 5 ml of solution in a closed cell. This is to be compared with the volumes of 125 to 200 ml (*1*, *21*) where stirring was required to maintain homogeneity. If the sample size is reduced, the intensity and physical size of the source can also be reduced.

Just as coulometric titrations depended on background knowledge regarding fundamental electrode processes, photochemical titrations will advance as new photochemical research reveals new methods and systems worthy of practical development as titrant generators. Since research activity in fundamentals of photochemical processes is currently at a very high level, one can hope for parallel advances in photochemical titrations.

ACKNOWLEDGMENT

The author acknowledges the support of Grant E-384 from the Robert A. Welch Foundation.

REFERENCES

1. C. E. Bricker and S. S. Schonberg, *Anal. Chem.*, **30,** 922 (1958).
2. T. Kuwana, *Anal. Chem.*, **35,** 1398 (1963).
3. W. A. Noyes, Jr., and P. A. Leighton, *The Photochemistry of Gases*, Dover, New York, 1966, p. 84.
4. J. J. Lingane, *Electroanalytical Chemistry*, 2nd edition, Wiley-Interscience, New York, 1958, pp. 488–495, pp. 511–513.
5. J. M. Fitzgerald, R. J. Lukasiewicz, and H. D. Drew, *Anal. Lett.*, **1,** 173 (1967).
6. H. C. Kessler, Jr., *J. Phys. Chem.*, **71,** 2736 (1967).
7. T. Kuwana, in *Electroanalytical Chemistry* (A. J. Bard, ed.), Dekker, New York, 1966, Vol. 1, pp. 217–219.
8. Y. Yokoyama and S. Ikeda, *Bull. Chem. Soc. Japan*, **42,** 2254 (1969).
9. R. J. Lukasiewicz and J. M. Fitzgerald, *Anal. Lett.*, **2,** 159 (1969).
10. D. M. Hercules, *Anal. Chem.*, **38,** No. 12, 29A (1966).
11. R. J. Lukasiewicz and J. M. Fitzgerald, *Anal. Lett.*, **1,** 455 (1968).
12. S. Udenfriend, *Fluorescence Assay in Biology and Medicine*, Academic Press, New York, 1962, Vol. I, p. 104.
13. H. D. Drew and J. M. Fitzgerald, *Anal. Chem.*, **38,** 778 (1966).
14. J. M. Fitzgerald and E. J. Smith, Abstracts, 154th National Meeting, American Chemical Society, Chicago, Illinois, Sept. 1967, No. B46
15. H. D. Drew, Doctoral dissertation, Seton Hall Univ., S. Orange, N.J., 1967.
16. W. M. Riggs and C. E. Bricker, *Anal. Chem.*, **38,** 897 (1966).
17. J. G. Calvert and J. N. Pitts, Jr., *Photochemistry*, Wiley, New York, 1966, pp. 686–700.
18. H. D. Drew and J. M. Fitzgerald, *Anal. Chem.*, **41,** 974 (1969).
19. D. D. DeFord, J. N. Pitts, Jr., and C. Johns, *ibid.*, **23,** 938 (1951).
20. E. J. Smith and J. M. Fitzgerald, Abstracts, 156th National Meeting, American Chemical Society, Atlantic City, N.J., Sept., 1968, No. B76.
21. J. L. Beck, J. M. Fitzgerald, and J. A. Bishop, *Anal. Chim. Acta*, **51,** 935 (1970).
22. H. V. Malmstadt and C. G. Enke, *Digital Electronics for Scientists*, Benjamin, New York, 1969, pp. 23–24.
23. M. Das, A. H. A. Heyn, and M. Z. Hoffman, *Talanta*, **14**, 439 (1967).
24. N. J. Turro, *Molecular Photochemistry*, Benjamin, New York, 1967, pp. 92–132.
25. S. R. Abbott and D. M. Hercules, *Anal. Chem.*, **42,** 171 (1970).
26. H. L. Pardue and S. N. Deming, *ibid.*, **41,** 986 (1969).
27. J. M. Fitzgerald and J. L. Beck, *Anal. Lett.*, **3,** 531 (1970).
28. J. M. Fitzgerald and D. E. Warren, *ibid.*, **3,** 623 (1970).

Part III

ANALYTICAL METHODS FOR THE STUDY OF PHOTOCHEMISTRY

CHAPTER 6

ANALYTICAL PROBLEMS IN THE STUDY OF PHOTOCHEMICAL REACTIONS

Earl L. Wehry

DEPARTMENT OF CHEMISTRY
UNIVERSITY OF TENNESSEE
KNOXVILLE, TENNESSEE

In this chapter, we consider some aspects of the application of analytical chemistry to elucidation of photochemical reactions in liquid solution. The reader will observe that this topic represents an inversion of the principal theme of this book. It is evident, however, that progress in analytical applications of photochemistry may be seriously inhibited by the present lack of fundamental understanding of photochemical processes in solution. The analytical chemist cannot remain unaware of activity in "mechanistic" photochemistry since such research will have important implications for analytical photochemistry. Furthermore, analytical chemists should be able to make important contributions to the study of photochemical reactions since, as we shall see, procurement of reliable analytical data is a sine qua non for understanding photoresponsive processes. We consider herein some important aspects of the analytical chemistry of systems undergoing photochemical change. Reference is made to Calvert and Pitts (*1*), who treat in detail the broad general problem of inferring mechanistic pathways in photochemical systems from analytical data.

I. General Considerations

The practice of chemistry has been revolutionized by the advent of modern, rapid, sensitive instrumental techniques; nowhere is the validity of this statement more evident than in the field of solution photochemistry. Regrettably, much of the pre-1950 literature in the area is of limited value owing to erroneous conclusions derived from incorrect or (very often) insufficient analytical data.

Analysis of a photoreacting system poses all the problems encountered in "ordinary" analytical chemistry. In addition, photochemical systems usually exhibit several unique characteristics which pose additional difficulties. The most fundamental problem encountered in the analysis of photoresponsive systems in liquid media is the fact that primary photochemical products are very frequently thermally unstable and quite labile. In general, the final, stable products of a photochemical reaction are formed via one or more thermal reactions involving the primary products, which may themselves have but a very transient existence. Hence, it is seldom that thorough qualitative and quantitative analyses of final, thermally stable reaction products are themselves sufficient to enable postulation of a mechanism for the primary photochemical process.

In addition to a search for thermally unstable ground-state photoproducts, a thorough photochemical study usually requires that the nature and eventual fate of one or more electronically excited species be ascertained. These are usually short-lived. It is obvious therefore that the analysis for transient species (many of which are not encountered in thermal chemistry) is an important component of photochemical research. Because the analysis for transient molecular species is itself a frontier in contemporary analytical chemistry, the difficulty of solving this problem in complicated photochemical systems is indeed formidable.

The usual thermal instability of primary photoproducts has another important consequence of concern to the analytical chemist. There exist many photochemical systems in which two or more competitive thermal reactions involving the primary product (or a subsequent transient species) can occur. If large concentrations of unstable photoproducts are generated, secondary reactions may occur which are capable of producing misleading distributions and quantities of final products. The ultimate product distribution may then depend on the fraction of starting material which has been photolyzed, as well as the rate with which it was consumed. In order to minimize such problems, one ordinarily limits the fraction of decomposition to 2% or less of the original quantity of reactant. Then, however, it becomes necessary to identify and determine small concentrations of products in the presence of a very large excess of starting material. It is therefore not surprising that chromatographic techniques are experiencing increasing use in the analysis of photoresponsive systems.

Since only light which is absorbed can be photochemically active, time considerations usually dictate that highly absorbing solutions be used in photochemical irradiations. Absorption of light by the reactant will then be geometrically inhomogeneous; consequently, it is possible that the rate of generation of primary photoproduct may be different in different portions of the photolysis cell. This situation is rather uncommon in thermal chemistry. Kinetic data obtained in any system in which the local rate of product formation is not constant throughout the reaction cell must be interpreted with considerable circumspection.

Elucidation of a photochemical reaction frequently proceeds by a series of three more or less well defined stages:

(A) First, the final (stable) products must be identified, and the rates or quantum yields of their formation determined.

(B) Next, information relating to the identity of excited states and transient intermediates is sought by relatively indirect means. Such experiments, which are described in a subsequent section, are usually easy to perform but require great care in their interpretation.

(C) Finally, the results of the first two stages of research suggest experiments by means of which reasonably "direct" observation and identification of excited states and transient intermediates are sought. The most popular techniques include electron spin resonance, photoluminescence, and flash photolysis. Most of these experiments are difficult, time-consuming, and require fairly elaborate apparatus; consequently, they are not usually attempted until more facile experiments have progressed sufficiently to suggest a tentative mechanism which can be used as a working hypothesis.

We consider below some important analytical difficulties encountered in each stage of a photochemical study. To provide an idea of what we may be up against, consider the photolysis, in aqueous solution, of $Co(dap)_3^{3+}$, where "dap" represents 1,2-diaminopropane. It is believed that the mechanism for this photoreaction is quite simple, corresponding to a homolytic cleavage of a Co–N coordinate bond. Yet the product distribution is formidable; among the identified products were $Co(aq)^{2+}$, ammonia, nitrogen, free 1,2-diaminopropane, 2-aminopropanol, ethylamine, and formaldehyde. At least six other products were noted by thin-layer chromatography but were not identified (*2*). In even a relatively simple photoreaction, a host of thermally stable products may be formed, to say nothing of the number of transients which may be produced.

II. Identification and Efficiency of Final Product Formation

In principle, identification of final photoproducts poses no special difficulties other than the circumstances described in Sec. I. Hence, any of the multitude of instrumental and classical analytical techniques may be used to accomplish this task. Since the products are usually present in much smaller quantities than the starting material, considerable trial and error may be required to choose suitably sensitive and specific techniques. With this constraint in mind, one can apply any analytical technique used in thermal chemistry to a photoresponsive system.

We now consider quantitative determination of the overall efficiency of a photochemical reaction. We choose the reaction sequence

$$A + h\nu \xrightarrow{\text{primary process}} X \xrightarrow{\text{one or more thermal steps}} Z \tag{1}$$

where X is the primary product (assumed to be unstable and labile) and Z the thermally stable product. One can measure the overall quantum yield for either appearance of Z or disappearance of A:

$$\Phi_Z = \frac{\text{Moles of } Z \text{ formed/unit volume/unit time}}{\text{Einsteins of radiation absorbed/unit volume/unit time}} \tag{2}$$

$$\Phi_A = \frac{\text{Moles of } A \text{ consumed/unit volume/unit time}}{\text{Einsteins of light absorbed/unit volume/unit time}} \tag{3}$$

Determination of an overall quantum yield thus requires measurement of three quantities:

(a) the intensity of the radiation source (Einsteins/unit time);

(b) the fraction of the source output absorbed per unit volume per unit time by A; and

(c) the quantity of Z formed, or of A consumed, during the experiment.

Measurement of source intensity, a problem of serious experimental complexity, has been discussed by Calvert and Pitts (*1*). This determination is often accomplished by chemical actinometry, a topic discussed elsewhere in this volume. Quantity (c), the measure of extent of reaction, can be a difficult matter, as outlined in Sec. I; with sufficient ingenuity, persistence, and luck, this measurement can usually be made with a high degree of accuracy. Rather surprisingly, evaluation of quantity (b), the rate of light absorption by the sample, can cause severe difficulty, failure to consider which has led to the publication of many incorrect quantum yields. Accordingly, we consider some aspects of this problem below.

If photolysis of A occurs with appreciable conversion of A to Z, it is probable that the absorbance of the solution will change significantly during the course of the experiment. In that case, assuming Beer's law to be valid, one must integrate the absorbed intensity, I_a, over the duration of the experiment:

$$\int_{I_a(0)}^{I_a(t)} dI_a = \int_0^t I_0(1 - e^{-\epsilon b[A]})\, dt \tag{4}$$

in which $[A]$ varies with time. The situation may be further complicated if it happens that concentration gradients are produced within the cell; in that case, one cannot assume adherence to Beer's law since these gradients will exhibit a time dependence if appreciable quantities of A are consumed. A mathematical treatment of this case has been published by Kessler (*3*). This factor can constitute an important source of error in overall yield determinations and should be explicitly considered in the evaluation of quantum-yield data.

To further complicate the situation, I_0, the incident intensity, is a function of the distance between an absorbing molecule and the front face of the cell, unless the absorbance of the solution is quite small. One normally assumes that the concentration of A is sufficiently constant such that Beer's law is valid, whereupon I_0 varies exponentially with x (distance from the front face of the cell). However, if $[A]$ is a function of time in a strongly absorbing solution, Beer's law may fail, in which case it becomes necessary to correct simultaneously for the time effect (see preceding paragraph) and for the distance factor (*3*).

Some of these problems can of course be minimized by suitable experiment design. A very common approach is to photolyze such a small fraction of A that $[A]$ does not change appreciably during the irradiation. This approach should in fact always be used when feasible, but it can cause complications in analyzing the final solution composition. A second approach which is occasionally followed is to utilize such an extremely large concentration of A that even after extensive photolysis virtually all incident radiation is absorbed. This procedure is quite simple from the experimental standpoint and eliminates the time dependence of the absorbance (*if* the inner-filter effect is of negligible significance; see below), but it limits the uses to which the measured yields may be put. In particular, one must be exceedingly cautious in attempting to convert yields so obtained to kinetic information. The reason, of course, is the geometric inhomogeneity of I_a in highly absorbing solutions.

We may usually view a primary photochemical process as a bimolecular reaction between an A molecule and a photon:

$$A + h\nu \rightarrow X \tag{5}$$

in which

$$-\frac{d[A]}{dt} = \phi I_a[A] \tag{6}$$

where ϕ is the primary-process quantum yield and I_a is the radiation intensity absorbed by A. It is generally correct to assume that ϕ and $[A]$ will be constant throughout the reaction vessel, but clearly this cannot be true for I_a. Thus the rate of the primary process will be greatest near the front of the cell, and appreciably slower near the rear. If $[A]$ is known as a function of time, and if it is assumed that there are no secondary reactions in which the primary product is generated or consumed, it is usually possible to make valid inferences concerning local reaction rates from measured overall quantum yields. Very frequently, however, the assumption stated in the preceding sentence is invalid. If the primary product engages in thermal reactions, their rates will also be different in different volume elements of the cell unless the thermal reactions are relatively slow or highly efficient stirring is employed. The relationship between the measured overall quantum yield and the local rate of the primary process then involves consideration of at least two reactions (the primary photochemical process plus one or more thermal reactions), all proceeding at rates which are a function of distance from the front face of the cell. This situation is mathematically intractable unless simplifying assumptions are made. Our conclusion, then, must be that attempts to relate overall quantum yields to local rates in totally absorbing solutions are frequently unsuccessful. It is obvious that if secondary thermal reactions are occurring at different rates in different volume elements within the cell, the measured overall quantum yield will be a function of I_0, the light intensity incident upon the front cell face. The variation of overall quantum yield with I_0 should therefore be routinely investigated in a photochemical study, especially since good (and relatively inexpensive) neutral-density filters are now commercially available.

Another important experimental problem in solution photochemistry is the "inner-filter effect," in which either an intermediate or final product absorbs radiation of that frequency used to excite the reactant. For simple cases, the influence of the inner-filter effect can be treated mathematically *if* one assumes that both the reactant and inner filter obey Beer's law at all times during the photolysis (*4*, *5*); such treatments have been applied to actual systems with some success (*5*, *6*). If, however, time-dependent concentration gradients of either reactant or inner filter occur, Beer's law does not obtain; correction for the inner-filter effect under these conditions is exceedingly difficult. Occurrence of an appreciable inner-filter effect is commonly accompanied by an apparent change of Φ_Z or Φ_A as a function of duration of photolysis of A; if such experimental observations are made, an inner filter should be sought.

Occasionally, inner-filter effects can cause dramatic increases in overall quantum yields with irradiation time, especially if the inner filter is itself photochemically active. An example of such behavior occurs in photolysis of the dinuclear iron(III)–1,10-phenanthroline complex,

$$[(\text{phen})_2(\text{H}_2\text{O})\text{FeOFe}(\text{H}_2\text{O})(\text{phen})_2]^{4+},$$

in aqueous media. This reaction proceeds according to the stoichiometry

$$3[\text{Fe}_2(\text{phen})_4\text{O}(\text{H}_2\text{O})_2]^{4+} \xrightarrow{h\nu} 4\text{Fe}(\text{phen})_3^{2+} + 2\text{Fe}(\text{aq})^{3+} \quad (7)$$

The photoreaction is commonly excited with light of about 28,000 cm^{-1}, which is also absorbed by the aquo Fe(III) ion. This by itself would be troublesome since the extinction coefficients for the various aquo Fe(III) species are rather large. The situation is further complicated by the fact that aquo Fe(III) ions undergo photoreduction to Fe(II) and OH radicals. This reaction is more efficient by at least a factor of 10 than the photoreduction of $[Fe_2(phen)_4O(H_2O)_2]^{4+}$. Consequently, an appreciable fraction of absorbed light is in fact "consumed" in this side reaction if an appreciable fraction of the dinuclear complex is destroyed. Furthermore, the aquo Fe(II) produced by the inner-filter reaction undergoes an exceedingly rapid thermal redox reaction with the dinuclear complex:

$$4Fe(aq)^{2+} + 3[Fe_2(phen)_4O(H_2O)_2]^{4+} \rightarrow 4Fe(phen)_3^{2+} + 6Fe(aq)^{3+} \quad (8)$$

We note that (a) the thermal reaction produces the same iron–phenanthroline complex as the photoreaction of interest, and (b) the thermal reaction generates two moles of $Fe(aq)^{3+}$, which can undergo further inner-filter photolysis, for every mole of dinuclear iron–phenanthroline complex consumed. The reaction system is a chain; one accordingly observes a drastic increase in the quantum yield for appearance of $Fe(phen)_3^{2+}$ with increasing exposure time (*7*). Under such conditions it is obviously difficult to interpret the significance of the observed quantum yields.

We have heretofore ignored what is probably the most common source of error in overall quantum-yield determinations. Solution photochemistry is normally performed with cutoff filters or, more desirably, narrow-band interference filters or a monochromator, for selection of exciting frequency. Bandwidths of light are usually on the order of 5 nm or larger. Obviously, these bandwidths are much larger than those encountered with a high-performance ultraviolet absorption spectrophotometer. Under these circumstances, ϵ values obtained from simple uv spectrophotometry (or from the literature) will not pertain to the conditions extant in the photochemical irradiation. The usual procedure is to determine an average ϵ, integrated over the photolysis bandwidth:

$$\bar{\epsilon} = \frac{\int_{\lambda_1}^{\lambda_2} \epsilon_{\lambda_i} I_{\lambda_i}^0 \, d\lambda_i}{\int_{\lambda_1}^{\lambda_2} I_{\lambda_i}^0 \, d\lambda_i} \quad (9)$$

where ϵ_{λ_i} is the extinction coefficient, and $I_{\lambda_i}^0$ the incident intensity, at wavelength λ_i. Clearly, one must know the intensity distribution of his source–monochromator or source–filter combination to effect this correction. This is most easily obtained by directing the incident light into a high-resolution (preferably 0.5 nm or greater) spectrograph and photographing the resulting spectrum. In the absence of a high-resolution grating spectrograph, one can use a recording uv-visible spectrophotometer for this purpose, provided that its owner does not object violently to dismantling of the cell compartments. This correction must be made whenever one desires to correct for inhomogeneous light absorption, inner-filter effects, or time-dependent absorption. If the photolysis beam is also split and used to monitor the course of the reaction (as by absorption, fluorescence, scattering, or other optical measurements), it may be necessary to correct for nonmonochromaticity to obtain meaningful kinetic data.

It should be evident that there are a number of experimental difficulties to be considered in performance and evaluation of quantum-yield determinations. The problem of interpreting such a yield, once a reliable value has been obtained, is considered in detail by Calvert and Pitts (*1*).

III. "Indirect" Analytical Techniques in Photochemistry

It is frequently desirable to learn something about the identity of excited states and unstable intermediates formed in photoreactions by relatively simple, indirect experiments before proceeding to more direct, but also more difficult, experimental techniques such as matrix spectroscopy or flash photolysis. Of the extant "indirect" techniques, two are of particular note: (a) chemical scavenging of free radicals; and (b) electronic energy-transfer studies. We consider each of these methods below.

A. *Chemical Free-Radical Scavenging*

Free radicals are produced as primary photoproducts in many organic and some inorganic systems. Most radicals are exceedingly reactive species; under certain circumstances, this extreme reactivity can be utilized to detect the formation and discern the identity of a radical formed in a photochemical reaction. A free-radical scavenger is a species which undergoes a rapid thermal reaction with a radical to yield a characteristic product which can be analytically detected. For example, photolysis of a number of organic compounds (e.g., aliphatic esters, alkyl halides, alkyl sulfides, some aliphatic ketones) is postulated to involve generation of alkyl radicals as an important primary process. An effective scavenger for alkyl (and other) free radicals is molecular iodine:

$$R\cdot + I_2 \rightarrow RI + I\cdot \tag{10}$$

Relative to most free radicals, the iodine atom is rather unreactive; an important secondary reaction, therefore, is recombination of two iodine atoms to re-form molecular I_2. The product alkyl halide can be detected easily by a number of common analytical techniques; its identity is easily established by nuclear magnetic resonance and mass spectrometry. When the identity of the alkyl halide has been established, the identity of the original alkyl radical is known. This experiment is usually quite easily performed and can be enormously informative in a qualitative sense.

One of the most useful forms of radical scavenger consists of a species which can undergo a free-radical-initiated polymerization (*8*). The polymer can be isolated and analyzed to determine the identity of the initiator. Palit and co-workers have described a series of spectrophotometric procedures for identifying various functional groups present in polymethyl methacrylate, polystyrene, and related systems (*9–11*). Such polymerization systems can be used to identify primary products such as hydroxyl radicals and halogen atoms in the photolysis of transition-metal complexes, and can also be utilized to estimate kinetic parameters in such processes (*12*).

Many other species can act as radical scavengers. Oxygen and NO, both of which are ground-state triplets, react very rapidly with many free radicals. Certain metal ions, most notably Ce^{4+} and Fe^{3+}, can be reduced by radicals. A number of polyfunctional organic compounds are effective scavengers; for example, *p*-nitroso-*N*,*N*-dimethylaniline is reported to act as a selective scavenger for OH radicals in alkaline aqueous solutions (*13*). Stable radicals such as 1,1-diphenyl-2-picrylhydrazyl (DPPH) often scavenge free radicals; they are especially advantageous scavengers because the scavenging reaction does not produce other radicals.

A desirable free-radical scavenger should exhibit four important properties:

(a) It should react sufficiently rapidly with the radical being trapped to minimize or completely suppress the normal secondary reactions of that radical.

(b) The scavenging reaction should be "clean." That is, there should be one predominant reaction path, and the products thereof should not engage in a complicated series of secondary reactions.

(c) It should be possible to analyze for the products of the scavenging reaction by relatively prosaic analytical techniques.

(d) The scavenger should have no effect upon the photochemical system other than the consumption of radicals.

With sufficient trial and error, the first three criteria can usually be satisfied reasonably effectively. However, mention should be made of the fact that free radicals are highly reactive species; analysis of a photochemical radical-formation process thus requires great care in purification of reagents and solvents and in replication of experimental conditions.

A very significant limitation of radical-scavenging techniques is criterion (d). It is rare indeed that a radical scavenger does not exert other effects upon the course of a photochemical reaction. Most free-radical scavengers are efficient quenchers of electronically excited molecules. Furthermore, they are often thermally (or even photochemically) reactive species; consequently, the product distribution of the photochemical reaction of interest may be altered by the radical scavenger. In extreme cases, the reaction may be suppressed completely by the sensitizer; a good example is the photochemical reduction of $Cu(dmp)_2^{2+}$ (dmp is 2,9-dimethyl-1,10-phenanthroline), which is completely suppressed by addition of methyl methacrylate. Presumably the scavenger in this case forms a weak complex (which is photostable) with the Cu(II) species (*14*).

Molecular oxygen is an especially effective free-radical trap. Unfortunately, O_2 is also an extremely efficacious quencher of excited singlet and triplet states of many organic molecules. For many organic molecules in fluid media, excited-state quenching by O_2 occurs at or near the diffusion-controlled rate (*15*). Since the oxygen molecule has a very high diffusion coefficient in most organic solvents, oxygen quenching can be very efficient even when its concentration is quite small. Hence, oxygen, when used as a radical scavenger, may effectively inhibit the photoreaction of interest by quenching the excited state(s) involved; under such conditions, the experiment is obviously uninformative.

We consider now an organic photoreaction which illustrates both the value and the ambiguity of radical-scavenging experiments. Photolysis of 4-iodobiphenyl in benzene solution is thought (*16*) to proceed via the primary process

$$p\text{-}C_6H_5C_6H_4I \xrightarrow{h\nu} p\text{-}C_6H_5C_6H_4\cdot + I\cdot \qquad (11)$$

The primary process is, as usual, followed by a series of thermal steps, including as possibilities the following:

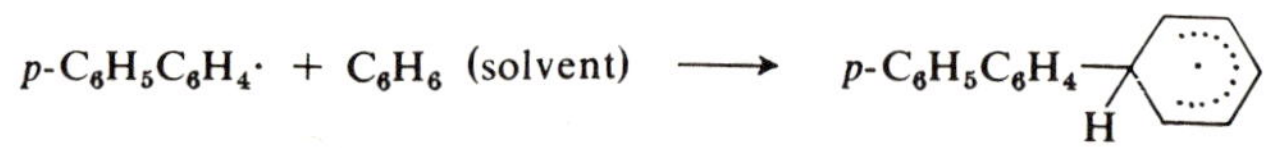 (12)

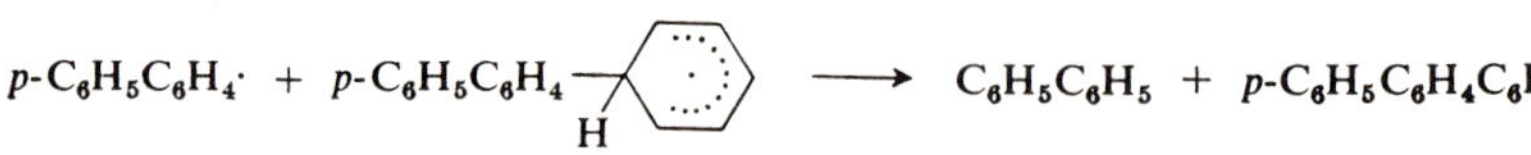 (13)

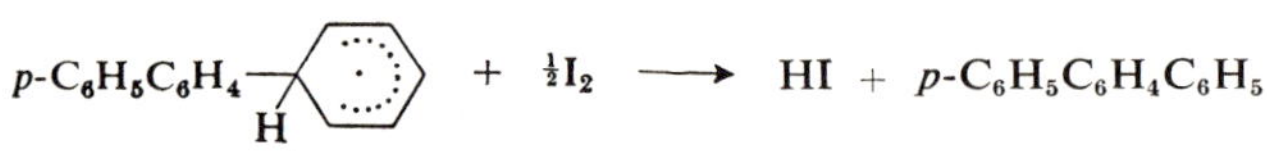 (14)

$$p\text{-}C_6H_5C_6H_4\cdot + HI \longrightarrow I\cdot + C_6H_5C_6H_5 \qquad (15)$$

$$I\cdot + I\cdot \longrightarrow I_2 \qquad (16)$$

Following the primary photoprocess, there are two important thermal pathways, leading to formation of two principal hydrocarbon products, biphenyl and *p*-terphenyl. In this system, oxygen can be utilized as a radical scavenger. It is observed that the overall quantum yield for I_2 evolution is significantly enhanced by oxygen (*16*). One rationalizes this observation by asserting that the aryl radical ($p\text{-}C_6H_5C_6H_4\cdot$) is scavenged by O_2, inhibiting recombination of the primary radicals, the reverse of process (11). By analogy with other systems, one then expects that trapping of $p\text{-}C_6H_5C_6H_4\cdot$ by O_2 will proceed through a hydroperoxide intermediate to yield *p*-hydroxybiphenyl as a final product. Indeed, if the initial solution is saturated with O_2, one observes formation of a significant yield of *p*-hydroxybiphenyl. The radical-scavenging experiment thus provides important prima facie evidence that (11) is indeed a reasonable mechanism for the primary process. It further indicates that, in the absence of scavenger, radical recombination must be important. The simple scavenging experiment therefore provides important qualitative information regarding mechanistic pathways in this system.

When one attempts to extract quantitative information from such an experiment, however, ambiguities are encountered. One would like to assert that O_2 is a sufficiently effective scavenger of $p\text{-}C_6H_5C_6H_4\cdot$ that the overall quantum yield for appearance of I_2 in the presence of scavenger is a good measure of the primary quantum yield for photolysis of *p*-iodobiphenyl. There are a number of serious uncertainties inherent in that assumption, including:

(a) Unless one can produce evidence that O_2 is indeed an exceedingly effective scavenger, the assumption described in the previous paragraph remains just an assumption. Such evidence is not easy to obtain.

(b) Little is known of the ability of O_2 to quench excited states of iodoaromatics. In the absence of such information, one cannot arbitrarily assume that O_2 quenching is unimportant in this specific system.

(c) Scavenging of p-$C_6H_5C_6H_4\cdot$ by O_2 leads to formation of p-hydroxybiphenyl. One must ascertain that this species, as well as the hydroperoxide assumed to be intermediate in its formation, does not engage in secondary reactions with the other transient species present. Again, this is not an easy task.

(d) One must be concerned with the possibility that the scavenger, even if present in small quantities, will influence the distribution of final products. Following the primary process, there are two important thermal paths, leading to formation of two principal hydrocarbon products. It is observed (*16*) that oxygen alters the *ratio* of biphenyl to p-terphenyl in the final product mixture. As oxygen is removed by the freeze–thaw technique, the *relative* yield of biphenyl increases. Apparently molecular O_2 interferes with the thermal hydrogen-transfer processes (13) and (15). One may draw incorrect inferences from the observed relative yields of stable products if he has not performed the experiment at several scavenger concentrations; indeed this is a wise precaution in all radical-trapping procedures.

We conclude that free-radical scavenging is a useful qualitative analytical tool for identifying transient radicals formed in photochemical reactions. Such experiments are easy to perform, and the voluminous literature in free-radical chemistry ensures that analogies with earlier work will be available. Such studies are thus very useful in formulation of tentative hypotheses, which are then tested by more direct means. As exemplified by the iodobiphenyl example, however, radical-scavenging experiments are of very limited quantitative utility. Even for relatively simple kinetic studies, ambiguities usually arise in treatment of scavenging data (*17*).

B. *Electronic Energy Transfer*

Use of electronic energy transfer as a probe for electronically excited states in photochemical systems is burgeoning. Electronic energy transfer in liquid solution has been the subject of several recent reviews (*18–22*); we accordingly restrict the present discussion to use of energy transfer as an analytical technique for electronically excited molecules in solution.

We consider as an example an organic photoreaction which is postulated to proceed via a triplet state:

$$A + h\nu \rightarrow {}^1A^* \xrightarrow{\text{intersystem crossing}} {}^3A^* \rightarrow \text{Products} \tag{17}$$

Accurate quantum yields for product formation are assumed available. One wishes to ascertain whether or not ${}^3A^*$ is indeed the photochemically active state. One also wishes to estimate the efficiency of intersystem crossing from ${}^1A^*$ to ${}^3A^*$, since intersystem crossing ordinarily proceeds with less than unit efficiency. A comparison of the quantum yield for intersystem crossing (hereafter denoted Φ_{ISC}) with the overall product yield may be informative with respect to the efficiency of primary recombination.

The most popular technique for utilizing energy transfer to measure Φ_{ISC} is that proposed by Lamola and Hammond (*23*), where the compound A is used as sensitizer of a well-characterized photoreaction known to proceed via the triplet state of a suitable substrate Q:

$$A + h\nu \rightarrow {}^1A^* \xrightarrow{\text{intersystem crossing}} {}^3A^* \tag{18}$$

$$ {}^3A^* + Q \xrightarrow{\text{energy transfer}} {}^3Q^* + A \tag{19}$$

$$ {}^3Q^* \xrightarrow{\text{photochemical reaction}} \text{Products} \tag{20}$$

If one measures the yield of photoproducts produced from ${}^3Q^*$, one can determine the efficiency of formation of ${}^3A^*$ by intersystem crossing from ${}^1A^*$ *if* several important conditions are satisfied:

(a) It is required that all triplets of A are quenched by Q—that is, triplet–triplet transfer must be efficient relative to the other modes by which ${}^3A^*$ can decay (including photochemical reaction, triplet–triplet annihilation, intersystem crossing to the ground singlet, and phosphorescence). It is also necessary that all triplets of Q be formed by quenching of ${}^3S^*$; i.e., that no ${}^3Q^*$ be formed directly by intersystem crossing from ${}^1Q^*$. In practice the former condition necessitates use of a rather large excess of Q, which, by the second condition, imposes rather severe constraints on the species which can be used as quenchers for this purpose. First, substrates with low extinction coefficients in the spectral region for $A \rightarrow {}^1A^*$ absorption must normally be used in order to minimize direct formation of ${}^3Q^*$ by intersystem crossing from ${}^1Q^*$. Second, due to the requirement that Q be present in large excess, its purity is a critical consideration; painstaking care must be exercised in that respect. Small quantities of impurities may act as efficient quenchers or sensitizers, completely vitiating the experiment. These two conditions severely restrict the choice of quenchers that can feasibly be used in organic systems. Olefins, which undergo a photosensitized cis–trans isomerization via a triplet mechanism, are the most popular substrates for such experiments because they absorb but weakly in the ultraviolet and are usually susceptible to purification by preparative gas–liquid chromatography.

We must concern ourselves further with the efficiency of the actual triplet–triplet energy-transfer step. Most kinetic treatments of triplet–triplet transfer assume that the process proceeds at the diffusion-controlled rate (i.e., every collision between $^3A^*$ and Q is effective in inducing excitation of the latter). In some cases triplet–triplet transfer can occur by a long-range transfer process at distances considerably greater than the normal encounter distance (*21*). In other cases it appears that there is sufficient inefficiency in the energy-transfer process that diffusion apart of $^3A^*$ and Q without energy transfer can be appreciable, especially in relatively nonviscous media (*24*). Furthermore, if the quencher is present in excess, "static" quenching may become highly significant. In other words, if an A molecule absorbs a photon while a molecule of Q is a nearest neighbor, that particular A molecule is quenched immediately and can therefore not enter into the normal competition between diffusional quenching and other modes of energy dissipation. This does not in itself appear detrimental since one desires the energy-transfer step to be highly efficient; however, static quenching may cause anomalies in "Stern–Vollmer plots" (*1*) which are often used for interpreting quenching data. It is clear then that one should ascertain whether or not Φ_{ISC} values obtained by this technique are independent of the concentration ratio of Q to A—Q, of course, always being present in excess.

(b) It is required that the lowest triplet state of the acceptor be lower in energy than that of the donor. It is desirable that the first excited singlet of the quencher be above that of the donor, to prevent significant sensitization of $^1Q^*$ by $^1A^*$.

(c) One desires that $^3Q^*$ react with known efficiency to yield products, that this reaction be relatively free of thermal complications and side reactions, and that analytical methods for determining the concentrations of products be rapid, sensitive, and reliable.

(d) The actual energy-transfer step should be well behaved. Since triplet–triplet transfer is an area of current research interest, with all aspects of the phenomenon not being well understood, this may presently be a rather difficult requirement to satisfy. For example, piperylene

(1,3-pentadiene) is a rather popular triplet quencher, and occasionally it has been assumed that quenching of a photoreaction by piperylene indicated that the reaction proceeded through a triplet state. It has been observed, however, that piperylene can also act as a singlet quencher (*25*, *26*), indicating that a few simple quenching experiments are not necessarily sufficient to establish the multiplicity of a photoactive excited state. Furthermore, considerable care must be exercised in the use of Stern–Vollmer plots (*27*), which are often obtained in the course of an energy-transfer experiment. Still another complication is the observation of triplet–triplet transfer despite the fact that the triplet energy of the acceptor, as determined from singlet–triplet absorption or triplet–singlet emission studies, is greater than that of the donor. The apparent endothermic transfer is usually rationalized by postulating the existence of a "nonspectroscopic" triplet state of the acceptor, lower in energy than that observed by spectroscopic techniques, which is populated by energy transfer (*28*). Similarly, conjugated dienes can effectively quench excited singlet states of aromatic hydrocarbons, despite the fact that the diene possesses no low-lying singlets (*25*). Thus, even when the energy-level criteria are satisfied, as in (b) above, the system may not exhibit "ideal" behavior. Since triplet–triplet energy transfer is not as well understood as one might like, it is highly desirable to check triplet-formation efficiencies so obtained with more than one substrate.

One may attempt to "count" triplets of A by sensitizing phosphorescence, rather than photochemical reaction, of Q. Unfortunately, triplet–singlet phosphorescence in solution is restricted to a very few organic molecules, most notably biacetyl (2,3-butanedione) and benzil ($C_6H_5COCOC_6H_5$), and their phosphorescence yields in solution are not large. An example of the use of sensitized biacetyl phosphorescence in solution has been presented by Backstrom and Sandros (*29*).

Still another approach is to sensitize "P-type" delayed fluorescence in solution. This form of delayed fluorescence involves triplet–triplet annihilation, and the following processes occur in its sensitization:

$$A + h\nu \rightarrow {}^1A^* \xrightarrow{\text{intersystem crossing}} {}^3A^* \tag{21}$$

$${}^3A^* + Q \xrightarrow{\text{triplet transfer}} A + {}^3Q^* \tag{22}$$

$${}^3Q^* + {}^3Q^* \xrightarrow{\text{triplet–triplet annihilation}} {}^1Q^* + Q \tag{23}$$

$${}^1Q^* \rightarrow Q + h\nu' \tag{24}$$

Measurement of the relative lifetimes and intensities of P-type delayed fluorescence of the same acceptor (*Q*) when sensitized by two different donors enables one to compute the ratio of triplet-formation efficiencies for the two donors (*30*). If Φ_{ISC} for one donor is known, that for the other can be computed. This technique offers as its principal advantage the relative simplicity and high sensitivity of the fluorescence measurement. Delayed fluorescence lifetimes are on the order of milliseconds, and accordingly can be determined much more easily than the nanosecond lifetimes characteristic of "prompt" fluorescence. The principal disadvantage of the sensitized delayed-fluorescence technique arises from the fact that relatively few organic compounds exhibit intense delayed fluorescence in solution; most of those which do are aromatics, which absorb strongly in the ultraviolet. Consequently, direct excitation of *Q* can be difficult to avoid.

It is sometimes possible to determine triplet-formation efficiencies by the sensitizing or quenching of phosphorescence in rigid, glassy media, where phosphorescence is common. However, as is discussed in Sec. IV, great care must be exercised in extrapolating experimental results obtained in such media to a solution photoreaction.

Electronic energy transfer is achieving increasing use as a tool for identifying electronically excited states involved in organic photochemical reactions. In certain cases, quantum yields for triplet formation can be obtained by quenching or sensitization studies. The only other common techniques which can be made to provide this information are electron spin resonance or flash photolysis, both of which require more elaborate equipment and are more difficult to perform than a simple energy-transfer experiment. While the technique has not yet been applied to the study of inorganic photochemistry, scattered observations are beginning to occur in the literature, and it is likely that quenching studies will eventually assume an important position in the photochemistry of coordination compounds (*31*, *32*). It cannot be overemphasized, however, that electronic energy transfer is an area of great current research interest; energy-transfer phenomena are incompletely understood, with the result that problems of interpretation are by no means uncommon in these experiments.

IV. "Direct" Analysis for Transients in Photochemical Analysis

A "complete" mechanistic study of a photochemical reaction in liquid solution requires not only that the nature of final products and quantum yields for their formation be determined but also that the identity of electronically excited states and unstable intermediate species be established. Since excited states and thermally unstable intermediates usually have very short lifetimes in fluid media at or near room temperature, special techniques must be employed for their "direct" observation. In this section we consider some of the most important "direct" techniques applied to photoresponsive systems.

A. *Matrix Spectroscopy*

Many organic and some inorganic photoreactions generate free radicals which can be identified by spectroscopic techniques. Unfortunately, the lifetimes of free radicals in liquid solution are usually so short that detectable concentrations cannot be produced. One must therefore attempt to accumulate appreciable quantities of radicals (or other intermediates) under environmental conditions which inhibit thermal reactions thereof. This can usually, though not always, be accomplished by trapping the unstable species in a rigid solid or glassy medium at low temperature. This technique, known as "matrix isolation spectroscopy," or more simply "matrix spectroscopy," was independently demonstrated feasible in 1954 by Norman and Porter (*33*) and Whittle *et al.* (*34*).

Matrix spectroscopy is normally performed by irradiating a frozen sample for a suitable time, then examining the specimen by the appropriate spectroscopic technique. An "ideal" matrix should possess the following properties:

(a) The most critical requirement for a matrix material is that it inhibit diffusion of the molecular fragments formed by photolysis; they must be "trapped" by the matrix. For very small or very reactive fragments, the matrix must accordingly be highly rigid and used at temperatures considerably below its melting point.

(b) It is usually, though not always, desirable for the matrix to be chemically inert. The high reactivity of free radicals imposes a severe constraint on the matrix materials which can be utilized successfully.

(c) The matrix should not appreciably perturb the primary photochemical process. In that regard, xenon may be in some cases an extremely unsuitable matrix despite its chemical unreactivity, due to its considerable heavy-atom effect upon singlet–triplet intersystem crossing rates in organic molecules (*35*).

(d) A matrix must be reasonably transparent in the spectral region of interest. Alternatively, it should exhibit optical properties conducive to the application of diffuse reflectance spectroscopy.

(e) The matrix must dissolve an adequate concentration of the starting material to be photolyzed at the desired temperature. This statement is a good deal less trivial than it may appear upon initial reading. Rigid, glassy matrices, consisting of frozen organic solvents, are commonly used as matrices for optical phosphorescence spectroscopy. The usual procedure is to dissolve the solute in the liquid and then freeze the solution into a glass at liquid-nitrogen temperature. In the freezing process, it is quite possible for the solubility of the species of interest to be so diminished that formation of solute crystallites will occur (*36*). Formation of microcrystalline aggregates of solute within a rigid matrix defeats the purpose of the experiment, which is to isolate the reactive molecules, and fragments formed therefrom, from each other. Unfortunately, it is very difficult to ascertain the significance of this effect in many matrix experiments.

A number of materials have been used as solvents for matrix spectroscopy, including:

(a) The frozen gases Ar and N_2 are highly rigid and (especially for Ar) chemically inert. They often tend to crystallize, rather than forming glasses, upon freezing; hence, especially in the ultraviolet, they scatter light efficiently and accordingly may be unsuitable for absorption spectroscopy. This difficulty is considerably less pronounced in the infrared. The solids are commonly used at temperatures of the order of 20°K in order to ensure rigidity; cryogenic equipment is therefore required.

(b) A number of organic solvents or mixtures thereof form glassy matrices when frozen at liquid-nitrogen temperature (*37–39*). These glasses are very widely used for optical absorption and phosphorescence studies. They are considerably less useful for photochemical matrix spectroscopy since many of them contain components (e.g., aliphatic alcohols) which scavenge free radicals.

(c) Plastics such as polymethyl methacrylate possess good optical properties and can, in principle, be utilized to immobilize relatively large molecular fragments at temperatures considerably above cryogenic. In practice, however, the polymeric matrices are not usually sufficiently inert for this purpose. Furthermore, they are very difficult to purify; accordingly, traces of residual monomer and/or polymerization inhibitor are usually present. Either of those components is likely to be highly reactive toward free radicals and may also undergo photolysis.

(d) Boric acid glasses, largely discredited in optical phosphorescence work, are sometimes used in matrix spectroscopy (*40*). For relatively bulky intermediates which diffuse slowly, boric acid glasses can be used at temperatures approaching room temperature. They appear especially useful for trapping high concentrations of intermediates in ionic systems, due to the high polarity of the solid boric-acid matrix, and are therefore of potential utility in inorganic photochemistry. Unfortunately, preparation of reproducible boric acid glasses is at best a black art. Since the glass-preparation procedure involves heating to about 250°C, the precursor solute species is likely to undergo appreciable thermal decomposition. Furthermore, boric acid is not especially inert in the presence of highly reactive intermediates.

(e) Photoreactive entities can sometimes be positioned within clathrates and photolyzed therein. If the fragments so produced are sufficiently large, diffusion is inhibited and the intermediates can be detected spectroscopically. For example, organic mercaptans can be induced to form clathrates in which water acts as the host. Photolysis of RSH yields RS· and H·. The hydrogen atoms are so small that they escape from the cage and eventually combine to form H_2. The RS· radical remains immobilized within the cage and can be detected by diffuse reflectance (*41*). Relatively large fragments can be trapped in this manner at relatively high temperatures. The generality of this approach is, however, severely limited by the paucity of well-characterized clathrate systems and by the chemical reactivity of many host materials.

(f) Since infrared spectroscopy is very useful for identification purposes, Pitts *et al.* (*41a*) have used KBr pellets as a matrix. It has been demonstrated that the photolysis of *o*-nitrobenzaldehyde can be used as an actinometer for KBr-pellet matrix photolysis. KBr matrices can be used at or near room temperature. The rate of diffusion of organic molecules in KBr pellets seems to be somewhat greater than one would expect for a solid matrix, which may limit the intermediates which can successfully be sought by this approach. There is also reason to be concerned about the possible heavy-atom effect of the Br^- ion on singlet–triplet intersystem crossing rates in organic molecules. More work is required before the utility of KBr pellets as a matrix can be discussed with certainty, but the technique appears quite promising.

Despite their unfavorable optical properties and the necessity of access to cryogenic apparatus, it will be apparent from the above discussion that the solidified "inert" gases are the most widely used matrix materials.

The principal caveat associated with matrix spectroscopy, from the viewpoint of the solution photochemist, is the possibility that information procured by the matrix experiment will not be related easily to mechanistic details of the reactions in liquid media. If the primary fragments produced in a matrix photolysis do not diffuse from their solvent cage (which presumably is the reason for performing the experiment), the probability may be high that the fragments will recombine to form the starting material, causing the experiment to fail in its principal purpose—production of relatively high concentrations of unstable intermediate. The original results of Norman and Porter (*33*) seemed to indicate that this "cage effect" was not especially significant. Subsequent work has demonstrated that in actuality the cage effect can constitute a serious difficulty. For example, in the photolysis of 4.7×10^{-5} M acetophenone in a mixed hydrocarbon solvent (2-methylpentane–3-methylpentane), benzene was produced as a product, with a quantum yield of the order of 1.0 at room temperature. In contrast, irradiation of the same solution at 77°K effected *no* detectable decomposition of acetophenone after ten times the exposure required for *complete* destruction of acetophenone in the liquid sample (*42*). One cannot expect, therefore, to obtain valid estimates of primary quantum yields via matrix spectroscopy unless the cage effect can be demonstrated inoperative in a particular system. In practice, one can be assured of fulfilling that condition only if the primary fragment is exceedingly stable—as, for example, the triphenylmethyl radical (*33*).

Additionally, one cannot ignore the disquieting possibility that not only the efficiency but also the nature of the primary process may be different in a rigid glass than in a liquid sample. We can do no better than to quote an important paper on this subject by Symons and Townsend (*43*): "In many instances the course of photolysis in rigid media seems to differ from that in fluid media or gas phase. Sometimes a substance which is readily photolyzed in solution is not decomposed at all when the medium is rigid. Possibly because of (recombination), processes which can follow both radical and non-radical paths in the fluid state seem to follow a non-radical path only in rigid media." For example, 1,1′-azoisobutyronitrile is a well-known initiator of liquid-state free-radical polymerization of vinyl monomers. Yet matrix photolysis of this substance produces no discernible electron spin resonance signal. While one must avoid the temptation to become overly philosophical about the absence of esr signals in such an experiment, the fact remains that this excellent source of free radicals in liquid-state photolyses undergoes photochemical decomposition by pathways which do not appear to involve radical production in rigid matrices. The implications of such examples for liquid-state photochemists are obvious.

Assuming that the matrix photolysis does in fact proceed in a favorable manner, there remain the problems of obtaining and assigning the spectra of the fragments. Any form of spectroscopy may be utilized; ultraviolet, infrared, nuclear magnetic resonance, electron spin resonance, photoluminescence, and reflectance techniques all have been used. The technology of obtaining a spectrum is not especially difficult if the optical properties of the matrix are reasonable. Identification of the spectrum is usually accomplished by comparison with "known" spectra of suspected intermediates; unfortunately, the comparison spectra will usually turn out to have been obtained in the gas or liquid phase. In that case, one must be concerned with the influence of the matrix on whatever form of spectrum has been obtained. Unfortunately, there is a severe shortage of information that may be applied to this problem.

Pimentel and Charles (*44*) have described some important matrix effects which must be considered. One factor which can be extremely significant occurs whenever a solute molecule forms part of the solvent cage of another solute species—i.e., whenever two solute molecules are nearest neighbors in the solid. The presence of the "cage" solute molecule may perturb the spectrum of any fragments resulting from photolysis of the solute molecule of interest. In effect, this "aggregation" phenomenon signifies that spectra of intermediates in various different microenvironments are observed; this may be manifested by broadening or even splitting of spectral bands, and by frequency shifts. Furthermore, the aggregation effect seems to be difficult to eliminate; it apparently persists at matrix-to-sample mole ratios as large as 500 (*44*). One can, of course, use exceedingly dilute solute solutions to surmount aggregation, but only at the risk of operating below the sensitivity limit of the desired detection method. Clearly, this difficulty will be most pronounced when relatively insensitive spectroscopic techniques (such as infrared and nuclear magnetic resonance) are employed.

Even if the solute concentration is sufficiently small to render solute aggregation negligible, there remains the possibility that the photolysis fragments will be trapped in microenvironments which are sufficiently unlike to lead to band broadening or splitting. For example, in the irradiation of 1% HI in solid Ar at 4.2°K, esr spectra of product hydrogen atoms were sought. Instead of the anticipated one doublet, an esr spectrum consisting of three doublets was observed; the three doublets decayed at widely varying rates as the sample was subjected to slow heating. These observations were attributed (*45*) to the presence of three "types" of trapping sites in the argon matrix. It is difficult to distinguish between this phenomenon and the aggregation effects described in the preceding paragraph. The conclusion, however, is clear enough: interpretation of matrix absorption spectra may be severely complicated by "matrix effects," the nature and significance of which cannot presently be predicted with certainty.

Many of the reported applications of matrix spectroscopy (*44*) have been directed toward elucidation of free radicals produced in "known" photochemical reactions rather than toward elucidation of photochemical mechanisms per se. Most of the reported "photochemical" work in this area has been concerned with photolysis of relatively "small" molecules such as H_2O_2, hydrogen halides, various inorganic cations, aliphatic alcohols, etc. A recent publication (*46*) demonstrates that the matrix technique can be of considerable utility in elucidating photochemical mechanisms in large organic molecules. By the use of both ultraviolet-visible and infrared absorption spectroscopy in organic glasses at 77°K, the course of seco–endo isomerizations in various conjugated hydrocarbons and ketones was investigated and correlated with the Woodward–Hoffmann rules. Such experiments should find increasing use in the photochemistry of large molecular systems.

The reader of the above paragraphs may develop a rather pessimistic attitude toward the value of matrix spectroscopy in mechanistic liquid-state photochemistry. A healthy skepticism in the performance of such experiments, and especially in the interpretation of results thereof, is justified. Despite the uncertainties, however, the popularity of the matrix technique continues to increase. The experiments are relatively easy to perform with equipment which, except for some exotic cryogenic items, is available in every research installation. The information obtained from such experiments, while not always conclusive, is often of great value, provided that the urge to overgeneralize the interpretation of the results is resisted.

B. *Electron Spin Resonance Spectroscopy*

In principle, electron spin resonance (esr) spectroscopy should be an exceedingly powerful tool for direct observation of free radicals and triplet states in organic photochemical systems. In practice, the number of liquid-state photochemical systems to which esr has been successfully applied is rather small. This is due, at least in part, to the relative unfamiliarity of many solution photochemists with the esr technique. It is, however, also true that "the complexities of esr can expedite the making of mistakes perhaps more so than other forms of absorption spectroscopy" (*47*). When the complexities of esr are superimposed upon those inherent in many photochemical systems, it becomes evident that successful application of esr to photoresponsive systems requires careful experiment design, persistence, and a certain amount of luck. Two recent books dealing with experimental techniques in esr spectroscopy (*47*, *48*) will be of considerable interest to photochemists wishing to utilize esr for the first time.

From the liquid-state photochemist's viewpoint, the ideal approach to esr consists of irradiation of a solution within the esr cavity, accompanied by monitoring of the free radicals formed. Such procedures can lead to relatively secure identifications of radicals formed in photoreactions, free from the vagaries of chemical scavenging techniques. Good estimates of rates for formation and thermal decay of radicals in solution can also be achieved by esr. However, certain practical difficulties arise. First, while the sensitivity of esr, especially in nonpolar liquid samples, is quite great, the rate of radical formation in inefficient photoreactions may be sufficiently slow so that (especially for short-lived radicals) adequate radical concentrations may be inaccessible. One way to combat this problem is brute force—i.e., use of a very intense light source (high-pressure xenon or mercury lamp, or possibly a laser). Aside from possible experimental complications and expense, such an approach may induce possible side reactions, including photolysis of the very radicals one is attempting to observe. Another possible approach is to use large volumes of sample, either in a static or flow system. In the case of nonpolar samples, use of large samples is not normally the source of any major difficulty (other than expense), but this approach can fail badly when polar solvents are employed, due to the cavity damping commonly produced by such species. Consequently, optimum conditions for radical detection by esr are frequently not identical with those under which many steady-irradiation photoreactions are performed. This fact should be kept in mind when designing both steady-state and esr experiments. An especially detailed description of an example esr study, that of the photodecomposition of pure liquid alkyl hydroperoxides by Piette and Landgraf (*49*), is worthy of consultation.

A new procedure which may prove useful for enhancing the sensitivity of free-radical detection by esr in liquid systems has been described by Janzen and Blackburn (*50*). One uses an efficient chemical free-radical scavenger which forms a long-lived, paramagnetic adduct with the trapped radical. In this manner, relatively large concentrations of paramagnetic adduct can be accumulated and then detected by esr. If it happens that the hyperfine spectrum of the adduct has characteristic features sensitive to the structure of the attached groups, the spectrum can provide an identification of the trapped radical. For example, phenyl *tert*-butyl nitrone is an efficient radical scavenger; its adducts with a number of common radicals exhibit characteristic hyperfine patterns. While this approach may not be especially amenable to the study of kinetics of formation or decay of radicals, it should prove quite useful in their identification.

Attempts to determine quantum yields for radical formation in photochemical reactions by esr lead one to the problem of standardization of quantitative esr measurements (*51*). It is not usually feasible to use internal standards in photochemical systems; a "fixed" standard, such as a synthetic ruby situated within the cavity, is usually most desirable. The problem of standardizing quantitative esr determinations is not so formidable as seems to be commonly assumed; it is anticipated that quantitative applications of esr to photochemistry in liquid media should appear with considerably greater frequency in the future.

An alternative to determining esr spectra of photoproduced radicals in liquid solution is to trap the radicals in a solid, thereby prolonging their lifetimes. Two approaches to such experiments are available. One is to freeze the sample before irradiation; this is merely a specific case of matrix spectroscopy, discussed in Sec. IV, A. Alternatively, a liquid solution may be frozen into a glass for esr measurements subsequent to irradiation outside the cavity. Other than the fact that optically transparent glasses are not required, this approach is of little significance; if the radicals are sufficiently inert that they can survive until the medium is frozen, they presumably can be detected in the liquid, so that no advantage accrues from the freezing procedure. On the other hand, if the radicals are short-lived, they will not survive until freezing is effected, and no signal will be obtained.

Difficulties can arise in either form of trapping procedure. First, esr spectra in glassy matrices are often less detailed than those in solution; line broadening in the former may be brought about by anisotropy of the g-factor and hyperfine coupling coefficients which, in liquid solution, is normally averaged out by rapid molecular tumbling. Second, if one wishes to assign an observed esr spectrum to a given radical, he will often attempt to do so by comparison with well-known gas- or liquid-phase spectra. Matrix effects may vitiate the use of such "assignment by analogy." Finally, the possibility of multiple trapping sites (see Sec. IV, A) rears its ugly head again.

One of the most important applications of esr to photochemistry is the study of triplet states of organic molecules. Some aspects of this topic have been reviewed by McGlynn *et al.* (*52*), Thompson (*53*), and Alger (*47*). Owing to the extreme importance of triplet states in organic photochemistry, a number of early but unsuccessful attempts were made to observe esr absorption in glassy media containing randomly oriented triplets. It was not until 1958 that esr absorption of triplets was reported; Hutchison and Mangum (*54*) were able to observe an esr spectrum for triplet naphthalene, present as a dilute sample in a single crystal of durene at liquid-nitrogen temperature. In that work, it was observed that the triplet state of naphthalene was split into three energy states, even in the absence of an applied magnetic field. The observed splitting is presumably due to unusually strong dipole–dipole interaction of the spin magnetic moments of the two unpaired electrons. The resulting anisotropy is so great that randomly oriented molecules (as in a rigid glass) would exhibit an exceedingly broad, and therefore not easily detectable, esr absorption. Despite this difficulty, $\Delta M_S = \pm 1$ esr absorption for randomly oriented triplets in rigid glasses were subsequently observed (*55*); resonance absorption is observed only for those triplets with a molecular axis approximately parallel to the applied magnetic field. Because of this restriction, the $\Delta M_S = \pm 1$ esr absorption for triplets in glassy media is usually a factor 10—100 less intense than the $\Delta M_S = \pm 2$ signal. Nonetheless, the glassy matrix is much easier to prepare than a single crystal, and it is also more reflective of conditions in fluid solution, a consideration of importance in liquid-state photochemistry.

One may well ask why the anisotropy referred to in the preceding paragraph would not be averaged out by rapid molecular tumbling in liquid media. Weissman (*56*) has estimated that in order to narrow the dipolar splitting to the point where $\Delta M_S = \pm 1$ esr absorption could be easily observed, tumbling frequencies on the order of 10^{12} sec^{-1} are required. Such tumbling frequencies can be attained in solvents of relatively low viscosity. However, at such tumbling rates, spin-lattice relaxation becomes appreciable; the resulting relaxation broadening negates the narrowing that would otherwise be anticipated. Hence, esr spectroscopy of triplets in fluid media has received very little attention. In glasses, on the other hand, successful esr work has been carried out with triplets of lifetimes ranging from 0.10 to 20 sec.

Despite the inapplicability of triplet-state esr to liquid solutions, the technique has a number of important implications in solution photochemistry, including:

(a) Lifetimes of nonphosphorescent triplet states can be determined in glasses by esr measurements. In fact, even for phosphorescent triplets, it appears that esr generally yields more reliable rise and decay curves than phosphorescence, due to problems associated with self-absorption of phosphorescence (*57*). While triplet lifetimes so determined will be appreciably greater than those which would obtain in liquid solution (*58*), they are nonetheless useful for mechanistic considerations.

(b) One of the most important parameters available from a triplet esr spectrum is the "zero-field splitting," i.e., the energy splitting in the absence of an external magnetic field. Analysis of the zero-field parameters of the esr spectrum of an organic triplet can yield important information concerning electron-density distributions and structures of triplets (*52*). Further, the zero-field parameters of different triplets are seldom identical. Consequently, esr is of value for distinguishing between different components in mixtures of triplets, especially if one or more of the constituent triplets is nonphosphorescent. In triplet–triplet energy transfer, in which at least two triplet species are simultaneously present, esr can serve as a powerful probe.

(c) Line-broadening effects can provide information regarding interactions of a triplet with its environment. With a certain amount of arm waving, such data can sometimes be used to make inferences about environmental effects on triplets in liquid media.

A number of recent esr studies of triplet states have involved use of a technique known as *magnetophotoselection* (*59*). The term *photoselection* refers to selective excitation of molecules with certain orientations, from among a collection of randomly oriented molecules in a glass matrix, by the use of monochromatic, plane-polarized incident light. The polarized radiation preferentially excites those sample molecules which are oriented such that their transition moments lie parallel, or nearly so, to the electric vector of the exciting light. *Magnetoselection* relies on the fact that the intensity distribution of the $\Delta M_S = \pm 1$ esr absorption spectrum for a randomly oriented collection of triplet molecules gives overwhelming preeminence to those molecules which have one of their principal magnetic axes oriented along that of the steady applied magnetic field. Hence, the $\Delta M_S = \pm 1$ esr spectra are essentially due to three oriented assemblies of molecules. Both the optical and esr experiments thus constitute approaches whereby one can look at selected groups of oriented molecules from among a random array. *Magnetophotoselection* consists of the measurement of esr spectra of triplets excited with plane-polarized incident radiation; molecules having all orientations will be seen to some extent, but orientations other than those described above will provide only a relatively insignificant background to the observed esr spectra. In this manner, orientation effects upon triplet-triplet energy-transfer processes in glasses have been examined (*60–62*). Results of such studies seem to indicate that the probability of triplet–triplet transfer in glasses is not strongly dependent upon relative orientations of donor and acceptor, in contrast with the results usually found for triplet–triplet transfer in liquid solution (*63*). Once again, therefore, we note the difficulty of extrapolating experimental observations in rigid matrices to fluid media.

The use of esr in liquid-state photochemistry should gradually increase in importance as familiarity with the technique on the part of photochemists increases. Eventually, one would anticipate that esr will largely supplant chemical-scavenging studies for organic photoreactions in which free radicals are produced. Since most interesting photoactive coordination compounds are themselves paramagnetic, esr spectra of irradiated solutions of metal complexes are often intractable; accordingly, it seems less likely that inorganic photochemists will routinely utilize esr.

C. *Mass Spectrometry*

While it has achieved considerable popularity as an analytical tool for detection and identification of free radicals formed in gas-phase photolyses (*64*), mass spectrometry is of little assistance for tackling the analogous problem in liquid-state photochemistry. The difficulty, of course, is that in the process of introducing the liquid sample into the mass spectrometer, secondary reactions of the unstable intermediates will almost certainly take place, owing both to the time required and to the processes by which the liquid sample is converted into a low-pressure gas. Hence, while mass spectrometry is exceedingly useful for identification of *stable* products formed in photoresponsive solutions, it is rarely suitable as a technique for "direct" observation of unstable species in such systems. Hence, what is potentially one of the most powerful tools available to the gas-phase photochemist is essentially denied to his counterpart dealing with fluid media.

D. *Photoluminescence Spectroscopy*

Experimental observation of radiation emitted by electronically excited states is one of the most powerful tools for "directly" investigating the behavior of such species, at least in the case of singlet states. Unfortunately, phosphorescence from triplet states of organic molecules is rarely observed in liquid solution, due to quenching of triplets by oxygen or traces of other impurities which are exceedingly difficult to remove (*58*); contrary to popular belief, phosphorescent triplets are not usually quenched by collisions with solvent molecules. Hence, *if* quenching impurities can be completely removed from a liquid solution, phosphorescence can be observed in liquid solution, even from relatively long-lived triplets. The italic "if" in the preceding sentence, however, is an especially big "if"; consultation of the studies of Tsai and Robinson (*58*) will provide an accurate impression of the heroic measures which normally must be taken to observe phosphorescence from organic triplets in fluid media. Unfortunately, therefore, phosphorescence studies on photochemically significant triplets must normally be performed in glasses (*37–39*), raising once again the difficulty of extrapolating matrix spectra to liquid media.

The theory and practice of photoluminescence spectroscopy, with specific reference to liquid-state processes, have been discussed in a number of places. Nobody planning to perform liquid-state photoluminescence experiments should do so without access to Parker's recent book (*65*), which contains thorough, authoritative discussions of both theory and experimental practice. Other general discussions of the theory of photoluminescence (*1*, *66–68*), experimental techniques in fluorescence and phosphorescence (*69*, *70*), luminescence of coordination compounds (*71–73*), luminescence of organic molecules (*74*), and environmental effects in luminescence spectroscopy (*75*) have been presented. Each of the above references contains material of interest to liquid-state photochemists; due to the wealth of published material already available, we shall not deal further with photoluminescence in the present discussion. This should not be construed as an implication that luminescence is less important than the other techniques discussed in Sec. IV.

E. *Flash Photolysis*

The most popular technique for detecting the presence of unstable intermediates and relatively long-lived electronically excited molecules in liquid-state photochemical systems is flash photolysis. In this method, which is essentially a relaxation technique, an intense pulse of electromagnetic radiation of short duration is utilized to produce a high concentration of transient species, which is then detected by a suitable analytical technique. A thorough, comprehensive, and authoritative review of flash photolysis has been presented by Porter (*76*). Accordingly, we restrict the present discussion to some considerations of special interest to the analytical chemist, and to some recent advances in the field.

The usual approach to flash photolysis is to analyze the system as soon as possible after the photolysis flash by single-beam optical absorption spectroscopy. Often, one follows the photolysis flash with a second flash, also of short duration, but of considerably smaller intensity than the photolysis flash. The second, monitoring flash lamp is usually situated such that its flash is directed in a direction perpendicular to that of the bank of photolysis flash lamps. Light from the monitoring flashtube passes through the reaction cell and into a spectrograph; the transmission spectrum of the material in the cell is photographed. This form of single-beam monitoring, the goal of which is to obtain spectra of intermediates present after the photolysis flash, is commonly termed *flash spectroscopy*. In principle, one could use a rapid-scanning spectrometer with photoelectric recording, in place of a spectrograph, provided that the kinetics of the reactions by which the transients decay are sufficiently slow that the bandwidth required for adequate spectral definition does not result in an unacceptable signal-to-noise ratio.

An alternative approach is to replace the monitoring flash lamp with a continuous source. Following the photolysis flash, the transmittance of the test solution is determined at a single frequency as a function of time. Photoelectric detection is normally used. Obviously this form of measurement is of use for studies of kinetics of formation or decay of transients; it is therefore termed *flash kinetics*. High-resolution monochromators are available commercially which can be operated as either spectrographs (photographic detection) or spectrometers (photoelectric detection); they can therefore be used in both flash kinetics and flash spectroscopy. A block diagram of a commercial flash-photolysis apparatus, suitable for both kinetics and spectra of transients, is shown in Fig. 6-1.

A number of difficulties of interest to the analytical chemist are encountered in this form of analysis for transient species, including:

(a) Very careful alignment and collimation of optics is mandatory. The output of most high-energy photolysis flash lamps includes a "tail"

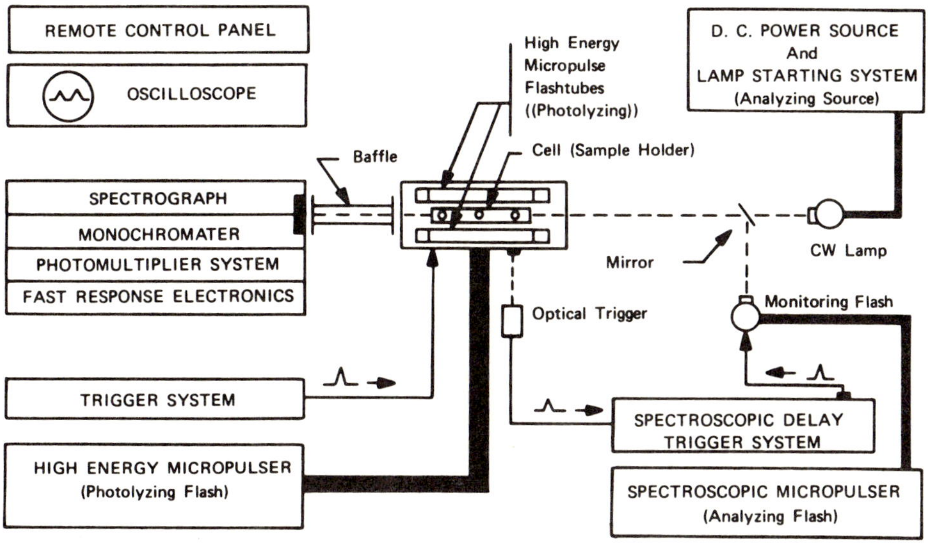

FIG. 6-1. Block diagram of commercial flash-photolysis apparatus for flash spectroscopy and flash kinetics (Xenon Corp., Medford, Mass.).

which extends well past the $1/e$ time of the lamp. Thus, scattered light from the photolysis flash can act as an important source of "background" in both flash spectroscopy and kinetics. It is absolutely necessary that the only light which enters the monochromator be that which has passed along the axis of the photolysis cell, and that no portion of the walls of the cell be visible at the entrance slit of the monochromator.

(b) For flash spectroscopy, it is necessary that the monitoring flash be synchronized with the photolysis flash, such that the interval between the two is known and reproducible. It is also highly desirable that this interval be capable of adjustment by the experimenter. Normally the photolysis flash serves as the master operation which triggers the monitoring flash; the forms of trigger and delay circuits commonly employed have been described (*76*, *77*).

(c) For flash kinetics, one requires a large short-term stability of the steady source. In most cases, automobile storage batteries are used as the power supply for this lamp, since the output of most commercial dc power supplies contains an intolerably large "ripple" component. Since the efficiency of a gas-filled lamp depends on the temperature of the surrounding air, it is critically important (1) that a carefully designed housing for the steady-state lamp be used such that, after a relatively short warm-up period, the temperature of the air within the housing becomes constant, and (2) that the rate of flow of cooling air over the lamp jacket not fluctuate significantly.

To further enhance the signal-to-noise ratio, the steady source is often chopped and used in conjunction with a tuned detection system.

(d) Since many photochemical intermediates have short half-lives, it is important to ensure that the detection system exhibit a rapid response. This is an important consideration in both flash kinetics and flash spectroscopy.

(e) Serious difficulty can arise in flash kinetics if the order of the decay reaction of an intermediate of interest is greater than unity, for then absolute rate constants for the decay reaction cannot be computed unless the concentration of the intermediate can be determined (*76*). This condition signifies that (1) either the absolute extinction coefficient for the intermediate at the wavelength of interest must be determined or (2) some other analytical technique must be applied to determine the concentration of stable products produced by the photolysis flash.

If the parent solute and the transient whose ϵ is desired are the only solute species present in appreciable quantity immediately following the photolysis flash, then alternative (1) can be accomplished relatively easily by simultaneously monitoring the formation of the transient and the depletion of the initial solute (*76*). Alternatively, one can attempt to work backward from the concentration of final products to estimate the concentration of intermediate which must have been present immediately following the photolysis flash. Utilizing this information, alternative (2) in the preceding paragraph, one can determine an ϵ for the intermediate if he has its flash spectrum available. This approach requires sensitive analytical techniques since it is normally undesirable for a high conversion of reactant to products to be produced in a photochemical kinetic study (*1*). It is also necessary to assume that only one important primary process obtains. In fact, both alternatives (1) and (2) tend to require that something be known of the mechanism of decay of the intermediate, as well as that of the primary photolytic process. Since acquisition of such information is the reason for performing the experiment, it is often necessary to proceed most of the way through the kinetic study using relative rate constants before ultimate determination of absolute concentration of intermediates. Considerable ingenuity may be required to tackle this problem since numerous complications can occur (*76*).

(f) The problem described in (e) becomes especially irksome in both flash spectroscopy and kinetics if it happens that the intermediate and the original solute, or two or more intermediates, exhibit partially overlapping absorption spectra. In flash spectroscopy using photographic detection, the approach described by Grossweiner and Matheson (*78*) can often be used to obtain an ϵ value for the desired species. In flash kinetics, this problem requires that kinetic studies be performed at several wavelengths, followed by solution of simultaneous equations, to obtain the desired ϵ values. Since the number of intermediates in a photochemical system is not usually known with certainty, performance of flash kinetics at several wavelengths is a useful routine precaution. The computations described above assume that Beer's law holds for the system [cf. (g)].

(g) Deviations from Beer's law will be noted in both flash kinetics and spectroscopy if the wavelength interval over which the absorption is measured is finite. This implies that high-resolution monochromators must be used. However, if the concentration of transient is small, the intensity difference between a nonabsorbing blank and the actual sample is also small, requiring the use of highly sensitive (low *f*-number) optics. These two requirements must be compromised, with the result that finite bandwidth corrections [Sec. II, Eq. (9)] must be applied. Fortunately, most transients obtained in flash photolysis of liquid systems exhibit broad, structureless spectra; accordingly, the error produced by the finite bandwidth is small unless very wide apertures are employed.

(h) Flash lamps produce intense pulses of heat, as well as light. In order to prevent establishment of thermal gradients within the reaction vessel, thermostating must normally be employed. Normally a water jacket is sufficient.

The influence of continuing advances in design of optical spectrophotometers is clearly reflected in improved optical systems for flash photolysis. An excellent example is a flash-kinetic spectrometer recently reported by Green and Tollin (*79*). In order to study flash kinetics of systems in which quantum yields for intermediate formation are small, or in which it is necessary to use a low-intensity photolysis flash, it is desired to have available a detection system which has a sensitivity of at least 0.010 absorbance unit, with a response time on the order of 10^{-3} sec. To achieve the specified sensitivity, it is necessary to modulate the steady-state beam incident upon the reaction vessel; the chopping frequency must be quite high if fast response times are to be achieved. A tuned detection system is of course utilized. The instrument designed by Green and Tollin, which is schematically represented in Fig. 6-2, is a double-beam, optical-null system, in which the steady light beam is chopped before splitting. Two photomultiplier tubes are utilized; the optical-null condition is achieved via filters in front of the photomultipliers. Since modulation occurs prior to beam splitting, the sample and reference signals are identical in frequency,

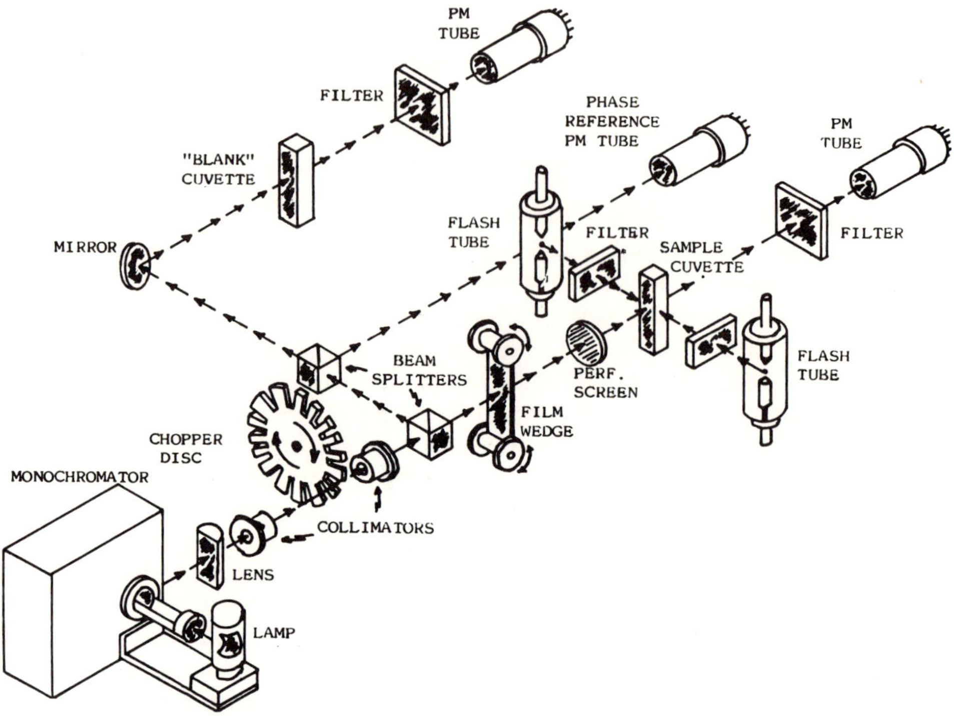

FIG. 6-2. Diagram of optics of double-beam differential spectrophotometer for flash kinetics. After Green and Tollin (*79*).

wave shape, and phase. Consequently, additional enhancement of the signal-to-noise ratio may be achieved by using a phase-sensitive detection system which rejects all signals not in phase or 180° out of phase with the reference signal, as well as all signals of frequency not equal to that of the reference signal. This is a useful precaution since scattered light from the photolysis flash will be an especially important difficulty at very high detector sensitivities. An instrument of this type can, if carefully constructed, have its sensitivity limited by shot noise from the photomultiplier tubes; considering the stray-light problem inherent in flash photolysis, this is no mean achievement.

A further technique, useful in many forms of spectroscopy as a method of enhancing the signal-to-noise ratio, is repetitive scanning of the spectrum followed by signal averaging. This technique is not normally applicable in flash photolysis, due to destruction of the sample solution by the photolysis flash. In certain cases this technique can be utilized to increase the sensitivity of flash kinetics or spectroscopy, especially for reversible photoprocesses in which the solution is not destroyed by the photolysis flash (*80*).

Optical spectroscopy is presently the most widely used detection method in flash photolysis, for (inter alia) the following reasons:

(a) Photochemistry and optical spectroscopy are complementary; the photochemist usually must have available detailed information concerning the electronic spectroscopy of excited states and intermediates in photochemical reactions. Thus, flash spectroscopy is an important component of a thorough study of a photoresponsive system. It is therefore usually most convenient to use optical detection for flash kinetics, especially since the required equipment (except perhaps for some of the electronics) is already available in any well-equipped photochemistry laboratory.

(b) There is a considerable body of data in the literature regarding electronic spectra of transients in photochemical systems; hence, assignment by analogy is facilitated.

Among the disadvantages of optical analysis of flashed samples are the following:

(a) Ultraviolet-visible spectra are not noted for their high information content. They are seldom used for "fingerprinting" in more prosaic forms of chemistry, and are not well suited for this purpose in most flash-photolysis systems. This is especially true in inorganic photochemistry, where the literature contains very few examples of transient absorption spectra which can be used as a basis for comparison. One would like to obtain a spectrum which is more informative from the standpoint of identification of transients.

(b) The design of rapid-scan spectroscopic apparatus, with adequate signal-to-noise ratios, is difficult, and is in fact a current research area of some importance in chemical instrumentation. Currently, such spectrometers must often be constructed in an individual's own laboratory, and are correspondingly expensive. Analytical techniques in which high time resolution can be achieved without serious adverse effects on the signal-to-noise ratio are accordingly of interest in flash photolysis.

(c) Optical absorption spectroscopy is limited in sensitivity, regardless of the time resolution desired. Analytical techniques which can offer very high sensitivities are therefore of considerable interest in flash photolysis.

It is obvious that disadvantage (a) could be alleviated by use of infrared, rather than uv-visible, spectrometry. Such procedures are, however, uncommon at present, due to the low intensities of ir sources, low sensitivities of ir detectors, and the slow response times of most ir detectors. Pimentel and Herr (*81*) have surmounted the last of these difficulties by using a zinc-doped germanium detector, which exhibits response times on the order of 1 μsec or less. Use of this detector, associated with properly designed optics, led to the development of an ir spectrometer for flash spectroscopy which could scan the 2–20-μm range in an interval of 500 μsec or less, with a spectral slit width of the order of 20 cm^{-1}. This instrument has been used (*81*) with advantage to investigate the identity and structure of intermediates formed in a number of gas-phase flash photolyses. Sensitivity problems, however, continue to seriously inhibit ir monitoring of flash photolyses in liquid media.

A second form of spectroscopy noted for its high information content is high-resolution mass spectrometry; time-resolved mass spectrometry has achieved notable success as an analytical tool in gas-phase flash photolyses (*82*) but, as discussed in Sec. IV, C, the technique is not especially amenable to liquid-state photochemistry. For the purposes of the liquid-state photochemist, disadvantage (a) has not been effectively surmounted in an appreciable number of cases.

Most efforts to surmount the response-time limitation of optical detection have involved the use of electrochemical techniques, which in some cases can offer sensitivities considerably greater than those available from optical spectroscopy. The application of classical dropping-mercury-electrode polarography to flash-irradiation studies has been discussed by Witt (*80*) and Berg (*83*). In principle, flash polarography should be amenable to the study of intermediates, but most reported applications have dealt with analysis of final products produced by flash photolysis. In any case, the technique is unlikely to achieve popularity; most photochemists (justifiably) cringe at the thought of a puddle of mercury collecting in their photolysis cell! There is also the distinct possibility that mercury electrodes may quench excited states in organic photoreactions.

Perone and Birk (*84*, *85*) have discussed the use of stationary-electrode techniques to the study of intermediates in flash photolysis. Using a time-delayed potentiostatic electrolysis technique at a hanging mercury drop, they were able to achieve time resolution in the microsecond range without undue difficulty. It is estimated that such techniques will exhibit a detection limit of 10^{-6} *M* or less, an improvement of at least a factor of 10^2 over optical spectroscopy. However, every silver lining has its cloud, and the potentiostatic stationary-electrode experiments exhibit several disadvantages. First, the charging current limits the time resolution of the method unless fairly sophisticated electronics is used. Second, and more serious, the photolysis flash produces an exceedingly complex, usually inhomogeneous, distribution of intermediate within the reaction cell. Electrochemical techniques respond only to the situation in the immediate proximity of the electrode surface. Consequently, the observed response may depend upon the relative position of the photolysis flash lamps and the electrode. This complication causes the mathematical relationships between observed current–time curves and decay rates for transient intermediates to become exceedingly complex, the extraction of kinetic data from which requires some mathematical skill. Third, the number of interesting photochemical reactions in which one can be assured of observing electrochemically reversible behavior is not as large as one might desire; while this may not be a deterrent to successful application of the technique, it will render the already formidable mathematics even more so. Finally, the information content of the experiment, in terms of parameters which can be used to identify transients, is not large, the more so as "hard" information concerning the electrochemical behavior of free radicals and other transient species is not abundant in the literature. It is nonetheless anticipated that stationary-electrode techniques should prove of considerable utility in flash photolysis; further developments in this area should be awaited with considerable interest.

From the standpoint of time resolution, flash-conductance experiments described by Witt (*80*, *86*) are noteworthy. For flash photolyses which produce solvated electrons or ions, a time resolution of at least 10^{-7} sec is claimed. Unfortunately, this technique is of little value for identification purposes, though it may be quite suitable for flash kinetics. The number of organic systems in which flash conductance will be useful is very small, but it may be of considerable interest in inorganic photochemistry.

Finally, an interesting series of experiments has recently been reported by Strong and Richtol (*87*), in which a standard flash-kinetic apparatus was used to obtain optical rotatory dispersion curves for photoexcited states of ketones, by introduction of a polarizer–analyzer system in the optical path of the steady-state monitoring source (a xenon lamp). In certain cases, this approach may provide considerably more specific information than standard optical absorption spectroscopy; further developments in this respect are to be anticipated.

One may conclude from the preceding discussion that, given the proper choice of analytical technique, time resolution is no longer a serious problem in flash photolysis. In fact, the factor limiting the time resolution of most flash systems is the duration of the pulse from the photolysis flash lamp. The usual flash lamps are gas-filled tubes which are fired by discharging an energy-storage capacitor across the electrodes of the lamp. Common photolysis flash tubes operate at an energy of the order of 1000 J dissipated in approximately 10 μsec. Much shorter durations can be achieved by reducing the inductance of the capacitor-bank–lamp combination and, for a given flash power, by increasing the charging potential while decreasing the capacitance (*88*). For the study of excited singlet states in organic photoreactions, one wishes to have flashtubes with $1/e$ times of the order of 10^{-9} sec; this feat cannot be accomplished simply by regulating the inductance and capacitance of the discharge circuit and the length of the current pulse. One must also attempt to increase the speed of the processes which occur within the gas-filled tube upon termination of the current pulse. Malmberg (*89*) has demonstrated that the electron gas which results from the applied current pulse can cool by collisions with neutral gas molecules, producing excited gas molecules which then emit, thereby producing a long "tail" on the flash. This electron gas must therefore be cooled as rapidly as possible; use of a high pressure of H_2 has been demonstrated successfully in that regard, and nanosecond-duration H_2 lamps have been fabricated (*89*).

It was subsequently demonstrated (*90*) that a commercially available mercury contact relay could serve as the basis for a flash lamp exhibiting a rise time of 4×10^{-10} sec and a fall time of about 5×10^{-10} sec. Still later, it was reported that gas-filled tubes, in which the gas (normally nitrogen or deuterium) is subjected to electrical breakdown by application of a step-function potential, exhibit decay times in the 10^{-9}-sec region (*91*). These lamps are now used in a commercially available apparatus for measuring nanosecond (and longer) photoluminescence lifetimes. All of these short-duration flash lamps can be used for luminescence-decay studies, provided that the system under study exhibits a relatively high emission yield (*91*, *92*), but they are not sufficiently intense for use in photolysis studies. Accordingly, there has been considerable interest in the development of alternative sources of intense, short-duration light pulses.

A high-intensity spark discharge can be used to attain pulse durations on the order of microseconds (*93*). The principal advantage of this source is that line emission is obtained, rather than a continuum, rendering the spark source of some interest for primary quantum-yield determinations, in which adherence to Beer's law is especially critical.

Another unique flash source is the exploding wire, described by Oster and Marcus (*94*). A capacitor bank, when discharged through a high-resistance alloy wire, causes the latter to explode, producing an intense light pulse (and a loud noise). The flash duration is on the order of 1 msec; the emission is relatively monochromatic, and the wavelength can be varied by changing the composition of the alloy from which the wire is fabricated. No jacket is required; the exploding-wire source is accordingly of interest in the high-frequency ultraviolet, where quartz jackets absorb appreciably.

The most recent development in flash photolysis is the use of the pulsed laser as flash source. A laser can combine high intensity with short duration; it is probable that the pulsed laser will eventually become the standard source for flash photolysis. Certain problems, however, are encountered in utilizing lasers in flash photolysis, including the following:

(a) The output of most lasers is at lower frequency than most photochemists would desire. However, the output of a laser can be passed through a crystal such as KH_2PO_4 which acts as a frequency doubler (*95*). Hence, with a ruby laser and a frequency-doubling crystal, light of 3471 Å can be obtained; while the intensity of the harmonic is much smaller than that of the "undoubled" ruby-laser line, it is adequate for most purposes.

(b) One must cause the laser to produce pulses of sufficiently short duration. The most common approach to this problem is termed "*Q*-switching." A laser tube with reflectors at each end is used. If a shutter is imposed in the optical feedback path, the laser cannot oscillate, and a large population inversion will result. If the shutter is suddenly removed, when the population inversion is maximal, the result will be an intense, short pulse (*96*, *97*). Various "shutters" can be employed, including a mechanical shutter or rotating disk, a Kerr cell, an ultrasonic-refraction shutter, or an absorbing species which can be photobleached (e.g., an organic dye). For more detail, the reader is referred to Lengyel (*98*). Pulses on the order of 10^{-8}-sec duration can be produced by *Q*-switching. The production of considerably shorter pulses (10^{-11} sec or less) from a laser can be accomplished by a technique known as *mode-locking* (*97*, *99*).

(c) In order to perform flash spectroscopy, one requires a short-duration source with a continuous intensity distribution, which is easily synchronized with the laser photolysis pulse. The laser output is anything but continuous, but it happens that a properly focused laser beam will produce a sufficient electric-field intensity to cause electrical breakdown of air. The result is an intense spark which, by proper choice of gas, will exhibit a continuum from 2000 to 8000 Å, with a duration not appreciably greater than that of the original laser pulse (*100*). Since the continuous spark is generated by the action of the primary laser pulse on the ambient gas, synchronization is not a serious problem.

(d) In a flash system with such short-duration pulses, one needs to be concerned with the response time of the detection and recording systems. Most photomultiplier tubes fail at times shorter than about 10^{-10} sec; Novak and Windsor (*100*) have used an image-converter camera, which should exhibit a time resolution on the order of 5×10^{-12} sec.

Components for laser flash photolysis are now available, and the technique is beginning to experience application. For example, a Q-switched ruby laser was used as photolysis-flash source, yielding 3×10^{-8}-sec pulses of 0.5-J energy output, in the flash photolysis of methylene blue; transients with lifetimes on the order of 2, 30, and 140 μsec were studied by flash kinetics (*101*). Novak and Windsor (*100*) have flashed solutions of aromatic hydrocarbons with a Q-switched ruby laser, using a frequency-doubling crystal, and have obtained transient spectra which they have interpreted as absorption from the lowest *excited* singlet state to higher singlets. If the interpretation is correct, this observation is very noteworthy since it could not have been recorded with any gas-filled flashtube sources extant.

A series of especially significant experiments has recently been reported by Rentzepis (*102*), in which a mode-locked, Q-switched ruby (or neodymium glass) laser was utilized to generate pulses of durations ranging from 2 to 15 ($\times 10^{-12}$) sec. The pulsed-laser system was utilized to measure *directly* the rates for vibrational relaxation and intersystem crossing in a number of aromatic hydrocarbons; likewise, the pulsed-laser apparatus was used to detect $S_1^* \rightarrow S_0$ fluorescence and $T_1^* \rightarrow S_0$ emission from azulene, the "anomalous" aromatic hydrocarbon which exhibits intense $S_2^* \rightarrow S_0$ fluorescence. These measurements deal with quantities which have been eluding direct experimental studies by photochemists for many years; it is anticipated that results of very considerable fundamental importance will be forthcoming from further experiments analogous to those described in Rentzepis' communication.

It should be clear from the preceding paragraphs that at least a few photochemists now have at their disposal apparatus which possess sufficient time resolution and sensitivity to probe deeply into almost any process of significance in photochemistry. In principle, therefore, time resolution is no longer an important component of instrument-design research in the field of flash photolysis. Further developments should thus concentrate upon enhancement of the reliability, specificity, and information content of flash-photolysis studies. It is conceivable that many of the analytical methods that lead to highly detailed structural information will be applied eventually to liquid-state flash photolysis (*103*). Of the various methods available for "direct" observation of excited states and transient intermediates in photoresponsive systems, flash photolysis has been, and appears likely to remain, overwhelmingly the most useful.

F. *Nuclear Magnetic Resonance Emission*

In a recent communication, Cocivera (*104*) has reported the observation of nmr emission in anthraquinone solutions subjected to steady-state irradiation by a 3500-W mercury lamp. Such emission was not observed in the absence of exciting light. The nmr emission is assumed due to an Overhauser effect which involves the protons and unpaired electrons of the anthraquinone triplet. The possibility that the observed emission is due to a free radical rather than a triplet state cannot be ruled out on the basis of presently available information. If the observations reported by Cocivera are indeed related to the presence of triplet states in the irradiated solutions, the implication is that nmr spectroscopy can, under certain conditions, be used to study electronically excited molecules in solution. Since the information content of nmr spectra is quite great, these recent observations may ultimately constitute the basis for a new, potentially very powerful, analytical tool for the "direct" study of photoexcited molecules. Further developments in this respect will be awaited, accordingly, with considerable interest.

V. Addendum

Since the completion of the original manuscript for this chapter, a number of publications have appeared that describe important work related to the analysis of photochemical reaction systems. Some of the more important recent work is summarized below, employing the same headings as those employed in the body of the chapter.

General Considerations. It should be clear from the above discussion that understanding of a photochemical reaction system requires experimental evidence obtained by a number of different approaches. An excellent example of the use of varied experimental techniques in the study of a complex photochemical reduction is a recent investigation of the photochemical reduction of $Co(NH_3)_5Cl^{2+}$. In this work (*105*), steady-state quantum-yield measurements, final product analyses, isotope tracer studies, chemical scavenging experiments, and flash photolysis were all employed. This paper represents an excellent example of how information taken from a variety of sources can be combined to yield a convincing mechanistic description of a photochemical reaction.

Identification and Efficiency of Final Product Formation. The gathering of meaningful photochemical kinetic data, under conditions of high initial light absorption, appreciable conversion of reactants to products, and polychromatic incident light, has been described by Burr, Gordon, and Park (*106*). The interrelationships between absorption spectra obtained under monochromatic irradiation and those obtained under conditions of polychromatic incident light have been described in detail by Morrey (*107*).

Matrix Spectroscopy. The formation of concentration gradients upon photolysis, and the problem of geometrically inhomogeneous light absorption, is even more critical in rigid matrices than in liquid solutions, due to the much lower diffusion coefficients in the former. The problem of correcting for such effects in matrix photolyses has recently been treated mathematically (*108*).

The fluorescence spectroscopy of matrix-isolated molecules, excited with a laser, has been described (*109*). Liquid crystals have recently employed as matrices for obtaining polarized absorption (*110*) and fluorescence (*111*) spectra of oriented molecules; in some cases, liquid crystals might prove interesting as matrices for photolyses of oriented molecules.

Electron Spin Resonance Spectroscopy. The detection and identification of transient free radicals by esr trapping techniques (*50*) has been described in more detail (*112*, *113*). The determination of reliable quantitative information from magnetophotoselection experiments (*59*) has been discussed by Judeikis and Siegel (*114*).

Photoluminescence Spectroscopy. A monograph dealing with the interpretation of luminescence spectra of organic compounds and metal complexes has appeared (*115*); it especially stresses structural and solvent effects on luminescence spectra and treats radiationless processes at some length.

Flash Photolysis. Procedures for obtaining quantitative absorption spectra from flash photolyses involving conventional equipment have been described (*116*).

There remains considerable interest in the development of new sources for flash photolysis. A spark source, with output concentrated in the ultraviolet, has recently been described. This source produces light pulses of approximately 7.5 nsec halfwidth with intensities of the order of 10^9 photons (*117*). A *z*-pinch arrangement, which produces 0.2-μsec pulses of intensities equal to about 3×10^{18} photons/pulse, and its associated optics have been reported (*118*). An especially interesting development has been reported by Callear and Hedges (*119*), who have produced a monochromatic flash source consisting of a mercury resonance lamp that is triggered by a microwave pulse. The light output is, of course, nearly monochromatic and is centered about 2537 Å.

By far the most active area in current flash-photolysis research is laser photolysis. Laser flash photolysis in the picosecond time domain is now a reality and has recently been reviewed by Rentzepis (*120*). The use of laser-flash apparatus for measurements of fluorescence decay times (*100*, *121*), for studies of orientational relaxation processes in liquids (*122*), and for rapid flash-spectroscopic studies of rapid, nonphotochemical, biological reactions (*123*) has been described. Emission esr spectra of free radicals in solution produced by laser flash photolysis have been described (*124*). Overviews of the use of laser flash photolysis to unravel photochemical kinetic schemes (including the rates of radiationless processes) have been given by Rentzepis (*120*) and Porter (*125*). It is clear that the use of lasers as sources for flash photolysis is destined to become widespread and will assist photochemists in making measurements that were undreamed of a relatively short time ago.

ACKNOWLEDGMENT

This work was supported in part by the Public Health Service (AM 11081) and the National Science Foundation (GP-8705).

REFERENCES

1. J. G. Calvert and J. N. Pitts, Jr., *Photochemistry*, Wiley, New York, 1966.
2. W. C. Taylor, Jr., and C. W. Moeller, *Inorg. Chem.*, **4,** 398 (1965).
3. H. C. Kessler, Jr., *J. Phys. Chem.*, **71,** 2736 (1967).
4. O. Kling, E. Nikolaiski, and H. L. Schläfer, *Ber. Bunsenges. Physik. Chem.*, **67,** 883 (1963).
5. G. M. Badger, R. J. Drewer, and G. E. Lewis, *Australian J. Chem.*, **19,** 643 (1966).
6. M. Schwartz, *Biochim. Biophys. Acta*, **131,** 548 (1967).

7. P. G. David, J. G. Richardson, and E. L. Wehry, *Inorg. Nucl. Chem. Lett.*, in press.
8. J. C. Bevington, *Radical Polymerization*, Academic Press, New York, 1961.
9. S. R. Palit, *Makrolmol. Chem.*, **38,** 96 (1960).
10. S. R. Palit and P. Ghosh, *J. Polymer Sci.*, **58,** 1225, 1243 (1962).
11. M. K. Saha, A. R. Mukherjee, P. Ghosh, and S. R. Palit, *ibid.*, **C16,** 159 (1967).
12. G. A. Delzenne, *ibid.*, **C16,** 1027 (1967).
13. I. Kraljic and C. N. Trumbore, *J. Am. Chem. Soc.*, **87,** 2547 (1965).
14. S. Sundararajan and E. L. Wehry, *Chem. Comm.*, 267 (1970).
15. W. R. Ware, *J. Phys. Chem.*, **66,** 455 (1962).
16. R. K. Sharma and N. Kharasch, *Angew. Chem. (Int. Ed.)*, **7,** 36 (1968).
17. G. Czapski and M. Ottolenghi, *Israel J. Chem.*, **6,** 75 (1968).
18. F. Wilkinson, *Advan. Photochem.*, **3,** 241 (1964).
19. R. G. Bennett and R. E. Kellogg, *Progr. Reaction Kinet.*, **4,** 215 (1967).
20. R. G. Bennett and R. E. Kellogg, *Photochem. Photobiol.*, **7,** 571 (1968).
21. C. A. Parker, in *Fast Reactions and Primary Processes in Chemical Kinetics* (S. Claesson, ed.), Almqvist & Wiksell, Uppsala, 1967, pp. 325–331.
22. C. A. Parker, *Photoluminescence of Solutions*, Elsevier, Amsterdam, 1968.
23. A. A. Lamola and G. S. Hammond, *J. Chem. Phys.*, **43,** 2129 (1965).
24. P. J. Wagner and I. Kochevar, *J. Am. Chem. Soc.*, **90,** 2232 (1968).
25. L. M. Stephenson, D. G. Whitten, G. F. Vesley, and G. S. Hammond, *ibid.*,**88** 3665, 3893 (1966).
26. S. D. Andrews and A. C. Day, *Chem. Commun.*, 477 (1967).
27. R. J. Campbell, E. W. Schlag, and B. W. Ristow, *J. Am. Chem. Soc.*, **89,** 5098 (1967).
28. R. S. H. Liu, N. S. Turro, and G. S. Hammond, *ibid.*, **87**, 3406 (1965).
29. H. L. J. Backstrom and K. Sandros, *Acta Chem. Scand.*, **14,** 48 (1960).
30. C. A. Parker and T. A. Joyce, *Photochem. Photobiol.*, **6,** 395 (1967).
31. D. J. Binet, E. L. Goldberg, and L. S. Forster, *J. Phys. Chem.*, **72,** 3017 (1968).
32. A. Vogler and A. W. Adamson, *J. Am. Chem. Soc.*, **90,** 5945 (1968).
33. I. Norman and G. Porter, *Nature*, **174,** 508 (1954); *Proc. Roy. Soc. (London)*, **A230,** 399 (1955).
34. E. Whittle, D. A. Dows, and G. C. Pimentel, *J. Chem. Phys.*, **22,** 1943 (1954).
35. A. R. Horrocks, T. Medinger, and F. Wilkinson, *Photochem. Photobiol.*, **6,** 21 (1967).
36. R. A. Keller and D. E. Breen, *J. Chem. Phys.*, **43,** 2562 (1965).
37. F. J. Smith, J. K. Smith, and S. P. McGlynn, *Rev. Sci. Instr.*, **33,** 1367 (1962).
38. D. R. Scott and J. B. Allison, *J. Phys. Chem.*, **66,** 561 (1962).
39. J. D. Winefordner and P. A. St. John, *Anal. Chem.*, **35,** 2211 (1963).
40. A. Treinin, in *Radical Ions* (E. T. Kaiser and L. Kevan, eds.), Wiley-Interscience, New York, 1968, Chapter 12.
41. P. Goldberg, *J. Chem. Phys.*, **40,** 427 (1964).
41a. J. N. Pitts, Jr., J. K. S. Wan, and E. A. Schuck, *J. Am. Chem. Soc.*, **86,** 3606 (1964).
42. R. G. Sowden and N. Davidson, *ibid.*, **78,** 1291 (1956).
43. M. C. R. Symons and M. G. Townsend, *J. Chem. Soc.*, 263 (1959).
44. G. C. Pimentel, in *Formation and Trapping of Free Radicals* (A. M. Bass and H. P. Broida, eds.), Academic Press, New York, 1960, Chap. 4. G. C. Pimentel and S. W. Charles, *Pure Appl. Chem.*, **7,** 111 (1963).

45. S. N. Foner, E. L. Cochran, V. A. Bowers, and C. K. Jen, *J. Chem. Phys.*, **32,** 963 (1960).
46. G. Quinkert, *Photochem. Photobiol.*, **7,** 783 (1968).
47. R. S. Alger, *Electron Paramagnetic Resonance: Techniques and Applications*, Wiley, New York, 1968.
48. C. P. Poole, Jr., *Electron Spin Resonance: A Comprehensive Treatise on Experimental Techniques*, Wiley-Interscience, New York, 1967.
49. L. H. Piette and W. C. Landgraf, *J. Chem. Phys.*, **32,** 1107 (1960).
50. E. G. Janzen and B. J. Blackburn, *J. Am. Chem. Soc.*, **90,** 5909 (1968).
51. A. J. Bard, in *Standard Methods in Chemical Analysis* (F. J. Welcher, ed.), Van Nostrand, Princeton, N.J., 1966, Vol. 3-A, pp. 630–632.
52. S. P. McGlynn, T. Azumi, and M. Kinoshita, *Molecular Spectroscopy of the Triplet State*, Prentice-Hall, Englewood Cliffs, N.J., 1969.
53. C. Thomson, *Quart. Rev.*, **22,** 45 (1968).
54. C. A. Hutchinson, Jr., and B. W. Mangum, *J. Chem. Phys.*, **29,** 952 (1958).
55. W. A. Yager, E. Wasserman, and R. M. R. Cramer, *ibid.*, **37,** 1148 (1962).
56. S. I. Weissman, *ibid.*, **29,** 1189 (1958).
57. J. S. Brinen and W. G. Hodgson, *ibid.*, **47,** 2946 (1967).
58. S. C. Tsai and G. W. Robinson, *ibid.*, **49,** 3184 (1968).
59. M. A. El-Sayed and S. Siegel, *ibid.*, **44,** 1616 (1966).
60. N. K. Chaudhuri and M. A. El-Sayed, *ibid.*, **42,** 1947 (1965).
61. S. Siegel and L. Goldstein, *ibid.*, **43,** 4185 (1965).
62. G. P. Rabold and L. H. Piette, *Photochem. Photobiol.*, **5,** 733 (1966).
63. W. G. Herkstroeter, L. B. Jones, and G. S. Hammond, *J. Am. Chem. Soc.*, **88,** 4777 (1966).
64. S. N. Foner, *Advan. Atom. Mol. Phys.*, **2,** 385 (1966).
65. C. A. Parker, *Photoluminescence of Solutions*, Elsevier, Amsterdam, 1968.
66. D. M. Hercules, in *Fluorescence and Phosphorescence Analysis* (D. M. Hercules, ed.), Wiley-Interscience, New York, 1966, Chap. 1.
67. J. N. Murrell, *The Theory of the Electronic Spectra of Organic Molecules*, Methuen, London, 1963.
68. C. Reid, *Excited States in Chemistry and Biology*, Butterworth, London, 1957.
69. C. A. Parker and W. T. Rees, *Analyst*, **85,** 587 (1960).
70. D. W. Ellis, in Ref. **66**, Chap. 2.
71. G. B. Porter and H. L. Schläfer, *Ber. Bunsenges, Physik. Chem.*, **68,** 316 (1964).
72. W. E. Ohnesorge, in Ref. **66**, Chap. 4.
73. F. Zuloaga and M. Kasha, *Photochem. Photobiol.*, **7,** 549 (1968).
74. E. L. Wehry and L. B. Rogers, in Ref. **66**, Chap. 3.
75. E. L. Wehry, in *Fluorescence* (G. G. Guilbault, ed.), Dekker, New York, 1967, Chap. 2.
76. G. Porter, in *Rates and Mechanisms of Reactions* (S. L. Friess, E. S. Lewis, and A. Weissberger, eds.), 2nd edition, Wiley, New York, 1963, Chap. 19.
77. D. N. Bailey and D. M. Hercules, *J. Chem. Educ.*, **42,** A83 (1965).
78. L. S. Grossweiner and M. S. Matheson, *J. Phys. Chem.*, **61,** 1089 (1957).
79. M. Green and G. Tollin, *Rev. Sci. Instr.*, **38,** 1316 (1967).
80. H. T. Witt, in Ref. **21**, pp. 81–97.
81. G. C. Pimentel and K. C. Herr, *J. Chim. Phys.*, **61,** 1509 (1964).
82. R. T. Meyer, *J. Chem. Phys.*, **46,** 967 (1967).

83. H. Berg, *Z. Anal. Chem.*, **216,** 165 (1966).
84. S. P. Perone and J. R. Birk, *Anal. Chem.*, **38,** 1599 (1966).
85. J. R. Birk and S. P. Perone, *ibid.*, **40,** 496 (1968).
86. H. Rüppel and H. T. Witt, *Z. Phys. Chem.* (*Frankfurt*), **15,** 321 (1958).
87. R. L. Strong and H. H. Richtol, in Ref. **21**, pp. 71–79.
88. S. Claesson and L. Lindqvist, *Arkiv Kemi*, **12,** 1 (1958).
89. J. H. Malmberg, *Rev. Sci. Instr.*, **28,** 1027 (1957).
90. J. T. D'Alessio, P. K. Ludwig, and M. Burton, *ibid.*, **35,** 1015 (1964).
91. R. C. Mackey, S. A. Pollack, and R. S. Witte, *ibid.*, **36,** 1715 (1965).
92. O. J. Steingraber and I. B. Berlman, *ibid.*, **34,** 524 (1963).
93. G. J. Mains, J. L. Roebber, and G. K. Rollefson, *J. Phys. Chem.*, **59,** 733 (1955).
94. G. K. Oster and R. A. Marcus, *J. Chem. Phys.*, **27,** 472 (1957).
95. G. Birnbaum, *Optical Masers*, Academic Press, New York, 1964, p. 214.
96. F. J. McClung and R. W. Hellwarth, *J. Appl. Phys.*, **33,** 828 (1962).
97. G. Magyar, *Nature*, **218,** 16 (1968).
98. B. Lengyel, *Introduction to Laser Physics*, Wiley, New York, 1966, p. 228.
99. A. J. DeMaria, D. A. Stetser, and W. H. Glenn, Jr., *Science*, **156**, 1557 (1967).
100. J. W. Novak and M. W. Windsor, *Proc. Roy. Soc.* (*London*), **A308,** 95 (1968).
101. R. M. Danziger, K. H. Bar-Eli, and K. Weiss, *J. Phys. Chem.*, **71,** 2633 (1967).
102. P. M. Rentzepis, *Chem. Phys. Lett.*, **2,** 117 (1968).
103. G. Porter, in Ref. **21**, pp. 473–476.
104. M. Cocivera, *J. Am. Chem. Soc.*, **90,** 3261 (1968).
105. G. Caspari, R. G. Hughes, J. F. Endicott, and M. Z. Hoffman, *ibid.*, **92,** 6801 (1970).
106. J. G. Burr, B. R. Gordon, and E. H. Park, *Adv. Chem. Ser.*, **81,** 418 (1969).
107. J. R. Morrey, *Anal. Chem.* **41,** 719 (1969).
108. P. G. Barker, M. P. Halstead, and J. H. Purnell, *Trans. Faraday Soc.*, **65,** 2404 (1969).
109. J. S. Shirk and A. M. Bass, *Anal. Chem.*, **41, (11),** 103A (1969).
110. R. A. Levenson, H. B. Gray, and G. P. Ceasar, *J. Am. Chem. Soc.*, **92,** 3653 (1970).
111. E. Sackmann, *Chem. Phys. Lett.*, **4,** 537 (1970).
112. E. G. Janzen and B. J. Blackburn, *J. Am. Chem. Soc.*, **91,** 4481 (1969).
113. K. J. Klabunde, *ibid.*, **92,** 2427 (1970).
114. H. S. Judeikis and S. Siegel, *J. Phys. Chem.*, **74,** 1228 (1970).
115. R. S. Becker, *Theory and Interpretation of Fluorescence and Phosphorescence*, Wiley-Interscience, New York, 1969.
116. H. Knibbe, D. Rehm, and A. Weller, *Ber. Bunsenges. Physik. Chem.*, **72,** 257 (1968).
117. D. K. Bedford, *J. Phys.*, **E3,** 77 (1970).
118. E.-G. Niemann, *J. Phys. Chem.*, **72,** 3766 (1968).
119. A. B. Callear and R. E. M. Hedges, *Trans. Faraday Soc.*, **66,** 605 (1970).
120. P. M. Rentzepis and C. J. Mitschele, *Anal. Chem.*, **42 (14),** 20A (1970).
121. H. Merkelo, S. R. Hartman, T. Mar, G. S. Singhal, and Govindjee, *Science*, **164,** 301 (1969).
122. K. B. Eisenthal and K. H. Drexhage, *J. Chem. Phys.*, **51,** 5720 (1969).
123. B. Chance, J. A. McCray, and J. Bunkenburg, *Nature*, **225,** 705 (1970).
124. P. W. Atkins, I. C. Buchanan, R. C. Gurd, K. A. McLauchlan, and A. F. Simpson, *Chem. Commun.*, 513 (1970).
125. G. Porter, *Chem. Brit.*, **6,** 246 (1970).

CHAPTER 7

ELECTROCHEMICAL AND OTHER ANALYTICAL METHODS FOR FLASH-PHOTOLYSIS STUDIES

S. P. Perone and *H. D. Drew*

DEPARTMENT OF CHEMISTRY
PURDUE UNIVERSITY
LAFAYETTE, INDIANA

FACULTY OF CHEMISTRY
SOUTHERN ILLINOIS UNIVERSITY
EDWARDSVILLE, ILLINOIS

I. Introduction and Scope of Chapter

Flash photolysis, involving the irradiation of photosensitive systems with a very short, very intense burst of uv-visible energy, is extremely useful for the direct observation of transient photolytic species. The intense burst of radiant energy produces a relatively high concentration of excited molecules and their immediate transient reaction products in the photolysis chamber, allowing the application of direct analytical measurements. It becomes possible, therefore, to study the chemistry of very-short-lived photolytic intermediates, particularly radical species. Moreover, the approach has been used to characterize and study directly excited triplet species in solution.

Much has been published regarding the principles and applications of the flash-photolysis technique. However, relatively little attention has been focused on the most crucial aspect of the approach, and that is the development of analytical techniques for transient systems. Another chapter in this monograph (Chapter 6 by Earl Wehry) provides complete coverage of the conventional analytical methods for flash-photolysis studies. It is the objective of this chapter to describe several less conventional analytical methods developed for flash-photolysis studies. The most detail will be devoted to electrochemical methods which have not been previously reviewed.

II. Basic Principles of the Flash-Photolysis Technique

A. *Description of the Method*

Flash-photolysis studies require three basic instrumental components:

1. photolysis flash to produce a short-duration, high-intensity pulse of radiation within the wavelength range to induce the desired photochemical reaction;
2. the reaction vessel, with any associated optical components for the most efficient utilization of the radiation and isolation of the desired radiation energy;
3. monitoring devices and associated instrumentation for estimation of the reactants, intermediates, and products of the photoinduced reaction.

Although the principle of the method is identical for all detection techniques, the designs of the units must be modified to achieve optimum conditions for each detection technique. By far, absorption spectroscopy is the most common monitoring system, but the other techniques discussed in this chapter offer complementary information not deducible from the now classical spectroscopic methods. The reaction-cell design and detection systems are discussed under their appropriate sections. The photolysis-flash assembly, which is basically identical for all detection techniques, is considered here.

B. *The Photolysis Flash and Triggering Methods*

The light source most common is flash photolysis is a gas-filled discharge lamp. The radiation is produced by a high voltage discharge through the flashtube filled usually with a rare gas (Fig. 7-1). Exploding-wire (*1*) and electric-spark (*2*) sources have received consideration, but for most applications these sources lack reproducibility for quantitative work. Increasing attention has been directed in recent years to laser excitation sources (*3*, *4*, *5*). The favorable characteristics of short-duration [20 nsec for 3371-Å laser (*5*)], high-intensity, and monochromatic emission are of considerable advantage. Undoubtedly with further development of lasers operable in the ultraviolet and vacuum-ultraviolet regions this course for photolysis flash will receive greater acceptance.

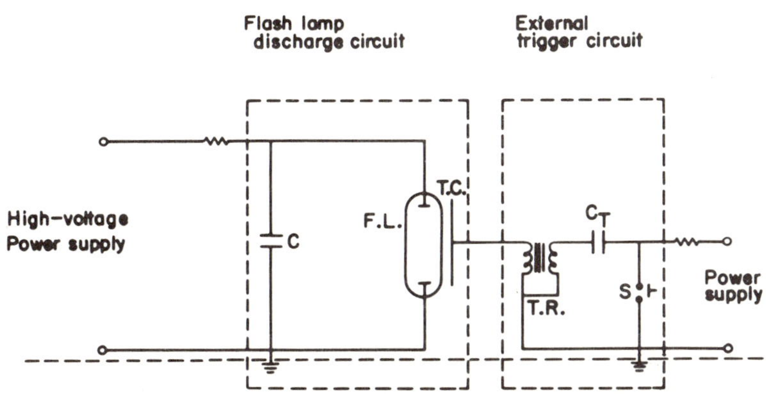

FIG. 7-1. Typical discharge circuit for external triggering of a high-pressure xenon flash lamp. C, Flash-discharge capacitor; F.L., xenon flash lamp; T.C., external triggering coil; C_T, trigger discharge capacitor; T.R., trigger transformer; S, trigger switch (often employed to initiate a time-delay sequence for the detection system).

A xenon-filled discharge lamp remains the most commonly used source; other rare gases are also suitable. Hydrogen- and deuterium-filled discharge lamps have been utilized but no advantages are apparent (*6*). Suitable discharge lamps are available commercially or if desired may be fabricated to meet specific needs.

The spectral distribution of high-pressure xenon flash lamps is a continuum which closely resembles the characteristics of sunlight. The high-energy (short-wavelength) cutoff is determined by the transmission of the envelope material; quartz, for example, absorbs wavelengths shorter than about 2000 Å. Xenon lamps can be operated to provide the vacuum-ultraviolet radiation with suitable modifications (*7–12*). Greatest radiant intensity is produced in the ultraviolet and visible regions with tailing into the infrared. The spectral characteristics depend on gas pressure and to some degree on the discharge parameters; higher voltage causes a slight shift toward the blue, whereas increased capacitance or inductance causes a slight red shift.

Of greater importance, especially with spectroscopic detection, are flash intensity and duration. Ideally, the flash should be of highest possible intensity and shortest duration. In practice, however, a compromise between the desirability of a short duration and the necessity of inducing sufficient photochemical change must be made. The intensity and duration of the flash are influenced by many factors. Among the more important factors are the flash-lamp parameters (tube diameter and arc length), the resistance, capacitance, and inductance of the discharge network, and the mode of discharge triggering. Usual discharge energies range from about 100 to 2000 J with flash durations, as measured by the half-peak width, less than 100 μsec. The conversion of electrical energy to radiant energy of greatest photochemical significance, between 2000 and 4000 Å, is relatively inefficient. Intensity within this wavelength range is typically 10^{-3} to 10^{-4} Einsteins.

Flash duration, measured by the half-peak width (Fig. 7-2), with no inductance in the discharge network, can be estimated from the following equation:

$$t_{1/2} = \tfrac{1}{2}RC \tag{1}$$

where $t_{1/2}$ is in microseconds when C, the discharge capacitance, is in microfarads and R, the flash-lamp resistance, is expressed in ohms. Likewise the energy per discharge can be estimated from the expression:

$$E\ (\text{Joules}) = \tfrac{1}{2}CV^2 \tag{2}$$

where the capacitance C is expressed in microfarads and the discharge

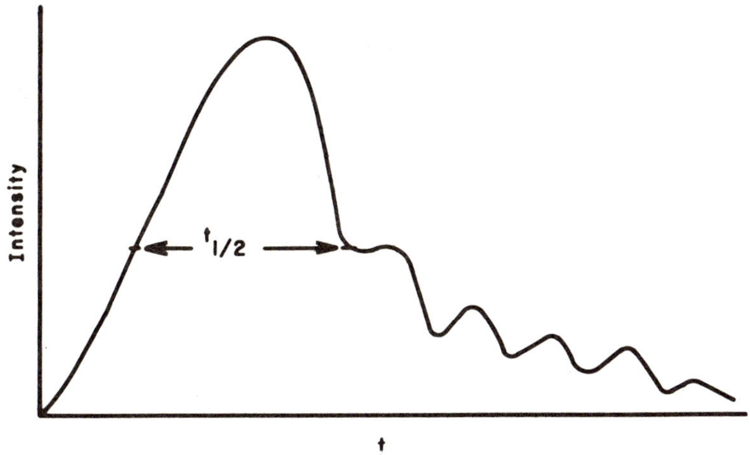

FIG. 7-2. High-pressure xenon flash-lamp intensity profile. The damped oscillations may lead to difficulty with spectroscopic detection techniques at short time intervals.

voltage V in kilovolts. If more exact information is necessary, flash duration can be measured directly with a photomultiplier detector, and the radiant flux can be determined by actinometry.

Several methods are available for increasing the discharge energy per flash without increasing the flash duration. Simply increasing the voltage while maintaining a low capacitance is not recommended since the efficiency of light production falls off at higher voltages and the flash lamp may be damaged or explode at current densities over the rated maximum. Several flash lamps with separate capacitor banks afford greater energy dissipation (*13*, *14*), but synchronous triggering offers some difficulty. Lindqvist (*15*) has reported a series triggering assembly which assures synchronous firing.

Tailing and oscillation of the flash output may persist well after $t_{1/2}$ (Fig. 7-2). With optical monitoring systems, this may offer considerable interference at short times. This effect can be minimized, with little loss in light output, by constructing the capacitor leads to the flash lamp out of a conductor with a positive temperature coefficient of resistance (*16*, *17*). The heat generated by the high initial current, often several thousand amperes, increases the resistance of the leads during the time of the flash itself and thus damps the discharge, which in turn reduces the flash time.

Two basic methods are generally utilized to trigger gas-filled discharge lamps. When the desired firing voltage is greater than the breakdown voltage of the lamp, a series thyratron or spark gap is required, which serves essentially as a switch. For voltages below the breakdown potential a third triggering electrode is necessary.

Both techniques incorporate a high-voltage pulse: the series triggering to initiate the thyratron or spark gap, and the external third electrode assembly to ionize the gas sufficiently to cause discharge. In this laboratory we have successfully triggered by connecting the low end of a floating charged capacitor to a trigger coil wrapped around the lamp. When the lamp is near to self-flash potentials, this method apparently induces ionization and discharge. Basic triggering circuits are described fully in the literature (*18*, *19*, *20*) and are commercially available.

Often it is necessary to initiate operation of the detection system at some short but definite time after the flash itself. This is usually accomplished by a time-delay circuit with a photomultiplier or light-activated solid-state switch used to synchronize the time delay with the flash. Time-delay methodology is considered later in this chapter.

III. Electrochemical Detection Methods in Flash Photolysis

A. *Similarities of Photoreductions and Electroreductions*

The similarity between photolytic reactions and electrolytic reactions is quite remarkable. For example, methane can be produced from acetic acid either by electrolytic oxidation (*21*) or by photolysis (*22*). Nitrosobenzene is the product of the photolysis of nitrobenzene (*23*) and is also formed as an intermediate in the electroreduction of nitrobenzene (*24*, *25*). Also, in the photolysis and electrolysis of benzophenone, the same radical intermediates, the ketyl radical and ketyl radical anion, and the same final product, benzopinancol, are produced (*26–29*). In addition, Stocker (*30*) observed that the photoreduction and electroreduction of acetophenone not only yielded the same products but also the same ratio of *dl* to meso forms.

With these similarities it is not surprising that photolytic intermediates are, in general, electroactive. This fact was first demonstrated in 1839 by Becquerel (*31*), who noted that the potential of an illuminated electrode was different than that of a nonilluminated electrode, both of which were immersed in the same solution of dilute acid. This phenomenon became known as the "Becquerel" or "photovoltaic" effect and is still the basis of much work that is being carried out in the area called *photopotentiometry*. Photopotentials have been shown to be the result of the altered oxidation–reduction properties of the solution in the vicinity of the irradiated electrode (*32–34*).

The combination of the facts that electroactive species are generated by photolysis and that many photolytically and electrolytically induced reactions are quite similar has prompted the application of electroanalytical techniques to the monitoring of photochemical reactions initiated by flash irradiation.

Electrochemical measurements have several analytical advantages for flash-photolysis studies: sensitivity is available for dilute concentrations of a wide variety of compounds; nearly the same detection limit exists for all electroactive species; and time resolution in the microsecond range is attainable (*35*, *36*). In addition, the qualitative characterization of photolytic intermediates can be simplified by purely electrochemical studies of the chemical processes in solution.

In this section a review and evaluation is made of the electrochemical techniques currently being used to follow photochemical reactions. The emphasis will be on the analytical aspects of the electrochemical measurements. For a good treatment of the fundamental photoelectrical aspects see the review by Kuwana (*37*).

B. *Potentiometry*

Potentiometry involves measurement of the electrical potential of an indicator (working) electrode in solution relative to a reference electrode (*38*). In zero-current potentiometry no net electrochemical reaction occurs at the working electrode, and the measurement is independent of mass transport to the electrode for homogeneous solutions.

Direct potentiometry utilizes a single measurement of cell potential for the analysis, concentration being related to potential through the well-known Nernst equation:

$$E = E^0_{O,R} - E_{\text{ref}} + \frac{RT}{nF} \ln \frac{\alpha_O[O]}{\alpha_R[R]} \tag{3}$$

Here E is the measured cell potential, $E^0_{O,R}$ is the standard electrode potential for the couple $O + ne^- = R$, E_{ref} is the potential of the reference electrode, R is the gas-law constant, F is the Faraday constant, n is the number of electrons transferred per molecule, T is the absolute temperature, and α is the activity coefficient of the respective species. The accuracy of the technique is limited by three major factors: (1) variations in liquid junction potential of the reference electrode; (2) uncertainties in activity coefficients; and (3) reversibility of the redox system, i.e., rate of electron exchange with the electrode.

C. *Photopotentiometry*

Photopotential, *photovoltaic effect*, *Becquerel effect*, and *photogalvanic effect* are terms applied to the experimental observation of a potential difference when one of two usually identical electrodes immersed in a photoactive medium is illuminated. The study of this effect has been described as *photopotentiometry*.

For almost four decades after the turn of the century there was considerable interest in this field. In fact, a review of the photovoltaic effect in 1942 (*39*) cited well over 300 references. However, at this time there is only scattered interest in the area of photopotentials.

Before the work of Surash and Hercules (*32–34*) there was considerable doubt as to the mechanism involved in the production of photopotentials; their work points to the conclusion that the electrode merely acts as an indicator for the altered oxidation–reduction properties of the solution in the vicinity of the illuminated electrode. However, little has been done to develop a good systematic approach for applying potentiometry to the quantitative study of photochemical processes. In fact, qualitative kinetic information is difficult to obtain for an unknown system since it is difficult to sort out the effects of electrode-surface problems, irreversibility of electroactive species, mixed potentials due to a combination of species, etc., from the phenomena of interest when using zero-current potentiometry. This is a direct result of the fact that a minimum of control is exercised over the electrochemical variables.

Thus, photopotential observations can demonstrate that the concentration(s) of electroactive species at the electrode are being changed as a result of irradiation of that electrode. Also, photopotentials can demonstrate that electroactive species are being produced in photolytic processes (*40*).

D. *Amperometric Measurements*

Perhaps the most sensitive electrochemical measurements on photolytic systems can be obtained from monitoring electrolysis *currents* at an inert working electrode in the region of the photochemical process in solution. The basic idea, as developed by Berg and co-workers (*41–57*) and by

Perone and co-workers (*58–60*), is to flash photolyze a volume of solution within an electrochemical cell (Fig. 7-3). By adjustment of the irradiating source, photolysis is caused to occur in the immediate vicinity of the monitoring electrode. In the most common approach, the electrode potential is held constant at a value insufficient to cause any electrolysis to occur before the flash. After the flash, any electroactive intermediates produced may be detected by the currents which will flow due to their electrolytic degradation at the electrode surface (Fig. 7-4). Selective monitoring of the intermediates can be accomplished by appropriate selection of the constant potential applied to the working electrode since the presence of oxidation–reduction couples results in sharp current excursions to new plateaus at potentials corresponding to their respective standard potentials. Thus, a current–voltage profile can be constructed which provides qualitative information regarding the initial, final, and intermediate species in the photochemical process. The current–voltage profile in Fig. 7-5 for example was obtained by plotting currents measured at a hanging-mercury-drop electrode 0.2 sec after a solution of 4×10^{-4} *M* benzophenone in 0.5 *M* NaOH was flash photolyzed in the electrochemical cell. Each point represents a separate experiment in which a different potential was imposed on the electrode. The resultant profile indicates the presence of at least three different electrode processes after the flash: the oxidation of the ketyl radical anion at potentials more anodic than -1.5 V

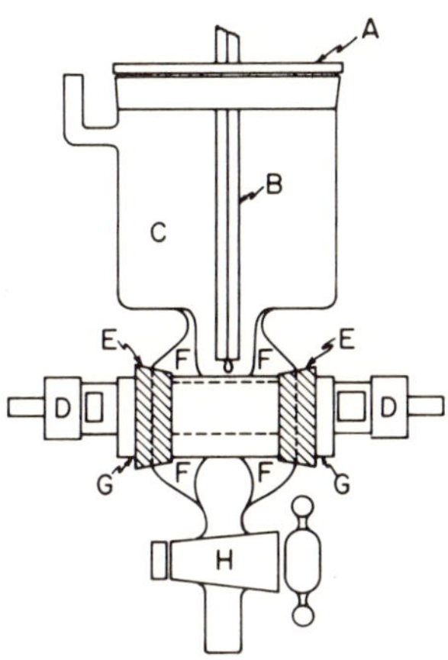

FIG. 7-3. Electrochemical cell for flash-photolysis studies. A, Machined Teflon lid; B, microburet-type HMDE working electrode; C, 200-ml borosilicate weighing bottle cell; D, flashtube; E, fitted rubber stopper; F, paraffin wax; G, quartz tubing, 28-mm o.d.; H, Teflon stopcock (other electrodes and nitrogen disperser not shown).

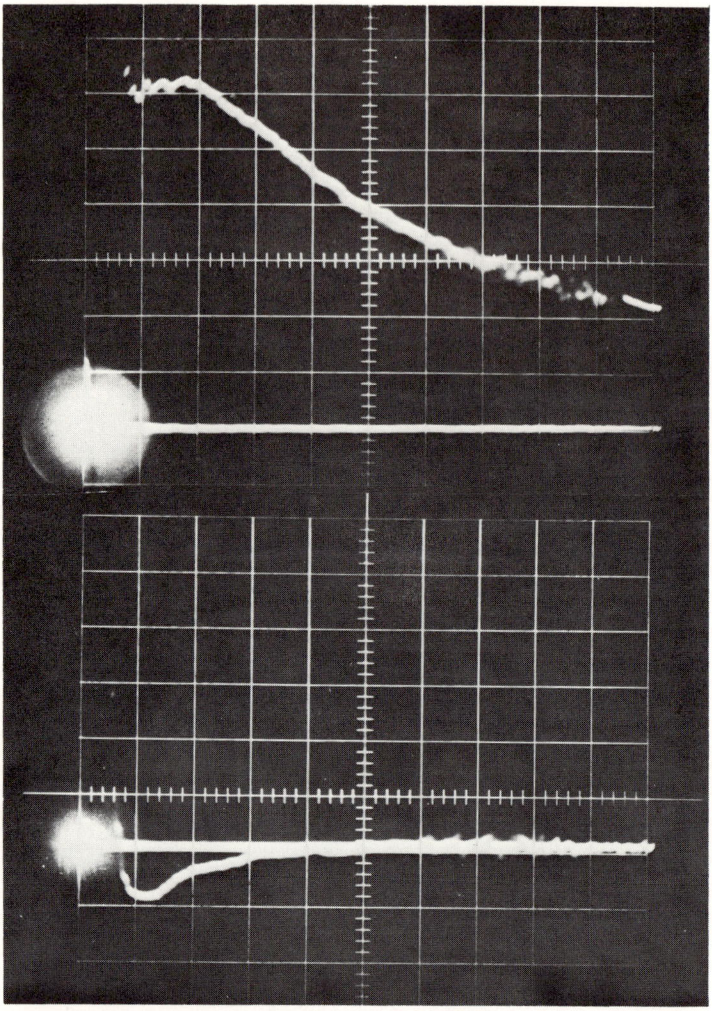

FIG. 7-4. Electrolytic currents observed in flash photolysis of benzophenone; continuous potentiostatic analysis following flash irradiation. *Top*, analysis of benzophenone solution (8×10^{-3} *M* in 0.25 *M* NaOH, 80% ethanol); *bottom*, analysis of a blank (0.25 *M* NaOH, 80% ethanol); *horizontal*, 50 μsec per division; vertical, 50 μA per division; potential, -0.700 V vs. SCE. The straight horizontal trace across each photograph is the zero-current line. The curved trace in the upper photograph is due to the oxidation of the ketyl radical and/or ketyl radical anion.

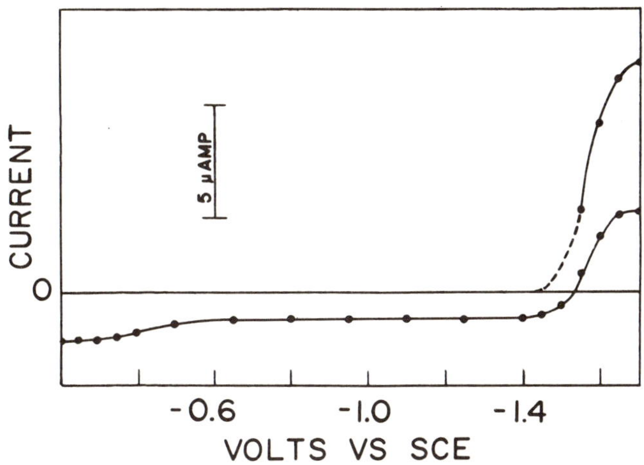

FIG. 7-5. Current–voltage profile from amperometric analysis of flash-photolyzed benzophenone. Currents measured at 0.2 sec after the flash from current–time curves are plotted vs. potential. The solution was 4×10^{-4} *M* benzophenone in 0.5 *M* NaOH, 80% ethanol. The upper curve at potentials more cathodic than −1.4 V is the current measured on the above solution without flash using the potential-step technique.

vs. SCE; the oxidation of benzopinancol at potentials more anodic than −0.4 V vs. SCE; and the reduction of unphotolyzed benzophenone at potentials more cathodic than −1.5 V vs. SCE (*58*). These assignments were made on the basis of previous electrochemical characterizations of the benzophenone reduction mechanism (*28*).

Kinetic information can be obtained from the current–time decay curves of Fig. 7-4. However, since the current decay is influenced by two interdependent processes—electrolytic depletion and chemical reaction—the extraction of kinetic data is possible only when the kinetic-diffusion differential equations are solved for the particular system involved (*61*). This is not difficult for a simple first-order kinetic step (*51*, *59*), but the equations for systems involving second-order kinetics are not amenable to closed-form solutions. Perone and Birk (*59*, *60*) have used several approximate approaches to obtaining useful theoretical relationships for current-decay curves involving a simple second-order dimerization of the electroactive species as in the scheme below:

$$\begin{array}{c} O \xrightarrow{h\nu} R \\ \uparrow \underline{\quad -ne^- \quad} \downarrow \\ 2R \xrightarrow{k} D \end{array} \tag{4}$$

The simplest form of the theoretical relationship (using the crudest approximations) is:

$$\frac{1}{i_t(t)^{1/2}} = \frac{\pi^{1/2}}{nFAD_R^{1/2}C_R^0} + \left(\frac{k\pi^{1/2}}{nFAD_R^{1/2}}\right)t \tag{5}$$

where i_t is the measured current at time t; n is the number of electrons transferred in the oxidation step; F is the Faraday constant; A is the electrode area; D_R is the diffusion coefficient for species R; and C_R^0 is the initial bulk concentration of R at $t = 0$, immediately after the flash. Equation (5) provides that a plot of $1/i(t)^{1/2}$ vs. t should be a straight line with slope dependent on k. However, the values of the other constants must be known. Except for D_R, the other constants are known or easily determined; D_R is generally not known because species R is a transient. However, it can usually be assumed that $D_R \simeq D_O$, and D_O can readily be determined for species O.

The relationship in Eq. (5) was used to analyze amperometric data obtained in the flash photolysis of benzophenone in the pH range of 10 to 14 (*59*). Table 1 presents the kinetic data obtained and compares these results to those obtained photometrically by Beckett and Porter (*26*). This table shows that satisfactory kinetic data were obtained when the half-life ($t_{1/2}$) of the chemical reaction was long compared to the measurement time.

TABLE 1

THE EFFECT OF PH ON THE RATE OF DIMERIZATION OF THE KETYL RADICAL

		Values of rate constants k, M^{-1} sec^{-1}		
pH	$t_{1/2}^a$	By amperometry[g]	By spectrometry[b]	Buffer
10.0[c]	90 μsec	$(0.9 \pm 0.3) \times 10^8$	1.8×10^8	0.1 M glycine/NaOH[d]
11.0	600 μsec	$(1.3 \pm 0.4) \times 10^7$	1.8×10^7	0.04 M Na_2CO_3 [d]
12.0	5.5 μsec	$(1.6 \pm 0.5) \times 10^6$	1.8×10^6	0.01 M NaOH[d]
13.0	64 msec	$(1.4 \pm 0.4) \times 10^5$	1.8×10^5	0.1 M NaOH[d]
13.0			1.3×10^5 [e]	0.1 M NaOH[e]
0.25 M OH^-		$(4.9 \pm 1.5) \times 10^4$ [f]	7.1×10^4	0.25 M NaOH
0.25 M OH^-	68 msec	$(5.3 \pm 1.6) \times 10^4$	7.1×10^4	0.25 M NaOH
0.50 M OH^-	0.21 sec	$(3.6 \pm 1.1) \times 10^4$	3.6×10^4	0.50 M NaOH

[a] Average value of $1/kC_R^0$ for series of experiments at each pH.
[b] Using Beckett and Porter's (*26*) results: $\log (k_{obs}) = -\text{pH} + 18.25$.
[c] Kinetic measurement made after the half-life for this pH.
[d] Enough tetramethylammonium chloride added to make ionic strength 0.25.
[e] From Berg and Schweiss (*51*).
[f] Time-delayed potentiostatic analysis.
[g] Taken from Birk and Perone (*59*).

Perone and Birk (*58*) also developed a generally applicable electroanalytical approach for kinetic studies in flash-photolyzed systems. This is a "time-delayed potentiostatic analysis" technique which involves operating the electrochemical cell at open circuit for an accurately specified period of time (delay time, τ) after the flash, not allowing any electrolysis to interfere with the normal photolytic processes in the vicinity of the indicator electrode. Then, at time τ the electrodes are switched in with a potential to the indicator electrode sufficient to cause electrolysis of the species whose concentration is to be measured. If the current is measured in a time interval after the original time delay, which is short compared to τ, it can be assumed that chemical decay is negligible during the measurement interval, and the measured current is proportional to the concentration of the unstable species at time τ. Thus, by repeating the experiment several times with different τ values, as shown in Fig. 7-6, a kinetic-decay curve can be obtained, which follows the homogeneous-solution kinetic scheme. The reader is referred to the original literature for details of the electronics and treatment of the electrochemical data obtained (*58–60*).

The advantage of the time-delayed potentiostatic approach is that the kinetic data are not distorted by mass-transfer effects, and therefore they can be treated with less complex rate equations based on homogeneous

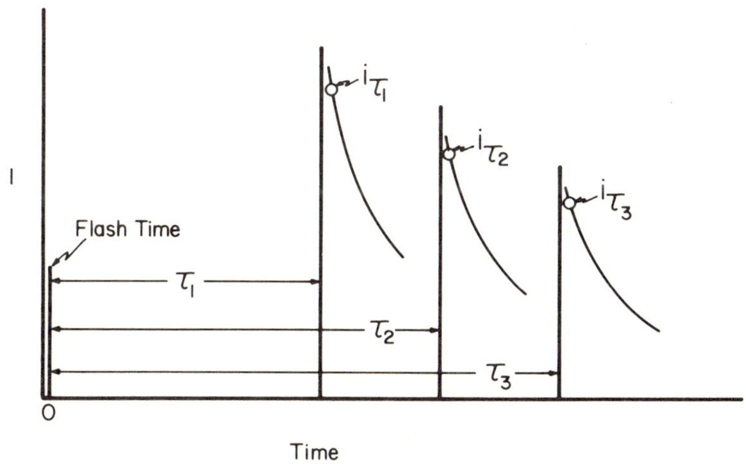

FIG. 7-6. Current–time functions observed with time-delayed potentiostatic measurements. τ_1, τ_2, and τ_3 refer to three different delay times used for three different experiments.

solution kinetics. Thus, the technique is applicable to much more complex kinetic systems than the analysis of continuous current–time curves after the flash.

On the other hand, the time-delayed technique cannot be applied to the study of very fast photochemical reactions because instantaneous application of the monitoring potentials results in large background currents needed to charge up the capacitive double layer at the electrode–solution interface. This limits the acquisition of meaningful faradaic current measurements to at least 50 μsec after τ. Thus, τ must be no shorter than about 500 μsec in order to make the assumption noted above.

Table 1 includes kinetic data obtained by the time-delay method. Results were comparable to those obtained by analysis of the continuous current–time curve according to Eq. (5). All results compared favorably with kinetic data for the same system obtained photometrically as mentioned above.

Some recent applications of the photoelectrochemical techniques have included the elucidation of the photochemical reduction mechanism of benzil in alkaline solution (*60*). This particular study pointed out the importance of a thorough understanding of the electrochemical characteristics of the system studied, and the final study was preceded by a complete electrochemical study in both acidic and alkaline solutions (*62*, *63*). In addition, Steipelfeldt and Perone (*60*) described the use of new photoactivated monitoring circuitry.

E. *Photopolarography*

Berg (*41–57*) has used polarography for investigating photochemical reactions and has termed the combination *photopolarography*. For slower secondary photochemical processes, conventional polarographic analysis with slowly varying potential has been employed. For the faster systems, constant-potential polarography was used. In the latter case time resolution down to 500 μsec was obtained for current–time studies at a single droplet formed from a conventional dropping-mercury electrode (*41*, *53*).

Berg has applied the photopolarographic technique to the study of several systems, these including benzophenone (*49–51*), fluorescein (*55*), and benzil (*42*), as well as several biologically important compounds (*44*, *45*). In his studies of quinones and ketones (*48*) he found a linear relation between the excitation energies required for the $n - \pi^*$ excited transition and the half-wave potentials. In another study (*53*) the effect of flash irradiation on solutions not containing a photoactive species was investigated at the dropping-mercury electrode. A small current, called the *photoresidual* current, was noted which Berg assigns to the electrons leaving the electrode and becoming solvated. This model is consistent with the observations that the current increases as the potential is made more cathodic and that the current is considerably higher in acid solution. In addition, his proposed mechanism explains the Becquerel effect noted in acid solutions and the lowering of the hydrogen overvoltage during irradiation. Other workers who have concerned themselves with the study of photoresidual currents include Delahay and Srinvasan (*64*), Barker *et al.* (*65*), and Heyrovsky (*66*).

F. *Summary*

As described above there are a variety of electrochemical measurements which might be used to monitor photolytic processes. The most promising of these appears to be amperometric measurements at a stationary electrode under potentiostatic conditions. These techniques provide measurement tools in flash photolysis to complement the commonly used photometric methods. The electrochemical methods are limited in applicability to solutions of relatively high conductivity and are sensitive only to electroactive materials. However, there are many photochemical studies which are conducted under such conditions. Moreover, where electrochemical techniques are applicable, they provide a very sensitive probe into photolytic processes.

One serious limitation on electrochemical monitoring was pointed out by Berg's work describing the photoresidual current (*53*). Electron emission from the electrode can be a serious interference to electrochemical measurements in flash-photolyzed systems. The interference is particularly great at low pH. The problem can be circumvented, to some extent, by a variety of approaches, including: (1) using a very short flash duration such that subsequent measurements are made at times when the photoresidual current and any side effects have been dissipated at the electrode surface; (2) using higher concentrations of the photosensitive components; and (3) adjusting the electrode potential to minimize the photoresidual current while still allowing detection of the electroactive photolytic products.

IV. Other Analytical Methods for Flash Photolysis

A. *Photoelectric Detection—Infrared Region*

The extension of optical detection for flash photolysis to the infrared region of the spectrum was delayed until the early 1960s primarily due to the lack of adequate detectors, both electronic and photographic, in this region. With the development of fast-rise, sensitive semiconductor devices with moderate wavelength selectivity, considerable impetus was given to rapid-scan infrared spectrophotometry. Emphasis directed toward infrared detection in flash photolysis was due largely to Pimentel and co-workers (*67–71*) and Hand, Kaufmann, and Hexter (*72*, *73*). Kasper and co-workers (*74–76*) investigated infrared emission processes initiated by flash photolysis which led to the first observations of chemical lasers.

Pimentel (*67*), in 1965, reviewed the advances of rapid-scan infrared-monitored flash photolysis and presented his prognosis of the method with regard to scan rate, flash noise duration, resolution, scan repetition rate, data acquisition, and detection limits. Since that time several improvements have been realized with the major limitations still remaining due to flash noise duration, a problem with all optical detection methods, and scan repetition rate (*71*, *72*). A description of the instrumentation is not presented here since excellent reviews of instrumental design features are readily available (*67*, *68*, *72*).

The kinetic study of chloroformic acid decomposition reported by Jensen and Pimentel (*71*) serves as an encouraging example of the utility of rapid-scan infrared-monitored flash photolysis. The spectrum is recorded by synchronous triggering of an oscilloscope and wavelength drive. Wavelength calibration is aided by the interference pattern (superimposed on formic acid spectrum, Fig. 7-7, spectrum *b*) generated by a silver chloride sheet positioned in the monitoring beam. Spectrum (*b*), Fig. 7-7, is the 768-cm^{-1} absorption band of formic acid recorded prior to photolysis. Spectrum *a*, Fig. 7-7, is the absorption band for the same vibrational mode of chloroformic acid. Assignment of this absorption mode was confirmed by band-contour analysis and deuteration studies.

The chloroformic acid was found to revert to starting material via first-order kinetics. Temperature effects on the kinetics were also studied.

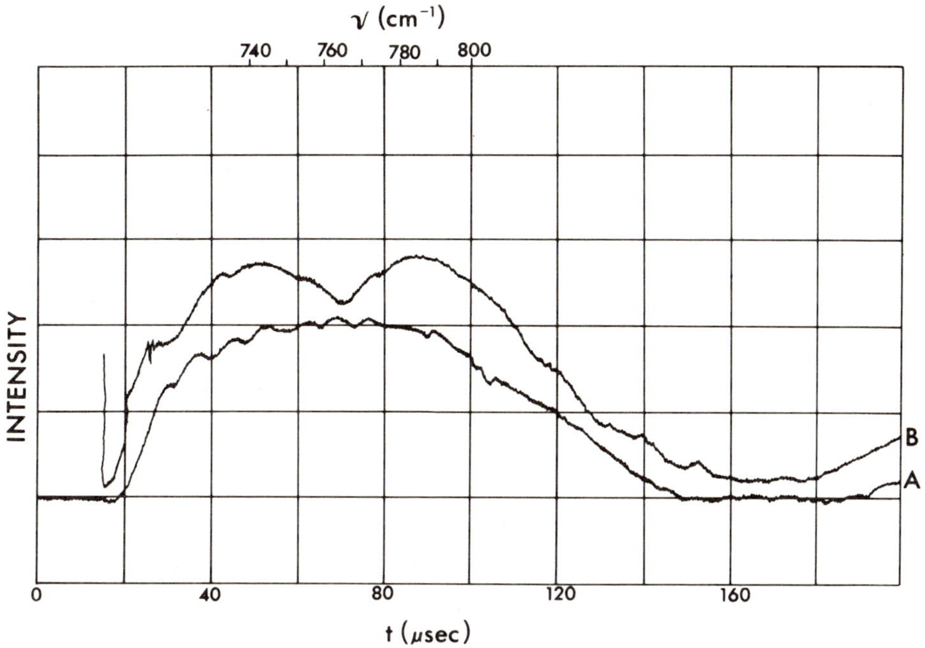

FIG. 7-7. Rapid-scan infrared spectra of photolysis of chlorine in the presence of formic acid. Band contour of 768-cm^{-1} band; spectral slit width, 8 cm^{-1}; *a*, 768-cm^{-1} band immediately after flash; time measured from flash trigger; *b*, before flash, formic acid and interference fringes from AgCl sheet; time measured from sweep trigger. After Jensen and Pimentel *(71)*.

Usually temperature studies are difficult, if not impossible, for gas-phase photolysis due to the difficulty of maintaining isothermal conditions during the flash. Jensen and Pimentel were fortunate with the formic acid system, however, because the temperature could be estimated by scanning the absorption band for the formic acid monomer (917 cm^{-1}) and dimer (1105 cm^{-1}). Employing the known temperature variation of the monomer–dimer equilibrium, the temperature, at any time after the flash, could be estimated.

The advantages of infrared investigation of reactive intermediates have already been illustrated with matrix isolation techniques. From the impressive initial studies utilizing rapid-scan infrared monitoring in flash photolysis indications are that this area of research will be fruitful in the near future.

B. *Mass Spectrometry*

Considerable effort has been directed toward utilizing time-resolved ("time-of-flight") mass spectrometry for detection of flash-photolysis processes. This combination is most suited for gas-phase reactions but in some instances has been employed when gaseous intermediates or products are involved (*77*). The initial adaptation of time-of-flight mass spectrometry for this purpose was reported by Kistiakowsky and Kydd (*77*). Berry *et al.* (*78*, *79*) have reported work in this area, but much of the recent improvements and research are attributed to Meyer and cohorts (*80–85*).

Adaptation of a time-of-flight mass spectrometer to flash-photolysis studies is relatively easy. The flash lamp is arranged to illuminate the reaction cell which is positioned in close proximity to the ion source of the spectrometer. The gaseous reaction mixture enters the spectrometer through a small aperture in the cell. By pulsing the ion source, the effluent gas is ionized, accelerated, and separated according to charge-to-mass ratio. The output of the ion-current detector is fed into a time-based oscilloscope which is synchronized with and triggered by the pulsed ion source. The entire mass spectrum can be displayed every 50 μsec, The record of the mass peak height versus time elapsed after the flash provides kinetic information about the system. Both adiabatic and isothermal conditions can be studied; the latter condition could be maintained by the addition of a nonultraviolet absorbing, preferably inert, diluent gas. Pressure changes under adiabatic conditions were also studied and suitable corrections incorporated when necessary (*82*).

Extensive studies have been reported, using mass-spectrometric analysis, which strongly support conclusions from earlier flash-photolysis investigations. In particular, the importance of the electronically excited iodine atom intermediate in flash photolysis has been confirmed by investigation of the kinetics of molecular iodine formation (*84*). The rate of formation of molecular iodine was found to be inconsistent with the reactivity of the ground-state atom, indicating the involvement of the more energetic excited-state atom. The significance of the hydroxyl radical in the combustion of oxygen and hydrogen in the presence of nitrogen dioxide, likewise, has been confirmed (*83*).

Difficulties in deciphering complicated reaction sequences are greatly simplified with mass-spectral detection of flash-photolysis processes. This is due in large degree to the ability to specify unambiguously the reactants, products, and intermediates by molecular weight. However, evidence for the primary photoproducts and excited-state intermediates can be only of indirect nature since the ionization process results in further reactions. Although the reproducibility of the technique is relatively poor; statistical analysis of the data has reduced this problem to some extent.

C. *Optical Rotatory Dispersion*

Carapelluci *et al.* (*86*, *87*) have combined flash photolysis with optical rotatory dispersion (ORD) to investigate possible structural changes and absolute configurations of excited states occurring on electronic excitation. Their apparatus was identical with conventional absorption-photometric monitoring systems with the addition of a polarizer and analyzer mounted within a centimeter of the cell windows. Appropriate corrections were made for residual rotation by the ground-state molecule while they scanned the triplet–triplet absorption region of the intermediate, recording the ORD spectrum 50 μsec after the flash.

The systems investigated were both *d*- and *l*-benzoin in benzene solvent. The observed ORD curves for the short-lived transient, assumed to be the benzoin triplet, were similar to the ground-state benzoin ORD curves indicating similarities in the Cotton effect between the ground and excited electronic states. The excited-state curves for *d*- and *l*-benzoin were roughly mirror images. A second, longer-lived transient exhibited no optical activity.

Although little work has been reported on the ORD detection technique, it appears to be a useful tool for investigating, in greater detail, the structure and symmetry of optically active excited states and intermediates of photochemical reactions.

D. *Electron Spin Resonance*

Transients observed in most photochemical reactions are radicals and it would seem that the electron spin resonance (esr) technique would lend itself to the study of these reactions. Indeed this technique has been employed for some time but the majority of these investigations have been at reduced temperature in frozen matrices. Radicals, although long-lived under these conditions, often exhibit spectra which are complicated by solvent and steric effects which would not be evident in free solution or the gaseous state. Such studies do not provide kinetic information about photoinduced chemical reactions either. The spectra of "free" radicals would be of greater importance, but, due to the reactivity of most radicals, such spectra are difficult to obtain.

Electron spin resonance has been utilized for investigation of triplet-state species generated by photolysis, but most of this work is restricted to low-temperature, matrix-isolation methods (*104*). The predominant difficulty is that in a nonviscous solvent the triplet-state molecule is generally very short-lived and not within the time resolution of present instrumentation. Detection of the triplet has been achieved during flash photolysis by optical absorption methods (*105*) by photolysis in a solvent of high viscosity which increases the lifetime by reducing collisional deactivation. Such a technique is not possible for esr detection, for under these conditions free rotation of the triplet is retarded, leading to a broadened, and in many cases undetectable, esr spectrum (*106*).

Long-lived radical spectra (where $t_{1/2}$ equals several hours) in solution have been observed with conventional esr methods (*88*), but such techniques would not be suitable for more reactive transients. Preliminary work employing a chopped ultraviolet light source and a C.A.T. (Computer of Average Transients) has provided esr kinetic data of short-lived intermediates (*89*). Many averaging cycles were required, however, due in part to the relatively low production of radicals by the light source.

The most promising approach thus far has been developed by Bennett and co-workers (*90, 91*). With this method a xenon flash lamp programmed for repetitive flashing is mounted near the resonance cavity of a conventional spectrometer. At a fixed magnetic field, the spectrometer output describes the radical decay curve produced by photolysis. After each flash the output of the spectrometer is monitored by six sampling channels, each channel gate set for a different time interval after the flash. The magnetic field is swept slowly through the spectral range and the resulting output of each channel is the "time-resolved" esr spectrum at the preset channel time delay. The method displays not only the kinetics of radical decay but also any spectral changes that may occur. Kinetic results for benzophenone photolysis agreed very well with spectrophotometric flash-photolysis results.

Estimation of the response characteristics indicates that transient spectra for species with rate constants as large as 10^5 sec^{-1} and 10^9 liters-mole^{-1} sec^{-1} for first- and second-order decay processes, respectively, can be obtained. Further modifications, as the authors suggest, would improve these response characteristics. Provisions for replenishing the solution during the investigation would be necessary if significant depletion occurred per flash. This could be accomplished easily by a flow-through cell assembly.

The high sensitivity (approximately 10^{-9} M) and favorable response time (5 μsec) indicate that esr investigation of rapid photolytic processes can reveal heretofore unavailable information of kinetic interest and, under favorable conditions, structural information as well. Presently, under high resolution, hyperfine structure is apparent in the transient spectrum of the benzophenone radical, but it is obscured to a large degree by noise. Improvement toward greater resolution would indeed be welcome.

E. *End-Product Analysis*

Product analysis offers no direct evidence about the reactions induced by flash photolysis. Some insight can be deduced, however, from the changes in product distribution as a function of the concentration of reactants and quenching additives. These studies may indicate a possible reaction mechanism and, in the case of competitive reactions, the ratio of the competing reactions' rate constants. Isotope exchange and distribution in the products can also shed considerable light on the reaction mechanism. Direct observation of the reaction kinetics, by one of the previously described methods, would be necessary for information about the order and rate of any reactions.

End-product analysis is most often accomplished by conventional analytical methods and equipment. One important advantage of flash-photolysis product analysis over low-intensity steady-state photolysis techniques is that the intermediates and products are not exposed to radiation. This precludes any possible complications that may arise due to further photochemical degradation of these species.

The most prevalent analytical techniques for determining the product distributions are vacuum line volumetric manometry (*92–95*) for gas-phase reactions, gas chromatography (*96–98*), and mass spectrometry (*99*, *100*). Most techniques involve transfer of the sample to the instrument for determination; however, the gas-chromatographic method of Martin and Ramstad (*96*) incorporates the flash-lamp assembly and reaction vessel in the sampling loop of a gas-sampling valve. The two-stage, two-column analysis was performed simply by sweeping the reaction mixture onto the column with the carrier gas, helium.

A combination of analytical techniques is often desirable—mass spectrometry and gas chromatography (*101*, *102*); volumetric monometry, mass spectrometry, and gas chromatography (*98*); and qualitative infrared spectroscopy (*103*).

The importance of end-product analysis should not be minimized. Far too often a proposed mechanistic interpretation of a photochemical reaction is reported without any mention of what products were finally recovered. The product analysis and distribution is a strong indication of the reactions involved and should be included in any thorough photochemical study.

ACKNOWLEDGMENTS

The authors are grateful for the contributions of Dr. J. R. Birk and Dr. H. E. Stapelfeldt to the preparation of this chapter.

This work was supported in part by Public Health Service Grant No. CA–07773 from the National Cancer Institute.

REFERENCES

1. G. K. Oster and R. A. Marcus, *J. Chem. Phys.*, **27,** 189 (1957).
2. G. J. Mains, J. L. Roebber, and G. K. Rollefson, *J. Phys. Chem.*, **59,** 733 (1955).
3. R. M. Danziger, K. H. Bar-Eli, and K. Weiss, *ibid.*, **71,** 2633 (1967).
4. G. Porter and J. I. Steinfeld, *J. Chem. Phys.*, **45,** 3456 (1966).
5. L. Lindqvist, *Compt. Rend. Ser. C*, **263,** 852 (1966).
6. I. B. Berlman, O. J. Steingraber, and M. J. Benson, *Rev. Sci. Instr.*, **39,** 54 (1968).
7. K. H. Welge, J. Wanner, F. Stuhl, and A. Heindrichs, *ibid.*, **38,** 1728 (1967).
8. E. Hayon and J. J. McGarvey, *J. Phys. Chem.*, **71,** 1472 (1967).
9. W. Braun, J. R. McNesby, and A. M. Bass, *J. Chem. Phys.*, **46,** 2071 (1967).
10. F. Stuhl and K. H. Welge, *Z. Naturforsch.*, **18A,** 900 (1963).
11. K. D. Bayes, K. H. Becker, and K. H. Welge, *ibid.*, **17A,** 676 (1962).
12. G. Black and G. Porter, *Proc. Roy. Soc.* (*London*), **266A,** 185 (1962).
13. G. Burnes and D. F. Hornig, *Can. J. Chem.*, **38,** 1702 (1960).
14. D. Kley, F. Stuhl, and K. H. Welge, *Z. Naturforsch.*, **18A,** 906 (1963).
15. L. Lindqvist, *Rev. Sci. Instr.*, **35,** 993 (1964).
16. L. Fischer, *Photochem. Photobiol.*, **2,** 411 (1963).
17. T. Ohno, S. Kato, and M. Koizumi, *Bull. Chem. Soc. Japan*, **39,** 232 (1966).
18. G. A. Theophanis, *Rev. Sci. Instr.*, **31,** 427 (1960).
19. G. Porter, in *Techniques of Organic Chemistry* (A. Weissberger, ed.), Vol. III, Pt. II, Wiley-Interscience, New York, 1963, Chap. XIX.
20. P. L. Goodfriend and H. P. Woods, *Rev. Sci. Instr.*, **36,** 10 (1965).
21. K. Clusius and W. Schanzer, *Z. Physik. Chem.*, **192,** 273 (1943).
22. L. Farkas and O. H. Wansbrough-Jones, *ibid.*, **B18,** 124 (1932).
23. S. H. Hastings and F. A. Matsen, *J. Am. Chem. Soc.*, **70,** 3514 (1948).
24. F. Haber, *Z. Elektrochem.*, **4,** 506 (1898).
25. F. Haber and C. Schmidt, *Z. Physik. Chem.*, **32,** 271 (1900).
26. A. Beckett and G. Porter, *Trans. Faraday Soc.*, **59,** 2038 (1963).

27. W. Kemula, Z. Grabowski, and M. Kalinowski, *Naturwiss.*, **47,** 514 (1960).
28. S. P. Perone, Ph.D. Thesis, Univ. of Wisconsin, Madison, Wis., 1963.
29. J. N. Pitts, Jr., R. L. Letsinger, R. P. Taylor, J. M. Patterson, G. Reckenwald, and R. B. Martin. *J. Am. Chem. Soc.*, **81,** 1068 (1959).
30. J. H. Stocker, private communication, Louisiana State Univ., New Orleans, New Orleans, La., April 1967.
31. E. Becquerel, *Compt. Rend.*, **9,** 561 (1839).
32. J. J. Surash, Ph.D. Theseis, Lehigh Univ., Bethlehem, Penna., 1960.
33. J. J. Surash and D. M. Hercules, Abstracts, 138th Meeting, American Chemical Society, New York, Sept. 1960.
34. J. J. Surash and D. M. Hercules, *J. Phys. Chem.*, **66,** 1602 (1962).
35. G. L. Booman and W. B. Holbrook, *Anal. Chem.*, **37,** 795 (1965).
36. S. P. Perone, *ibid.*, **38,** 1158 (1966).
37. T. Kuwana, *Electroanal. Chem.*, **1,** 197 (1967).
38. N. H. Furman, in *Treatise on Analytical Chemistry* (I. M. Kolthoff and P. J. Elving, eds.), Pt. I, Sec. D-2, Wiley-Interscience, New York, 1963, Chap. 45.
39. A. W. Copeland, O. D. Black, and A. B. Garrett, *Chem. Rev.*, **31,** 177 (1942).
40. K. M. Joshi and A. B. Karnik, *J. Ind. Chem. Soc.*, **42,** 127 (1965).
41. H. Berg, *Z. Elektrochem.*, **65,** 710 (1961).
42. H. Berg, *Z. Chem.*, **2,** 237 (1962).
43. H. Berg, *Rev. of Polarography* (*Japan*), **11,** 29 (1963).
44. H. Berg, *Abhandl. Deut. Akad. Wiss. Berlin, Kl. Med.*, 213 (1964).
45. H. Berg, *Abhandl. Deut. Akad. Wiss. Berlin, Kl. Chem., Geol. Biol.*, 128 (1964).
46. H. Berg, *Z. Anal. Chem.*, **216,** 165 (1966).
47. H. Berg and H. Kapulla, *Z. Elektrochem.*, **64,** 44 (1960).
48. H. Berg and K. Kramarczyk, *Ber. Bunsenges. Phsik. Chem.*, **68,** 296 (1964).
49. H. Berg and H. Schweiss, *Naturwiss.*, **47,** 513 (1960).
50. H. Berg and H. Schweiss, *Monatsber. Deut. Akad. Wiss. Berlin*, **2,** 546 (1960).
51. H. Berg and H. Schweiss, *Nature*, **191,** 1270 (1961).
52. K. E. Reinert and H. Berg, *Monatsber, Deut. Akad. Wiss. Berlin*, **4,** 26 (1962).
53. H. Berg and H. Schweiss, *Electrochim. Acta*, **9,** 425 (1964).
54. H. Berg, H. Schweiss, and D. Tresselt, *Exp. Tech. Physik*, **12,** 116 (1964).
55. F. A. Gollmick and H. Berg, *Ber. Bunsenges. Physik. Chem.*, **69,** 196 (1965).
56. H. Berg, *Z. Anal. Chem.*, **216,** 165 (1966).
57. H. Berg, H. Schweiss, E. Stutter, and K. Weller, *J. Electroanal. Chem.*, **15,** 415 (1967).
58. S. P. Perone and J. R. Birk, *Anal. Chem.*, **38,** 1589 (1966).
59. J. R. Birk and S. P. Perone, *ibid.*, **40,** 496 (1968).
60. H. E. Stapelfeldt and S. P. Perone, *ibid.*, **41,** 628 (1969).
61. P. Delahay, *New Instrumental Methods in Electrochemistry*, 2nd edition, Wiley-Interscience, New York, 1954.
62. H. E. Stapelfeldt and S. P. Perone, *Anal. Chem.*, **40,** 815 (1968).
63. H. E. Stapelfeldt and S. P. Perone, *ibid.*, **41,** 623 (1969).
64. P. Delahay and V. S. Srinvasan, *J. Phys. Chem.*, **70,** 420 (1966).
65. G. C. Barker, A. W. Gardner, and D. C. Sammon, Symposium, Electrode Process, Theoret, Div., Electrochem. Soc., Spring Meeting, Cleveland, Ohio, 1966, preprint papers, p. 435.
66. M. Heyrovsky, *Nature*, **206,** 1356 (1965).

67. G. C. Pimentel, *Pure Appl. Chem.*, **11**, 563 (1965).
68. K. C. Herr and G. C. Pimentel, *Appl. Opt.*, **4**, 25 (1965).
69. G. C. Pimentel and K. C. Herr, *J. Chem. Phys.*, **61**, 1509 (1964).
70. G. Carlson and G. C. Pimentel, *ibid.*, **44**, 4053 (1966).
71. R. J. Jensen and G. C. Pimentel, *J. Phys. Chem.*, **71**, 1803 (1967).
72. C. W. Hand, P. Z. Kaufmann, and R. M. Hexter, *Appl. Opt.*, **5**, 1097 (1966).
73. C. W. Hand and R. M. Hexter, Eighth European Congress on Molecular Spectroscopy, Copenhagen, Denmark, August 1965.
74. J. V. Kasper and G. C. Pimentel, *Phys. Rev. Lett.*, **14**, 352 (1965).
75. J. V. Kasper and G. C. Pimentel, *Appl. Phys. Lett.*, **5**, 231 (1964).
76. J. V. Kasper, J. H. Parker, and G. C. Pimentel, *J. Chem. Phys.*, **43**, 1827 (1965).
77. G. B. Kistiakowsky and P. H. Kydd, *J. Am. Chem. Soc.*, **79**, 4825 (1957).
78. R. S. Berry, J. Clardy, and M. E. Schafer, *ibid.*, **86**, 2738 (1964).
79. R. S. Berry, J. Clardy, and M. E. Schafer, *Tetrahedron Lett.*, **15**, 1003 (1965).
80. J. M. Freese and R. T. Meyer, U.S. Atomic Energy Comm., SC-RR-185 (1965). SC-R-65-956 (1965); through *Chem. Abstr.*, **63**, 17286b, 14204h (1965) and *Chem. Abstr.*, **63**, 14204h (1965).
81. R. T. Meyer, *Rev. Sci. Instr.*, **35**, 1064 (1964).
82. R. T. Meyer, *J. Sci. Instr.*, **44**, 422 (1967).
83. R. T. Meyer, *J. Chem. Phys.*, **46**, 967 (1967).
84. R. T. Meyer, *ibid.*, **46**, 4146 (1967).
85. R. T. Meyer, *J. Phys. Chem.*, **72**, 1583 (1968).
86. P. A. Carapelluci, H. H. Richtol, and R. L. Strong, *J. Am. Chem. Soc.*, **89**, 1742 (1967).
87. P. A. Carapelluci, *Diss. Abstr.*, **B28**, 2785 (1968).
88. M. Sofue and S. Nagakura, *Bull. Chem. Soc. Japan*, **38**, 1048 (1965).
89. L. H. Piette, Sixth Internatl. Symposium on Free Radicals, Cambridge, 1963 (cf. Ref. **3**).
90. T. J. Bennett, R. C. Smith, and T. H. Wilmhurst, *Chem. Commun.*, **513** (1967).
91. T. H. Wilmhurst, T. J. Bennett, and R. C. Smith, *Proc. Roy. Soc.* (*London*), **302A,** 305 (1967).
92. G. K. Oster and R. A. Marcus, *J. Chem. Phys.*, **27**, 472 (1957).
93. N. Slagg and R. A. Marcus, *ibid.*, **34**, 1013 (1961).
94. B. C. Roquitte and J. H. Futrell, *ibid.*, **37**, 378 (1962).
95. R. A. Back, *ibid.*, **40**, 3493 (1964).
96. S. B. Martin and R. W. Ramstad, *Anal. Chem.*, **33**, 982 (1961).
97. A. Shilman and R. A. Marcus, *J. Chem. Phys.*, **39**, 996 (1963).
98. W. Braun, K. H. Welge, and J. R. McNesby, *ibid.*, **44**, 2650 (1966).
99. D. W. L. Griffiths and R. A. Back, *ibid.*, **46**, 3913 (1967).
100. R. A. Back and D. W. L. Griffiths, *ibid.*, **46**, 4839 (1967).
101. L. C. Fischer and G. J. Mains, *J. Paths. Chem.*, **68**, 2522 (1964).
102. L. C. Fischer and G. J. Mains *ibid.*, **68**, 188 (1964).
103. B. C. Roquitte, *ibid.*, **70**, 2699 (1966).
104. C. Thomson, *Quart. Rev.*, **22**, 45 (1968).
105. G. Porter and M. R. Wright, *J. Chem. Phys.*, **55**, 705 (1958).
106. S. I. Weissman, *ibid.*, **29**, 1189 (1958).

CHAPTER 8

ANALYTICAL TECHNIQUES FOR THE SOLID-STATE PHOTOLYSIS OF COORDINATION COMPOUNDS

W. W. Wendlandt

DEPARTMENT OF CHEMISTRY
UNIVERSITY OF HOUSTON
HOUSTON, TEXAS

I. Introduction

The photochemistry of transition-metal coordination compounds in aqueous solutions has been rather extensively investigated during the past

50 years. The types of photochemically induced reactions, their reaction stoichiometries, kinetics, and mechanisms have been characterized by numerous investigators. From these studies there appear to be four distinct types of photochemical reactions of these compounds:

(a) reduction of the central metal ion by either a ligand or an anion in solution;
(b) oxidation of the central metal ion by either a ligand or an anion in solution;
(c) removal or substitution of a coordinated ligand or anion;
(d) isomerization or racemization of the complex.

Of the above reactions, types (a) and (c) have been perhaps the most extensively investigated.

In contrast to the solution photochemical studies, there have been relatively few investigations on the solid-state photochemical reactions of coordination compounds. The types of reactions reported are essentially the same as the above with the possible exception of the photochemical oxidation of the central metal ion [type (b)]. The solid-state photolysis reactions differ from those in aqueous solution in the following manner:

(1) Since the reactions involve only the surface layers of the sample, detection of the reaction intermediates and stoichiometries is difficult, if not impossible, to obtain.
(2) There is no "solvent effect" in solid-state reactions; hence, type (c) reactions may be difficult to observe. In some cases, however, hydrate-bound water may take part in the reaction.

The few studies that have been reported are mostly qualitative in nature. Generally, only the color changes of the sample have been reported with little data as to the reaction intermediates or products, reaction kinetics, mechanism, or thermochemistry involved. There have been, however, several comprehensive investigations in this area and these are reported in the latter part of this chapter.

II. Experimental Procedures

Since only the surface layers of the sample undergo photochemical reactions, the technique of diffuse reflectance spectroscopy is used to determine the reaction intermediates and/or products. The sample may be exposed to mercury or xenon lamp (*1*), monochromatic, or xenon flashtube irradiation (*2*); at periodic intervals, the diffuse reflectance spectrum of the sample is recorded. From the shift in wavelength of the curve maximum and/or changes in the reflectance intensity, the extent of the photochemically induced reaction may be determined.

The solid sample is placed in a 1- by 10-mm diameter circular indentation in a 5-cm square aluminum plate. The sample, as shown in Fig. 8-1, is then irradiated by one of the three means as previously discussed. After irradiation, the sample and plate are placed on the sample port of the integrating sphere of the spectroreflectometer and the reflectance spectrum recorded. A schematic diagram of the integrating sphere–sample holder configuration is shown in Fig. 8-2. The sample is irradiated at periodic intervals; after each irradiation, the sample's reflectance spectrum is recorded.

The effect of ultraviolet irradiation on a solid sample of

$$K_3[Co(C_2O_4)_3]\cdot 3H_2O$$

is shown in Fig. 8-3 (*1*). The irradiation source was a 200-W high-pressure mercury lamp with irradiation at 1-min intervals, at ambient room temperatures, for a total of 90 min. The initial dark-green-colored complex changed to violet on exposure to the radiation. This is quite evident from the curve as the original spectrum contained peak maxima at 415 and 590 nm, respectively, which changed to a curve containing a peak maximum at 540 nm with a shoulder peak at 590 nm after 90 min of irradiation. The intermediate curves, at 15- and 45-min irradiation periods, showed the presence of mixtures of the original complex with the final irradiation product, which in this case was $K_2[Co(C_2O_4)_2]$.

For irradiation of the sample with monochromatic radiation, the experimental arrangement shown in Fig. 8-4 may be employed (*4*). The system utilizes a high-pressure mercury lamp and power supply, grating-type

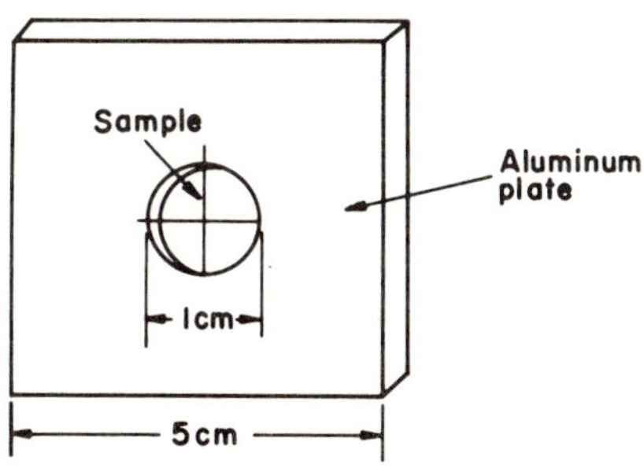

FIG. 8-1. Sample holder for irradiation of solid samples.

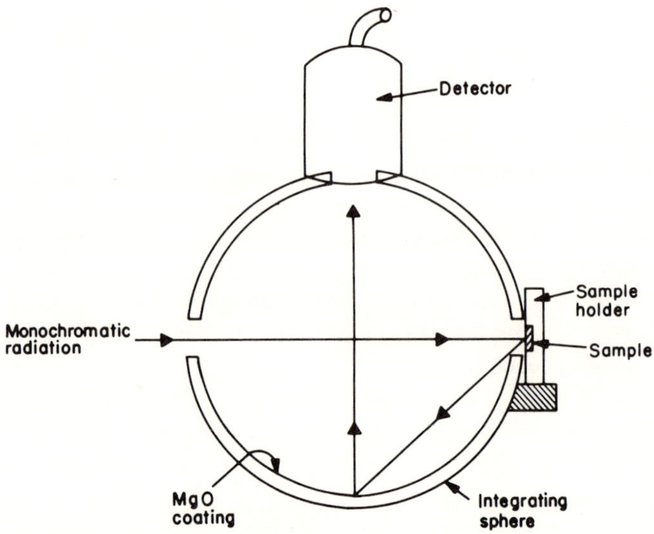

FIG. 8-2. Scheme illustrating the integrating-sphere–sample-holder configuration (*3*).

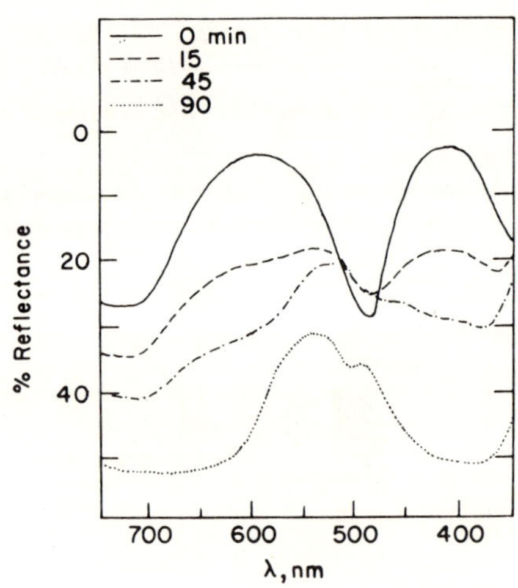

FIG. 8-3. Effect of ultraviolet irradiation on $K_3[Co(C_2O_4)_3]\cdot 3H_2O$ (*1*); irradiation intervals indicated.

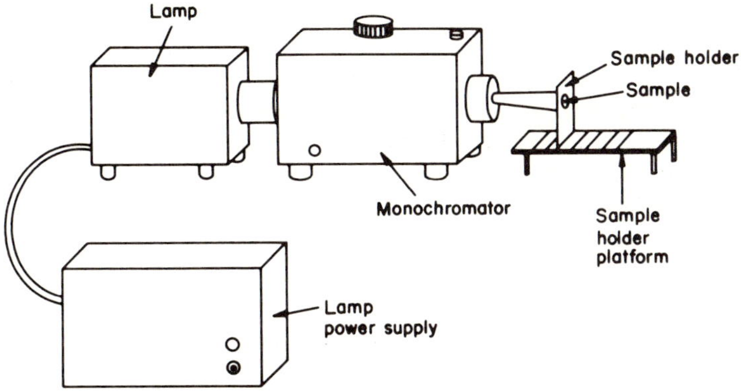

FIG. 8-4. Irradiation of sample using monochromatic radiation (*4*).

monochromator, and sample holder plate. The distance between the sample holder and the monochromator is adjustable by positioning the sample plate on the sample holder. At periodic intervals, the sample plate is removed from the apparatus and the reflectance spectrum of the sample recorded on the spectroreflectometer.

The experimental arrangement for flashtube irradiation is illustrated in Fig. 8-5. The system employed an Edgerton, Germeshausen and Grier Co.

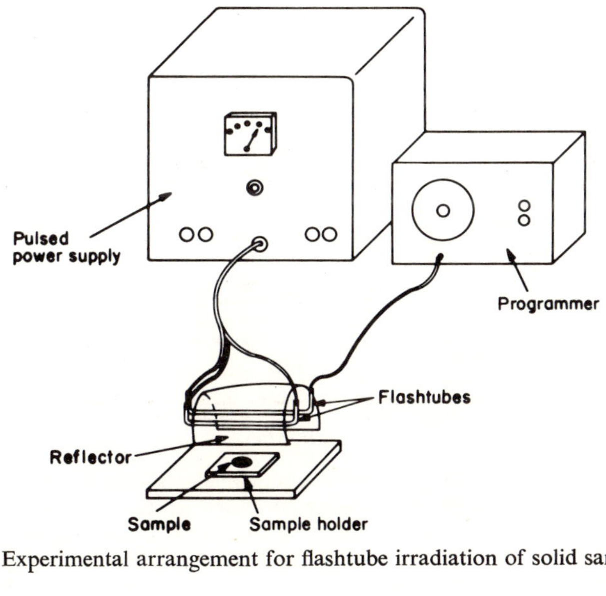

FIG. 8-5. Experimental arrangement for flashtube irradiation of solid samples (*2*).

Model 531 pulsed power supply equipped with two FX-100 xenon flashtubes placed 5 cm from the sample holder. The tubes were programmed to flash at 30-sec intervals at a radiation output of 100 Wsec (with a flash duration of 150 μsec). At periodic intervals, the sample plate was removed from the apparatus and the reflectance spectrum of the sample recorded. Usually, from 50 to 850 flash discharges were necessary for the conversion of the surface layer of a sample of $[Cr(en)_3]X_3$ to the bis(ethylenediamine) complex, $[Cr(en)_2X_2]X$.

III. Examples of Photolysis Reaction

To illustrate the use of the above irradiation techniques more fully, several examples of solid-state photolysis reactions of transition-metal coordination compounds will be presented.

A. *The Deamination of* $[Cr(en)_3]X_3$ *to* $[Cr(en)_2X_2]X$ *Complexes*

To illustrate the use of the flashtube technique, the photochemical deamination of $[Cr(en)_3]X_3$ complexes, as shown in Fig. 8-6, will be discussed. The irradiation of the yellow-colored $[Cr(en)_3]Cl_3$ and $[Cr(en)_3]Br_3$ complexes caused a darkening of their color to red and

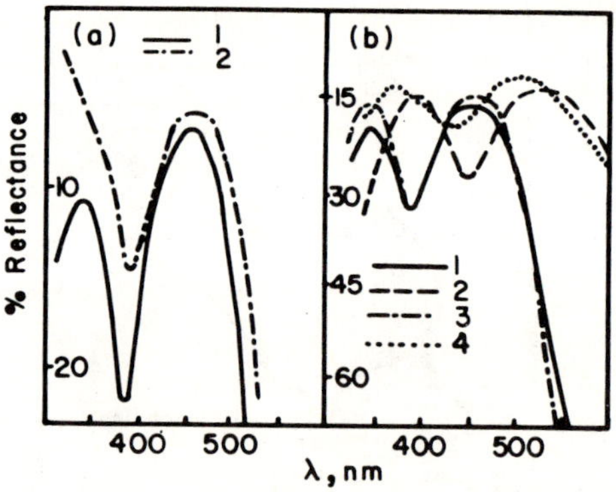

FIG. 8-6. Flashtube irradiation of $[Cr(en)_3]X_3$ complexes. (a) 1, $[Cr(en)_3](SCN)_3$; 2, $[Cr(en)_3](SCN)_3$ after exposure to xenon flashtube irradiation {identical to *trans*-$[Cr(en)_2(SCN)_2]SCN$}. (b) 1, $[Cr(en)_3]Cl_3$; 2, $[Cr(en)_3]Cl_3$ after exposure to xenon flashtube irradiation {identical to *cis*-$[Cr(en_2Cl_2]Cl$}; 3, $[Cr(en)_3]Br_3$; 4, $[Cr(en)_3]Br_3$ after exposure to xenon flashtube irradiation {identical to *cis*-$[Cr(en)_2Br_2]Br$} (*2*).

eventually purple. In the case of the chloride complex, this purple-colored product was identified from its reflectance spectrum to be *cis*-$[Cr(en)_2Cl_2]Cl$; for the bromide complex, *cis*-$[Cr(en)_2Br_2]Br$ was the ultimate product. The deamination reaction is then given by Eq. (1).

$$[Cr(en)_3]X_3\ (s) \xrightarrow{h\nu} cis\text{-}[Cr(en)_2X_2]X\ (s) + en(g) \tag{1}$$

The effect of irradiation on the $[Cr(eb)_3](SCN)_3$ complex was the color change from an original yellow to an orange-colored product. The latter compound was identified as *trans*-$[Cr(en)_2(SCN)_2]SCN$. The thiocyanate complex required the least number of flash discharges for the deamination reaction while the bromide required the most. The approximate order of photochemical deamination was SCN $\gg$ Cl $>$ Br. No effect was noted, however, on $[Cr(en)_3]I_3$.

B. *The Nitro → Nitrito Transition in Nitropentamminecobalt(III) Complexes*

Of the many studies on the linkage isomerism of the nitro group (NO_2^-) in the cobalt(III) ammine complexes, most of them have been concerned with the nitro → nitrito transition. For example, heating $[Co(NH_3)_5ONO]X_2$ complexes at 70°C results in the formation of the $[Co(NH_3)_5NO_2]X_2$ compounds. Wendlandt and Woodlock (*5*) found that on irradiating $[Co(NH_3)_5NO_2]X_2$ (X is Cl^-, Br^-, I^-, and NO_2^-) with ultraviolet radiation, the nitrito complexes, $[Co(NH_3)_5ONO]X_2$, were formed. The irradiation of $[Co(NH_3)_5NO_2]Br_2$ is illustrated in Fig. 8-7.

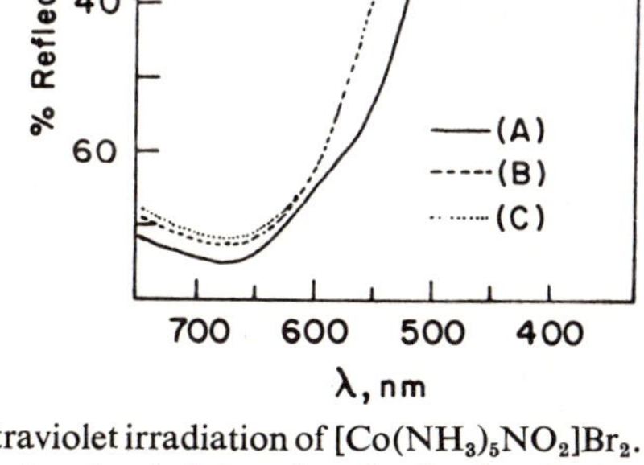

FIG. 8-7. Effect of ultraviolet irradiation of $[Co(NH_3)_5NO_2]Br_2$. A, $[Co(NH_3)_5NO_2]Br_2$; B, after irradiation for two 1-min intervals; C, after irradiation for two 1-min intervals.

The 460-nm peak in the reflectance curve shifted to 485 nm after an irradiation period of two 1-min intervals. Continued irradiation did not further change the curve's peak-maximum position.

C. *Potassium Tris(oxalato)metal Complexes*

A rather extensive series of investigations has been carried out on the photolysis of the tris(oxalato) complexes, $K_3[M(C_2O_4)_3]\cdot 3H_2O$, where M is cobalt (*1*), manganese (*6*), and iron (*7*).

The reflectance curves of $K_3[Co(C_2O_4)_3]\cdot 3H_2O$, showing the effect of irradiation by ultraviolet radiation, are shown in Fig. 8-3. The curves only indicate the formation of $K_2[Co(C_2O_4)_2]$ on irradiation. Further investigations, using magnetic susceptibility, mass loss (TGA), and mass-spectrometric analysis of the gaseous decomposition products, indicated the stoichiometry for the photolysis reaction shown in Eq. (2).

$$\begin{aligned} 2K_3[Co(C_2O_4)_3]\cdot 3H_2O(s) &\xrightarrow{h\nu} 2K_2[Co(C_2O_4)_2](s) \\ &\quad + K_2C_2O_4(s) + 2CO_2(g) + 6H_2O(g) \end{aligned} \tag{2}$$

Similarly, the solid-state photolysis of $K_3[Mn(C_2O_4)_3]\cdot 3H_2O$ was investigated using the same techniques (*6*). The dark-red-violet-colored complex changed to white on irradiation, as shown in Fig. 8-8. The curve for the original complex contained a single peak maximum at about 525 nm, which decreased in intensity as the time of irradiation increased. The stoichiometry of the photolysis reaction, as determined by various analytical procedures, was that shown in Eq. (3).

$$\begin{aligned} 2K_3[Mn(C_2O_4)_3]\cdot 3H_2O(s) &\xrightarrow{h\nu} 2K_2[Mn(C_2O_4)_2](s) \\ &\quad + K_2C_2O_4(s) + 2CO_2(g) + 6H_2O(g) \end{aligned} \tag{3}$$

An attempt was made to evaluate the rate of the photolysis reaction, using the expression shown in Eq. (4).

$$k = \frac{d(I_0 - I)^2}{dt} = \frac{dN^2}{dt} = 2N\left(\frac{dN}{dt}\right) \tag{4}$$

where I_0 is the initial percentage reflectance at 525 nm, I is the percentage reflectance at time t, k is the rate constant, and $N = (I_0 - I)$. The value for k was 52.7 min^{-1} and it was found that the reaction obeyed minus

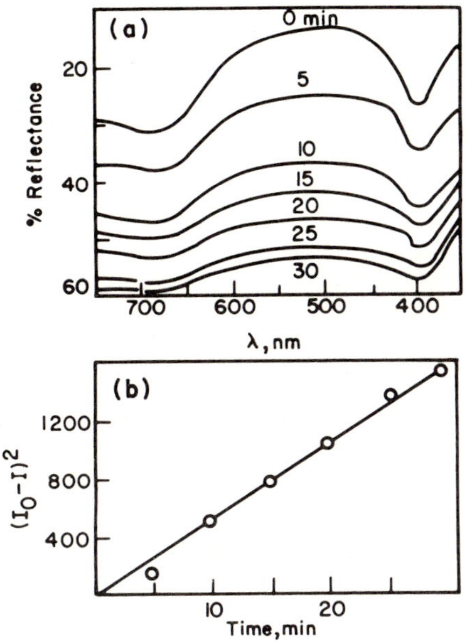

FIG. 8-8. (a) Change of reflectance of $K_3[Mn(C_2O_4)_3]{\cdot}3H_2O$ on irradiation; (b) kinetics plot (at 525 nm).

first-order kinetics. The latter appears reasonable since the reaction is inhibited by the photolysis product formed.

The solid-state photolysis of the green-colored $K_3[Fe(C_2O_4)_3]{\cdot}3H_2O$ showed that it changed to a yellow-colored compound according to the stoichiometry (*7*) of Eq. (5).

$$2K_3[Fe(C_2O_4)_3]{\cdot}3H_2O(s) \xrightarrow{h\nu} 2FeC_2O_4(s) + 3K_2C_2O_4(s) + 2CO_2(g) + 6H_2O(g) \tag{5}$$

Besides reflectance spectroscopy, magnetic susceptibility studies were employed to characterize the products formed. In the case of $K_3[Fe(C_2O_4)_3]{\cdot}3H_2O$, a μ_{eff} of 5.96 BM was obtained for the original sample (see Table 1) which changed to 5.07–5.19 BM after irradiation. The iron ion in the product is thus in the iron(II) state as iron(II) oxalate.

It is of interest to speculate on the possible mechanism for the photolysis of these complexes. The reduction of the M^{3+} ion must take place by some type of photoinduced electron transfer from the oxalate ligand to the

TABLE 1
MAGNETIC MOMENTS OF IRON COMPOUNDS

Compound	μ_{eff}, BM Found	μ_{eff}, BM Theory[a]
$K_3[Fe(C_2O_4)_3]\cdot 3H_2O$	5.77	5.92
	5.96	
	5.96	
$K_3[Fe(C_2O_4)_3]\cdot 3H_2O$ irradiated for 105 min	5.19	4.90[b]
	5.12	
	5.07	
FeC_2O_4	5.00	4.90
	5.11	
	5.08	

[a] Calculated from spin-only formula, $\mu_{eff} = [n(n+2)]^{1/2}$.
[b] Calculated for FeC_2O_4.

metal ion. In the case of the $[Co(C_2O_4)_3]^{3-}$ ion, free-radical mechanism is feasible, as shown in Eq. (6).

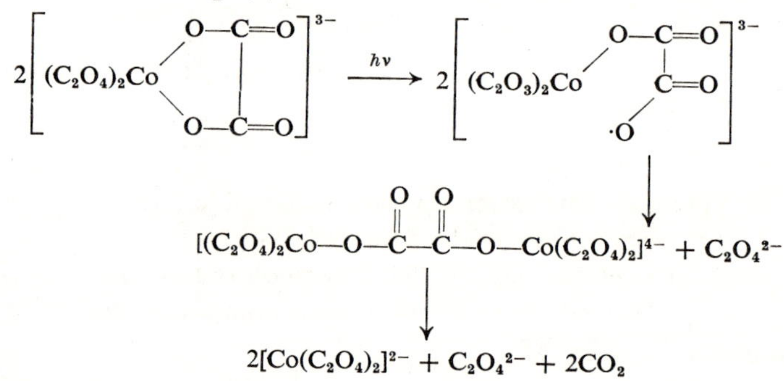

(6)

The manganese(III) complex had the same type of mechanism. Another mechanism which would explain the photolysis of $K_3[Fe(C_2O_4)_3]\cdot 3H_2O$ is shown in Eq. (7).

$$2K_3[Fe(C_2O_4)_3]\cdot 3H_2O \xrightarrow{h\nu} K_2[Fe(C_2O_4)_2] + 2K_2C_2O_4 + 3H_2O$$

$$\downarrow$$

$$[(C_2O_4)Fe{-}O{-}\overset{\overset{\large O}{\|}}{C}{-}\overset{\overset{\large O}{\|}}{C}{-}O{-}Fe(C_2O_4)]^{2-} + K_2C_2O_4 \qquad (7)$$

$$\downarrow$$

$$2FeC_2O_4 + 2CO_2$$

ACKNOWLEDGMENT

The generous support of this work by the Robert A. Welch Foundation, Houston, Texas, is gratefully acknowledged.

REFERENCES

1. W. W. Wendlandt and E. L. Simmons, *J. Inorg. Nucl. Chem.*, **27**, 2317 (1965).
2. C. H. Stembridge and W. W. Wendlandt, *ibid.*, **27**, 129 (1965).
3. W. W. Wendlandt and H. G. Hecht, *Reflectance Spectroscopy*, Wiley-Interscience, New York, 1966, Chap. IV.
4. W. W. Wendlandt, unpublished results.
5. W. W. Wendlandt and J. H. Woodlock, *J. Inorg. Nucl. Chem.*, **27**, 259 (1965).
6. E. L. Simmons and W. W. Wendlandt, *ibid.*, **27**, 2325 (1965).
7. W. W. Wendlandt and E. L. Simmons, *ibid.*, **28**, 2420 (1966).

CHAPTER 9

ANALYTICAL METHODS FOR STUDY OF PHOTODEGRADATION OF POLYMERS

Norma Z. Searle

AMERICAN CYANAMID COMPANY
CENTRAL RESEARCH DIVISION
STAMFORD, CONNECTICUT

I. Introduction—Survey of Problems and Approaches

A polymer consists of varying sizes of macromolecules composed of one or more relatively small repeating units. Polymers are chemically similar to ordinary organic compounds only in that they contain the same type of atomic groupings. The abnormally large size of the molecules (molecular weights ranging from 10^3 to 10^7) and the repetitive nature of the structure make them unique in their physical properties. The particular type of molecular architecture, i.e., whether the molecules are randomly coiled with little mutual interaction or whether they lie parallel to one another and interact strongly with some degree of branching and cross-linking, determines the physical properties and also the type of reactions they undergo under the influence of ultraviolet radiation.

Polymers generally undergo the same kind of photochemical reactions, with some modification, as the small repeating units of which they are constructed. However, because of the abnormal size and repetitive nature of the polymer molecule, an extremely small extent of reaction results in a significant change in physical properties. The photochemical aging of a plastic material may be defined as an irreversible change of physical and chemical properties in a direction which is usually to the detriment of its usefulness.

The term plastic is often used synonymously with synthetic polymer although in the strict sense of the word a plastic is either a rigid polymer or a rigid material containing a polymer along with other components such as cross-linking agents or plasticizers. In this chapter no sharp distinction is made between a plastic and a polymer although it is recognized that the presence of additives influences the effect of radiation and may interfere with the technique used to study the photodegradation.

The increasing use of polymers under deteriorative conditions has made the need for materials which will maintain their physical properties for long periods of time and the investigation of their photochemical stability increasingly important. Accelerated tests are carried out for the purpose of estimating the rate of aging under actual service conditions. A considerable effort has been expended in development of accelerated weathering devices which will produce in a short time the effect that natural exposures have over prolonged periods. Since the interrelationship of the various atmospheric factors in producing photochemical degradation is not fully understood, such an undertaking has been a difficult task. Nevertheless, it is realized that the most essential factor is adequate simulation of the radiation characteristics of sunlight or other source to which the materials will be exposed in practice. The adequacy of various radiation sources in satisfying this requirement is reviewed.

Radiation, heat, oxygen, etc., by modifying chemical structure and thus the interaction between chains, can affect the mechanical properties and appearance of polymers and plastics. Tests of physical properties such as tensile strength, elongation, and flexibility as well as of retention of gloss and color are made to evaluate the relative photochemical stabilities of the polymers. Molecular weight changes and spectroscopic measurements are also used as probes to assess the effects of ultraviolet radiation on chemical structure. The methods selected to evaluate photochemical stability are based on the type of material under study as well as on the ultimate application of the material. Emphasis is placed on the technique which allows the earliest detection of degradation.

While evaluation of the retention of the physical and chemical properties is a matter of practical concern, a basic understanding of the mechanisms of photodegradation is required for development of polymers which are not adversely affected by the environment. Therefore, studies of photodegradation of polymers include not only evaluation of rate and extent of degradation but also determination of the mechanism of degradation. The analytical methods used for the latter emphasize those techniques which will reveal the products and mode of degradation. Some of the spectroscopic techniques for identifying chemical structure are the same as those used in evaluating relative stabilities. In addition, mass spectrometry and gas chromatography are specifically used for analysis of the volatile degradation products while electron spin resonance is used in investigating free-radical mechanisms of degradation. Model compounds have often been employed in an attempt to confine the studies to simple chemical structures representative of the polymer backbone. They have also been used in some cases as an aid in identifying degradation products. The rate of oxygen consumption under the influence of ultraviolet radiation can be used for quantitative evaluation of the photooxidative instability of the polymer and also provides a means of investigating the degradation mechanism. Radioactive and stable isotopic studies have occasionally been used in conjunction with one of the other analytical techniques to help establish the mechanism of degradation. For mechanistic studies, irradiation sources are not restricted to those which most closely duplicate natural weathering conditions. Monochromatic line sources such as those used for general photolysis studies have merit in these studies also. Although investigations of high energy electron radiation are numerous, the present chapter deals exclusively with photon radiation in the ultraviolet region from about 220 to 400 nm.

II. Evaluation of Stability of Polymers

A. *Test Sources and Exposure Devices*

Ultraviolet sources used in evaluating the photochemical stability of polymers have the unique requirement of simulating and accelerating the radiation effects of the source to which the polymers will be exposed in practice. Ideally, the spectral energy distribution of the test source should match that which will be used in practice in the entire actinic region, but have a higher intensity. The necessity of matching the spectral emission characteristics is due to the fact that the extent and type of reactions are often determined by the wavelengths of irradiation. For polymers used outdoors, the test source generally required is one that simulates sunlight. If the polymer is to be used in fixtures for street lighting containing mercury lamps, the emission characteristics of the latter should be present in the test source. Simulation of other outdoor factors such as temperature, humidity, and rainfall is also important. One of the most common indoor applications of plastics is for fluorescent lighting fixtures. Evaluation of the relative stability of the plastics for this end use requires an accelerated test source that simulates the emission of the fluorescent lamps. Since this review is concerned primarily with clear, colorless plastics, the ultraviolet region is the one of major interest.

While the variability of weathering factors in nature precludes an exact duplication of the actual conditions, the least that should be achieved is simulation of some average radiation characteristics of the sun. This has been approached in recent years by the Pyrex glass-filtered high-pressure xenon arc. Its emission characteristics are compared with the radiation of the sun in Fig. 9-1. Several types of xenon arcs have been described in detail along with the commercial weathering units in which they are used (*1*). In most of these devices there is some control of temperature, humidity, and water spray on the samples and, in some cases, some control of atmospheric contaminants. Studies are being carried out to determine the interrelationship of the weathering factors. Because of the complex nature of the weathering process, it is realized that intensification of one factor alone, as for example the radiation source, can distort the results. Even acceleration on the basis of time only, i.e., constant exposure without the light–dark cycle present in nature, could prevent correlation with natural conditions.

Much of the early work on polymer stability was carried out with sources

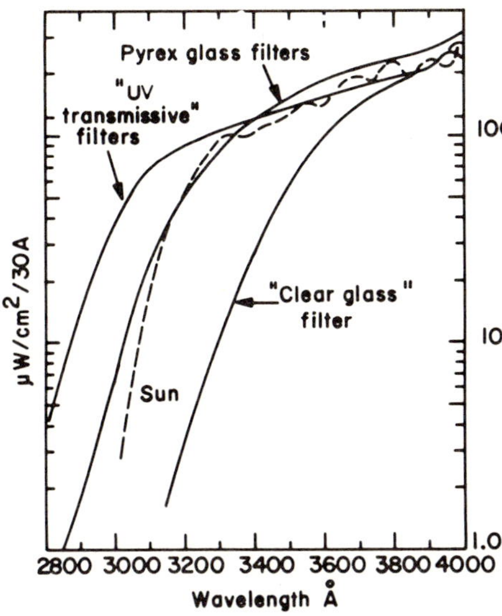

FIG. 9-1. Spectral energy distribution of 6000-W Osram xenon arc through different types of filters compared with sunlight at the earth's surface. Reprinted from Hirt and Searle (*1*, p. 74), by courtesy of John Wiley & Sons.

which had very little resemblance to sunlight in the ultraviolet region. For example, the S-1 sunlamp, a low-pressure mercury arc (*2*), was used extensively prior to the advent of the newer type sources. A typical emission spectrum of a low-pressure mercury arc is reproduced in Chapter 1. The radiation is confined almost exclusively to the mercury emission lines with very little background or continuum radiation. For this reason it is very useful in photochemical studies in which it is necessary to employ monochromatic radiation. However, it is a very poor substitute for the continuum exhibited by sunlight in the ultraviolet region. With increase in mercury pressure, the continuum radiation increases. The ultra high-pressure mercury arc gives relatively good simulation to the ultraviolet radiation of sunlight if properly filtered to remove some of the short-wavelength energy.

The fluorescent sunlamp, which has also been used extensively in accelerated testing of the lightfastness of plastics, is similarly a poor substitute for sunlight. It contains phosphors to convert the 254-nm mercury emission to longer wavelength ultraviolet radiation with peak emission at 313 nm. Its emission spectrum is shown in Fig. 9-2. It has considerably less intensity than sunlight in the long-wavelength ultraviolet

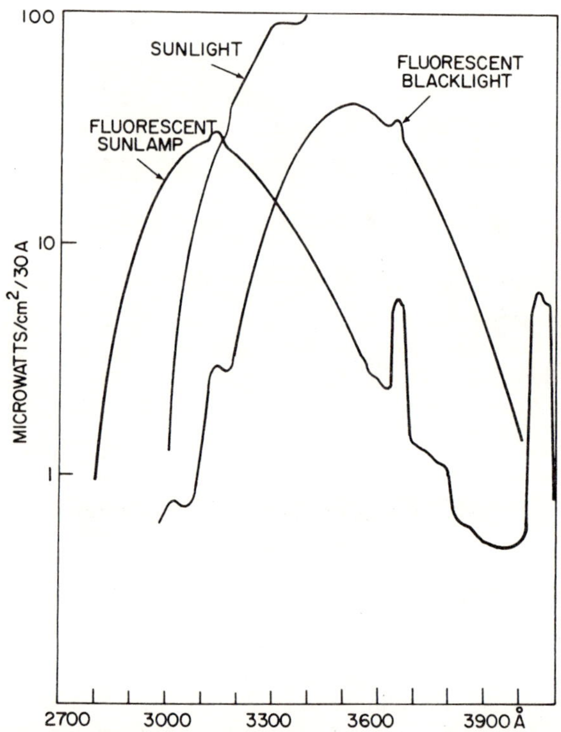

FIG. 9-2. Spectral energy distributions of the fluorescent sunlamp and fluorescent blacklight. Reprinted from Hirt and Searle (*1*, p. 70), by courtesy of John Wiley & Sons.

region and much more below 313 nm. A source with excessive short-wavelength energy should have an accelerating effect over sunlight. The more energetic, shorter wavelengths are not only absorbed to a greater extent, but have the ability to cause a greater amount of photochemical damage. However, the ability of such a source to give good correlation with outdoor exposure in ranking the stabilities of a series of materials is questionable. The fluorescent blacklight has an emission band peaking at 350 or 365 nm, depending on the method of manufacture. The use of this type of lamp along with the fluorescent sunlamp has served to increase the longer wavelength intensity of the radiation source and bring the spectral curve somewhat closer to that of sunlight. Although the energy still extends to shorter wavelength than is present in sunlight, the combination has given good correlation with outdoor exposure for some materials. It should be noted that lack of intensity in the visible region makes this source a particularly poor one for simulating sunlight when colored materials are involved. Exposure units containing these lamps, which are of tubular construction, generally incorporate an equal number of fluorescent sunlamps and blacklights alternately mounted inside a circular rotating drum (*3*). A combination of two blacklights and one sunlamp would further improve the match to the ultraviolet radiation of sunlight. Uniform illumination of the entire height of the sample rack offers an advantage over the arc-type sources in which the radiation emanates from a small central area.

Carbon arcs have been the standard laboratory test source for many years. The emission of two types is shown in Fig. 9-3. The enclosed carbon arc gives very poor simulation to sunlight in the ultraviolet region. It has excessive intensity in the region above 350 nm and negligible intensity below 350 nm, compared with sunlight. The open flame "Sunshine" carbon arc with the Corex "D" filter gives a much closer approximation to sunlight in the short-wavelength region and has considerably less intensity than the enclosed carbon arc in the region above 350 nm. It has been

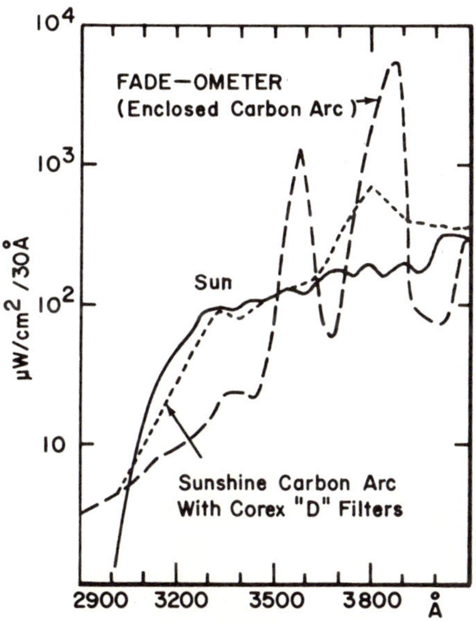

FIG. 9-3. Spectral energy distributions of the enclosed carbon arc Fade-Ometer and Sunshine carbon arc compared with sunlight. Reprinted from Hirt and Searle (*1*, p. 69), by courtesy of John Wiley & Sons.

shown to give relatively good correlation with outdoor exposure for many materials.

Comparative source intensities (*1*) and weathering studies under several of the sources (*1*, *4*) have been summarized. The high-pressure xenon arc, when properly filtered, gives the closest match to sunlight and best simulation of outdoor exposure. The enclosed carbon arc shows the greatest divergence from sunlight in the ultraviolet region. It is severely deficient in short-wavelength energy while the mercury arcs and fluorescent sunlamp have a higher proportion of short-wavelength energy than is present in sunlight. It should be recognized that the intensity as well as the spectral distribution of a source will change with use as a result of photolysis of the envelope and associated filters, contamination, and change in pressure. Aging characteristics of xenon arcs have been studied in detail (*5*), but no source is immune to these effects. The energy output of the weathering unit containing the fluorescent sunlamps and blacklights can be maintained most easily at a relatively constant level by a program of staggered replacement of the tubes.

Accelerated testing of polymers which are used primarily in fluorescent lighting fixtures requires a radiation source that simulates the actinic emissions of the cool white, warm white, or daylight fluorescent lamps. The emission spectra of these lamps are shown in Fig. 9-4 (*6*). Both the carbon arc and filtered high-pressure mercury arc have been recommended for such studies (*7*, *8*), but their emission characteristics have very little resemblance to the emission of the fluorescent lamps in the ultraviolet region. Recently, high-output fluorescent lamps (*9*) having the same

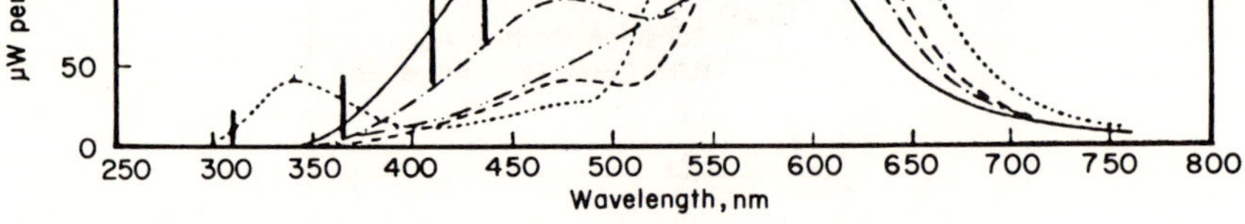

FIG. 9-4. Spectral energy distributions of several types of fluorescent lamps used in fluorescent lighting fixtures and of an incandescent lamp. (The relative intensities of the mercury lines are indicated by the black bars.) Reprinted from Withrow and Withrow (*6*, p. 180), by courtesy of McGraw-Hill Book Co. and General Electric Co.

energy distribution but higher intensity than those commonly used have been under study as test sources.

B. *Techniques for Determining the Extent of Photodegradation*

1. Spectroscopic Determinations

Photodegradation of polymeric materials proceeds from the irradiated surface inward. With the exception of surface effects such as crazing and cracking, changes in physical properties become evident only when the degradation has penetrated a sufficient distance into the plastic. In contrast, chemical changes of the surface can be detected early in the degradation process by spectroscopic measurements.

Infrared spectroscopy is particularly suited for characterization of chemical groups formed or destroyed by irradiation. The infrared multiple attenuated total reflectance (MATR) technique is a recent innovation. It is valuable for detecting surface chemical changes in materials which are opaque to the analytical wavelengths. Ultraviolet spectroscopy is more sensitive than infrared to some chemical changes and can readily detect changes occurring in surface layers by transmission measurements since most polymers are transparent in the near-ultraviolet region. Measurements in the visible region are used to evaluate the extent of degradation in terms of yellowness index, a factor which correlates with visual appearance of the sample. The sensitivity, however, is less than in the ultraviolet region. In some cases, ultraviolet transmission measurements are used to give an early indication of the relative stability to yellowing although they cannot be directly related to visual appearance.

Another technique which has been useful for determining chemical changes on the surface is that of complexing the oxidation products formed on the surface and quantitatively analyzing for the complexes by ultraviolet or visible spectroscopic techniques. Luminescence, which is the basis of a very sensitive analytical technique, is exhibited by many polymers and is frequently associated with the formation of photodegradation species. The measurement of changes in luminescence is a potentially valuable technique for early detection of structural changes produced by exposure to ultraviolet radiation.

a. Infrared Spectroscopy. Infrared spectral bands can be correlated with specific functional groups in a polymer and thus are useful for detecting and quantitatively determining certain chemical structures formed or destroyed on irradiation of the polymers. In the presence of oxygen, radiative degradation of organic polymers results in the formation of oxygen-containing groups such as acids, aldehydes, ketonic carbonyls, hydroxyls, or peroxides. Carbonyl or hydroxyl functions are most readily identified and estimated, particularly in hydrocarbon polymers since these do not already contain oxygenated compounds with interfering absorption bands as structural components. However, very little information on degradation can be gained from infrared studies of polymers such as polyester, polyurethane, or polyamides (*10*, *11*).

The strong carbonyl absorption (near 1720 cm^{-1}) is a very sensitive probe for detecting degradation of polyenes, poly(vinyl chloride), and polystyrene. The application of infrared analysis to photooxidation of polyethylene and polypropylene based on carbonyl absorption was demonstrated by Luongo (*12*), Rugg *et al.* (*13*), and Heacock *et al.* (*14*). It was shown by means of high-resolution infrared spectrophotometry that the carbonyl groups in photooxidized samples of polyethylene are well distributed among aldehydes, ketones, and carboxylic acid structures. Cotten and Sacks (*15*) also identified an acid anhydride group by its absorption at 1765 cm^{-1}. Photooxidation reactions in poly(vinyl chloride) give rise to carbonyl bond vibrations characteristic of ketones (*16*, *17*) plus aldehyde (*16*) or peracid (*18*) groups while in cellulose-type polymers ketone and acid groups are formed (*19–21*).

The main absorption band in the carbonyl region in photodegraded polystyrene has been reported by different investigators to occur between 1685 and 1745 cm^{-1} (*22–28*). It is generally agreed that the band can be ascribed to an aromatic carbonyl structure similar to that in acetophenone (*22*, *27*, *28*) although it has also been ascribed to an aromatic conjugated carbonyl similar to benzal acetophenone (*29*) or to a dicarbonyl compound similar to 1,2-propanedione (*22*). Based on a combination of evidence from infrared, ultraviolet, and nuclear magnetic resonance spectroscopy along with chemical tests of functional groups, Beachell and Smiley (*27*) ruled out the latter structures. The band can be shown to consist of several components (*22*), particularly after prolonged irradiation, thus indicating the presence of several types of carbonyl structures. Achhammer and co-workers (*26*) postulated degradation mechanisms which account for the presence of aldehyde, ketone, acid, and quinoid carbonyl structures. However, Beachell and Smiley (*27*) concluded that all carbonyl groups in degraded polystyrene are ketonic.

The molar absorptivity of the hydroxyl group absorption is weaker than that of the carbonyl group. In addition, changes caused by irradiation in the region of the OH stretch fundamental (between 3300 and 3600 cm^{-1})

are often masked by moisture or by absorption of other groups in the polymer. However, the hydroperoxide group (OOH) can frequently be detected near 3555 cm^{-1} through its OH stretching vibration. Measurement of the unbonded OOH group vibration at 3550 cm^{-1} was used by Rugg and co-workers (*13*) and by Luongo (*12*) for detecting hydroperoxide groups in the early stages of autoxidation of polyethylene. Baum (*30*) and Beachell and Tarbet (*31*) studied the kinetics of polyethylene oxidation by measuring the rate of formation of both carbonyl and hydroperoxide groups. Hydroperoxide groups were detected as precursors of the carbonyl groups in poly(vinyl chloride) (*16*) and postulated as the initial products of degradation in polystyrene (*22*, *26*, *32*) based on infrared absorption changes in the hydroxyl region. The hydroperoxide in polystyrene, which is on the tertiary carbon atom in the chain, subsequently breaks down and rearranges to carbonyl and hydroxyl groups. The latter were identified by Achhammer and co-workers (*26*) as a hydrogen-bonded type which could indicate intermolecular bridging through this group. Absorption bands attributed to unbonded and hydrogen-bonded OH groups of both the inter- and intramolecular types could be detected in both films (*33*) and fibers (*17*) of irradiated poly(vinyl chloride). In irradiated polyethylene, the increase in absorption at 3450 cm^{-1} was assigned by Heacock and co-workers (*14*) predominantly to alcoholic hydroxyl groups while in cellulose-type polymers, a decrease in absorption in the hydroxyl region was associated with photooxidation of alcoholic groups to form carbonyl groups. Mitchell and Perkins (*34*) showed that the sensitivity as well as the specificity of infrared analysis of hydroperoxide groups could be enhanced by exposing the oxidized polymer to sulfur dioxide vapors and measuring the much stronger absorption of the sulfate groups at 1195 cm^{-1}.

In some cases the type and content of vinyl groups aid in characterization of degradation. In polyethylene, for example, there is evidence for changes in concentration of various types of vinyl groups absorbing in the 890–1000-cm^{-1} region as irradiation progresses (*13*, *14*). It has also been shown by infrared spectroscopy that the first stage in the photodegradation of poly(vinyl chloride) is formation of the unsaturated linkages which are considered to be responsible for the yellowing of the polymer (*16*, *17*). In polystyrene the formation of a shoulder at 825 cm^{-1} on the 840-cm^{-1} band has been attributed to the out-of-plane vibration of the C—H bond in a trisubstituted ethylene (*35*). Other infrared bands present in photodegraded polystyrene have also been attributed to vibrations of C═C groups (*25*).

Caution is indicated in attempting to rely on a single criterion such as formation of functional groups as the only measure of breakdown. These are often related to physical changes as in the case of poly(vinyl chloride) in which the formation of C=C bonds, indicative of rupture of the macromolecular chains, is accompanied by decrease in viscosity, tensile strength, and elongation (*17*). However, the C=O and C—O groups initially formed have been shown to decompose on further irradiation as cross-linking occurs (*16*). It was also shown by Tamblyn and co-workers (*36*) that carbonyl formation in polyethylene did not always correlate well with stress-cracking resistance or loss in elongation.

(1) *Transmission Measurements.* Thin films, either unsupported or cast on rock-salt plates are most suitable for infrared transmission studies of polymer stability. If the polymers are too insoluble to be cast as films, they can often be investigated in KBr pellets provided they can be ground into a fine powder. The infrared spectrum is obtained before and after each irradiation period and the changes in absorbance of the bands are measured. In the films, the changes on irradiation may be determined more precisely by differential spectrophotometry. The spectrum of the irradiated specimen is compared instrumentally with that of an unirradiated one of the same thickness placed in the reference beam of a double-beam spectrophotometer. In this way, the absorption of the original polymer is automatically eliminated. The difficulty of preparing KBr discs having identical thickness and concentration precludes the use of differential spectrophotometry with such samples.

Estimation of the concentration of the carbonyl and hydroxyl groups formed is based on the molar absorptivities of these functional groups measured in simpler model compounds. For example, di-*n*-butyl ketone in cyclohexane has been used (*37*) as a standard the the carbonyl absorption in polyenes. A better estimate of the concentration can be obtained if the standard is added to the polymer. The ketone used for this purpose is stearone (dioctadecyl ketone) because of its higher molecular weight and lower volatility. For the calibration of hydroxyl concentration, simple alcohols are used as standards. The molar concentration of the degradation product is found by dividing the Absorbance of the absorption band of interest by the thickness of the film and the molar absorptivity of the equivalent band of the standard. In a study of polyethylene degradation (*14*), extinction coefficients of 190, 55, 150, and 500 liters/mole-cm for ketone, hydroxyl, vinyl, and carboxyl, respectively, were used to calculate the approximate molar concentration of these groups in relation to the number of carbon atoms in the polymer. Frequently the concentration is expressed in percentage by weight of the functional group. Since the film thickness is generally difficult to measure accurately, in many studies only the relative extent of degradation of different samples is determined. In order to correct for the variations in film thickness the ratio of the Absorbance in the carbonyl or hydroxyl region to that in a region of the polymer absorption is used to follow the degradation. For example, in copolymer film of styrene and butadiene the Absorbance ratio 1720 cm^{-1}/1925 cm^{-1} is an index of the carbonyl content corrected for film thickness via the Absorbance of the phenyl ring in polystyrene at 1925 cm^{-1} (*38*).

(2) *Attenuated Total Reflectance.* The transmission measurements discussed above involve the entire sample whereas the chemical properties of only a small portion, that of the surface layers, have probably undergone changes. Often the surface changes cannot be detected by transmission measurements because of interference by strongly absorbing structural groups of the polymer. A technique for abrading the surface and determining the chemical composition of the removed material has been reported (*39*). However, this is a destructive test and can give only a qualitative indication of the products formed. Attenuated total reflectance (ATR), also known as internal reflectance spectroscopy (IRS), has recently been developed (*40*) primarily for infrared analysis of surface layers of a material. Multiple attenuated total reflectance (MATR) has come into widespread usage because of its greater analytical sensitivity.

The method depends on the measurement of reflectivity as a function of frequency at an interface between the sample and a nonabsorbing material of high refractive index. The incoming radiation penetrates the high refractive index material, the internal reflection element, first. It then strikes the sample, placed against the reflecting surface of the element, at an angle greater than the critical angle, i.e., the smallest angle of incidence giving rise to total reflection. The beam is totally reflected back into the denser medium unless the sample has absorption bands at the frequencies being used. In this case, the radiation interacts with the absorbing medium at the interface, leading to a reduction or "attenuation" of the totally reflected radiation. For multiple internal reflection, the design of the internal element or plate is such as to permit the incoming radiation to be reflected many times, permitting multiple sampling of the specimen. As many as 50–100 reflections can be obtained easily. In addition, increased sensitivity makes the requirement for intimate contact between the sample and plate somewhat less stringent.

The spectrum traced out by measuring the change in reflection as a function of frequency resembles qualitatively the transmission spectrum of the sample of interest. Since the depth of penetration of the radiation into the sample is small (usually less than the wavelength of light used) the technique is ideal for nondestructive measurements of chemical changes in the polymer surface. Under conditions of total reflection the electric-field intensity at the interface can be considerably greater than exhibited during transmission measurements (*41*). Therefore, in the case of surface effects in the range of monomolecular layers, ATR measurements can give greater sensitivity of molecular absorption even with the single reflection technique. However, the greatest usefulness of the technique is for material that does not lend itself readily to transmission measurements.

The ATR technique has been employed successfully to study the ultraviolet degradation of several types of polymers. The degradation rates of polycarbonates were determined by monitoring the absorption of the carbonyl group along with other polymer bands in the infrared region (*42*). It was shown that this method was able to detect initial changes in structure about 400 hr before the initial change was detected by measurements of yellowness index. Chan and Hawkins (*43*), using a reflector element giving about 25 reflections, showed that the MATR technique is capable of detecting degradation in black as well as in clear polyethylene compositions. The extent of oxidation was determined from the ratio of the increasing C=O absorption at 1710 cm^{-1} to the CH_2 absorption at 1450 cm^{-1} which is unaffected by initial degradation reactions. The clear sample oxidized more than the sample protected with channel black. Comparison with brittleness temperature tests showed that the latter technique required twice the length of exposure that was required for infrared ATR measurements to detect the effect of irradiation. The photochemical degradation of poly(vinyl chloride) (PVC) and acrylonitrile-butadiene-styrene (ABS) resins were also investigated by the MATR technique by Chan and Hawkins. While the PVC formulation showed only slight changes in the carbonyl region, ABS resin showed very rapid surface degradation as evidenced by increases in hydroxyl and carbonyl content along with a decrease in the amount of trans unsaturation. Due to the weak penetration of 254-nm radiation in poly(ethylene terephthalate), degradation is confined to a very thin surface layer. While transmission measurements were found to be adequate to detect the formation of a band near 775 cm^{-1} following exposure to 313-nm radiation, the MATR technique was required in the case of the shorter wavelength radiation (*44*).

ATR and MATR accessories for infrared spectrophotometers are available from manufacturers of the instruments as well as from specialty manufacturers (*45*, *46*). Barr and Flournoy (*47*) have given an excellent review of the theory, instrumentation, and application of attenuated total reflectance in relation to macromolecules. The major requirements of the method are a transparent material of considerably higher refractive index than the sample, which should have a smooth surface in order to ensure intimate contact between the two media when pressed together. A KRS-5 (thallous bromide–iodide) prism having a refractive index of 2.2–2.4 is routinely used as the reflection element. Polycrystalline intrinsic germanium having a refractive index of 4.0 is frequently more useful since it allows a wider range of useful angles of incidence and a larger number of internal reflections.

b. ULTRAVIOLET SPECTROSCOPY. The use of ultraviolet spectroscopy to follow chemical changes in polymers caused by ultraviolet irradiation dates back to the work of Coblentz and Stair (*48*) on cellulose-type polymers. Although measurements in the ultraviolet region are not as specific as those in the infrared region in identifying the structural components produced, they generally provide a more sensitive means of detecting radiation effects. Studies of several polymers by both methods have shown evidence of chemical changes by ultraviolet measurements long before they were observed by infrared techniques (*10*, *11*, *14*, *49*, *50*). This is due both to the higher molar absorptivities in the ultraviolet for many chemical groups and to the fact that most polymers show no characteristic absorption spectra of their own in the near ultraviolet region (300–400 nm). Therefore, small changes in opacity in the ultraviolet region can be detected easily. Unsaturated structures, particularly in conjugation with carbonyl groups, have considerably higher molar absorptivities in the ultraviolet than in the infrared region. Although the absorption peaks are frequently hidden under the absorption of the main polymeric structure, the band is broad because of the variety of degradation structures present and extends well into the transparent ultraviolet region. It is observed as an increase in opacity in this region. It was demonstrated in the case of polystyrene (*51*, *52*) that a linear relationship exists between increase in opacity in the ultraviolet region and increase in carbonyl content as measured by infrared spectroscopy at 1720 cm^{-1}.

The tail absorption of the ultraviolet band extends into the blue region of the visible, imparting yellow color to the material. However, the increase in opacity is much more pronounced in the ultraviolet region and thus the latter region is also more sensitive than the visible region to chemical changes. For many polymers, ultraviolet spectroscopy is able to predict after relatively short exposure times the relative ranking of a series of polymers with respect to the yellowing which would be observed only after prolonged exposure. Ultraviolet transmission measurements of polystyrene (*53*) invariably showed the effects of degradation long before changes in the visible region were apparent. Complete ultraviolet spectra or change in transmission at either 340 or 365 nm have been used to assess the relative stabilities of different formulations of this material (*23*, *49–55*). However, no attempt has been made to relate such measurements to yellowing. For this reason, the ultraviolet technique has not gained as widespread acceptance as the visible technique which can be directly related to visual appearance.

Ultraviolet transmission curves of polymer films and castings are generally structureless. The aliphatic polymers, containing predominantly saturated structures, show negligible absorption and no characteristic bands at wavelengths longer than 250 nm. The characteristic monomer absorption in this region is lost when, as a result of polymerization, the unsaturated groups are converted to saturated structures. The aromatic polymers, such as polyester and polystyrene, generally start absorbing light at wavelengths near 390 nm but the absorption only becomes significant at about 340 nm, the exact wavelength depending on the film thickness. The absorption then increases steadily in the direction of shorter wavelengths and takes a sharp increase near 300 nm. The thinnest unsupported poly(ethylene terephthalate) film available (0.006 mm) is opaque to wavelengths shorter than about 303 nm (*44*). However, a 500-Å-thick film of the material cast on a quartz plate revealed a strong absorption peak near 240 nm, characteristic of the phthalate ester, and a weaker band near 290 nm. Spectra of diluted solutions will also reveal the characteristic bands of aromatic polymers in this region. Owing to the lack of suitable solvents with adequate ultraviolet transparency, little work has been carried out in solution. The advantages of carrying out photochemical degradation studies in solution is that the light absorbed by the sample can be measured with greater accuracy since the sample absorption is more uniform and more accurately known. The disadvantage is that the solvent may play a large part in the degradation process, sometimes forming a complex with the polymer. This is thought to be the case, for example, for cellulose in cadoxen solution (*56*). In addition, the mechanism and rate of degradation is very likely to be different in solution than in the solid state.

Generally, the greatest differences among a series of similar polymers following ultraviolet irradiation are in the wavelength region on the long-wavelength side of the characteristic absorption bands of the polymer.

Therefore, it is not necessary in evaluating stabilities to work with solutions or very thin films which allow observation of the strong absorption band. Changes in chemical structure are usually observed as a general increase in opacity in the longer wavelength region. The opacity advances progressively with length of exposure and is larger with decreasing wavelength of measurement. In some polymers, particularly those of the aliphatic type, the characteristic peak(s) of the degradation product(s) can be observed. For example, in both poly(vinyl chloride) (*18*, *50*) and cellulose (*19*) an absorption band in the region of 270–285 nm attributable to carbonyl groups and conjugated unsaturation is superimposed on the broad general increase in absorption. In cellulose, a shorter wavelength band, at 212 nm, ascribed to COOH groups is also evident. The increase in its intensity with irradiation corresponds to the increase at 1740 cm^{-1} in the infrared region.

In poly(vinylidene chloride) the band at 285 nm formed on irradiation is attributed to a chlorinated ring compound having three double bonds (*57*). A carbonyl group conjugated to a ring would also give the strong absorption peak in this region, but it is not a reasonable assumption considering that the band is stronger when the irradiation is carried out in vacuum rather than in air. The fact that no single substance is known which absorbs maximally at 285 nm and for which the absorption falls off so gradually in the visible region indicates it is an unusual species. The possibility of a carbonium ion species was also suggested.

Ultraviolet spectra of polyacrylonitrile irradiated in the absence of oxygen show a strong absorption band at 295 nm (*58*) probably due to conjugated double-bond formation involving the nitrile side groups of the polymer chains. On further irradiation in the presence of oxygen, the band is gradually eliminated, indicating a photooxidation reaction of the group responsible for the absorption. If the film is thin enough, a band at 216 nm is also observed to increase on irradiation in the absence of oxygen. In polyamides, a band at 280 nm, characteristic of the carbonyl group originally present in the peptide structure, disappears prior to the general increase in absorption (*10*). In poly(methyl acrylate) spectral changes in the 270–300-nm region are similar to those found in poly(methyl methacrylate). The increased absorption in this region was considered to be indicative of the formation of aldehyde groups along the polymer chain (*59*). Poly(vinyl acetate) shows only an overall increase in absorption in the ultraviolet region on irradiation (*60*).

Ultraviolet measurements of an irradiated film of radical polymerized, branched polyethylene initially showed a group of peaks in the vicinity of

273 nm which disappeared on continued exposure (*14*). The transient multiplet was ascribed to a triene which is formed under the shorter wavelength radiation (about 275 nm) but is not observed if radiation in the region of 390 nm is also present. The latter produces cross-linking which becomes the predominant reaction and is inhibited to some extent by the shorter wavelength radiation. The exposure time required for complete disappearance of the multiplet corresponds roughly to the induction time for onset of formation of OH, C=O, COOH, and vinyl groups as evidenced by infrared measurements. Following disappearance of the transient peaks of the multiplet, only a general increase in opacity in the 260–360-nm region is observed with increasing exposure.

In aromatic polymers, frequently an increase in transparency is observed in the region of strong polymer absorption following ultraviolet irradiation. In poly(ethylene terephthalate) this has been attributed to destruction of the phthalate ester caused by the ultraviolet radiation below 300 nm which it absorbs (*44*). Voigt (*61*), however, ascribed the increase in transparency in polyesters to radiation wavelengths longer than 300 nm. These cause partial elimination of unsaturated groups in the polymer by cross-linking reactions according to Voigt's theory. The particular photochemical reaction depends on the type of polyester. For example, cross-linking does not occur following rupture of the polymer chain in poly(ethylene terephthalate) (*62*). The condition favoring cross-linking results from opening of the lactone ring in terephthalic and isophthalic acid polyarylates with subsequent elimination of CO and CO_2. The wavelength of irradiation will also determine the type of reaction that occurs in any one material. Voigt (*61*) claims that in polyesters radiation of wavelengths shorter than 300 nm is required to form the unsaturated conjugated carbonyl structures which absorb in the near ultraviolet region. Other experimental evidence using a different source of radiation and different experimental conditions (*63*) shows that the increase in opacity of polyesters is caused by wavelengths as long as 340 nm.

Ultraviolet absorption measurements of an extremely thin film of poly(ester-urethane) (*11*) showed the growth of a broad absorption band (possibly a doublet) in the 270–310-nm region believed to be due to the formation of a quinoid structure. The simultaneous decrease in the 245-nm band characteristic of the unirradiated polymer indicated the loss of aromatic moiety. Studies of thicker films of polyurethane prepared from tolylene diisocyanate and a polyoxypropylene diol (*64*) showed only a shift in the absorption edge to longer wavelengths which was assumed to be due to a mixture of breakdown products.

As a consequence of the increased opacity in the long wavelength region and decreased opacity in the short wavelength region, subtraction of the curve of the unirradiated polymer from that of the irradiated material either manually or by differential spectrophotometry produces an absorption band even with the thicker samples that show only a shift in absorption edge by the normal measurement procedures. In aromatic polymers the peak of the band is in the region of 300 to 340 nm. The width of the band suggests that the degradation products consist of a mixture of structures with different degrees of conjugation.

The secondary reactions promoted by heat frequently persist after the irradiation has been discontinued. This is evidenced by the continuing increase in opacity of some polymers on storage in the dark in the absence of oxygen following irradiation (*54*, *61*, *65*). The postirradiation effect may be ascribed to decay of the relatively stable free radicals, produced under the influence of light, to compounds that absorb more strongly in the ultraviolet region. The degradation products formed in the dark show a narrower absorption band than those formed under the immediate influence of light. The increase in absorption produced in the dark can generally be eliminated by further irradiation. The reversible dark reaction in polystyrene has been attributed to cis–trans isomerism (*54*). However, it is usually not a simple reversible reaction, but is accompanied by chemical changes. On storage, in the presence of oxygen, a decrease in opacity has been noted in certain polyesters (*63*). It proceeds from the outer surface inward and can be retarded by decreasing the temperature over ambient. Oxygen can serve to destroy free radicals, some of which are highly colored. Oxygen also exerts a bleaching action on the degraded polymers by reaction with the conjugated double bonds, thus breaking up the conjugated unsaturation known to be color-bearing structures.

Because of the possibility of dark reactions, the temperature and length of time between cessation of irradiation and measurement should be kept relatively constant in comparing the stabilities of a series of materials. In evaluating the stability of cellulose (*19*), it was found that continuous, uninterrupted irradiation of the sample and periodically interrupted irradiation of the sample gave different results. Therefore, all runs were carried out in a continuous manner and the spectra were obtained immediately after irradiation.

Studies have been carried out on films ranging in thickness from 0.01 to 0.25 mm and occasionally on thicker molded disks. Where practical, attempts are made to eliminate ultraviolet-absorbing impurities such as plasticizers which would interfere with the detection of degradation products. The resins are further purified by dissolution and precipitation techniques. It must be recognized, however, that the impurities may play a major role in the degradation of the material as used in practice. The film thickness at which the extent of degradation becomes independent of thickness is determined by the absorption of the polymer for the particular wavelength region. According to Grassie and Weir (*66*), this is about 10 μm for polystyrene if the irradiation is carried out with monochromatic light of 254 nm. At the longer wavelengths, the absorption capacity of the polymer decreases and the light can penetrate further. Since it is not desirable to work with thicknesses in which all the actinic light is absorbed, an appropriate thickness is selected and maintained for all samples in a series of similar materials. It is preferable to work with a thickness in which the diffusion of small molecules out of or into the film does not play a major role in degradation. With formation of cracks and crazes as exposure progresses, reflection losses increase and should be corrected for in order to accurately evaluate the increase in opacity. The change in reflection can be determined only at a wavelength at which neither the unirradiated nor the irradiated polymer absorbs. For example, Absorbance changes at 570 nm are considered to be due only to reflection losses. The latter increase with decreasing wavelength in a nonpredictable manner. Therefore, subtraction of the Absorbance at 570 nm from the absorbance in the ultraviolet region only minimally corrects for reflection losses.

c. Yellowness Index. For many applications of polymeric materials, the ultimate usefulness of these materials is determined by their retention of appearance. Yellowness is one of the objectionable factors which appears early in the degradation of colorless plastics. It is defined as deviation in chroma from whiteness or water-whiteness in samples which have maximum reflectance or transmittance in the region of 570–580 nm after correction for eye response (*67*). Yellowness is characterized by an increase in absorption in the visible region of the spectrum, being most pronounced in the short-wavelength, blue-violet region. Color in organic compounds is usually associated with unsaturated groupings such as C=O, C=C, C=N, etc. Approximately four to six of these groups in conjugation are required for appearance of a yellow color in aliphatic compounds. In aromatic compounds, color can be observed when less than four of these conjugated unsaturated groups are present. The absorption peak of these structures is generally in the ultraviolet region, the tail end extending into the visible region and thus producing the color.

The determination of yellowness index (YI) is applicable to transmittance or reflectance measurements of nonluminescent, nearly white, transparent, translucent, or opaque plastics. The yellowness values can be related to visual observation as long as distinctly reddish or greenish samples are avoided. For this reason, measurement of degree of yellowness is commonly used to evaluate the changes caused by exposure of plastics to ultraviolet radiation. A positive YI describes the presence and magnitude of yellowness; a sample with a negative YI will appear bluish. Since the yellowness index is a function of thickness (except in the case of opaque materials), comparisons of the extent of yellowing can be made only between samples of comparable thickness. It is permissible to correct the YI value of the unexposed sample, only, for thickness. Correction for thickness cannot be made for the irradiated sample since the yellowness produced on exposure is not uniformly distributed throughout the sample but is confined to a thin surface layer. The change in visible absorption can be detected most readily when the sample is measured by the differential reflectance technique using a standard white backing for nonopaque materials.

Various mathematical expressions exist for numerically characterizing the yellowness of plastics. Determinations of yellowness as observed under daylight or simulated daylight illumination (source C) are described by the American Society for Testing Materials (ASTM) (*67*). The magnitude and direction of the yellowness index relative to a magnesium oxide standard can be calculated from Eq. (1).

$$\mathrm{YI} = \frac{100(1.28X_{\mathrm{CIE}} - 1.06Z_{\mathrm{CIE}})}{Y_{\mathrm{CIE}}} \tag{1}$$

where X_{CIE}, Y_{CIE}, and Z_{CIE} are CIE tristimulus values of the sample related to source C and the response curve commonly associated with human vision (*67*). The designation CIE stands for Commission Internationale de l'Eclairge (International Commission on Illumination). The tristimulus values, designated X, Y, and Z in the CIE system, are the amounts of three stimuli, i.e., three points in color space, required to match a color. The X_{CIE} and Z_{CIE} tristimulus values are associated with transmittance or reflectance in the red and blue regions, respectively. The Y_{CIE} tristimulus value is the luminosity factor. It gives the fraction of visible light transmitted or reflected and has occasionally been used by itself for evaluating optical stability (*68*). The coefficients of the X_{CIE} and Z_{CIE} tristimulus values in Eq. (1) are empirical adjustments for correlation with visual observation.

The tristimulus values are most accurately determined by numerical integration from recorded spectral data. The integral includes the products obtained by multiplying at each wavelength of the visible spectrum from 380 to 780 nm the spectral irradiance of the source, E_λ, the spectral reflectance of the sample, R_λ, and the tristimulus values of the spectrum for unit density of radiant flux of infinitely narrow wavelength range $\bar{x}(\lambda)$, $\bar{y}(\lambda)$, $\bar{z}(\lambda)$. The computational method generally used is the "weighted ordinate" method using wavelength intervals of 5 nm for approximating the value that would be obtained by true integration. Automatic and semiautomatic computers, either digital or analog in operation, are available for performing these calculations during spectrophotometric operation (*69*).

Direct calculation of the yellowness index can be made with the use of tristimulus colorimeters according to Eq. (2) (*67*).

$$\mathrm{YI} = \frac{100(A_0 - Z_0)}{Y_0} \tag{2}$$

The A_0, Y_0, and Z_0 filter photometer readings are reflectance values in the amber, green, and blue regions, respectively, relative to an instrumental reference standard. By approximating the CIE functions with a source–filter–photometer combination, the need for calculating tristimulus values is eliminated and the process is much simpler than the spectrophotometric method of color evaluation. It is adequate in cases where high accuracy is not essential. A calculation utilizing reflectance values in the blue and green regions only is also described (*70*).

The IES-SPI-NEMA yellowing factor (*71*) is a mathematical expression for calculating the change in yellowness of transparent plastics. In a comparison of this method with the ASTM method for yellowness index given above, Billmeyer (*72*) concluded that the latter was preferred because it more accurately ranked samples in accordance with visual experience, particularly samples containing ultraviolet light absorbers. The yellowing factor is misleading in that it does not properly distinguish among samples with differently shaped spectrophotometric curves. The ASTM method is also preferred because it is capable of determining the absolute yellowness of a single sample, while the yellowing factor is concerned only with change in yellowness on exposure. The change in yellowness has also been determined by measurement of loss in reflection or transmission at a single wavelength in the blue region between 420 and 440 nm (*28*, *65*, *73*). This appears to correlate well with observed discoloration when changes in reflection in the red end of the visible spectrum are comparatively minor. Hill (*74*) recently compared seven different methods of characterizing the yellowing of clear urethane coatings based on spectrophotometric curves in the visible region. These included calculation of the change in total area under the curve, equations utilizing individual wavelengths in the blue and green, or blue, green, and red regions (including the IES-SPI-NEMA yellowing factor), and measurement of the change in optical density at a single wavelength. An equation using selected wavelengths of 420 and 560 nm was found to correlate best with visual estimation of yellowing in samples both with and without protective ultraviolet absorbers. However, none of the techniques evaluated gave complete agreement with the visual ratings.

The rate of yellowing is generally more rapid initially and decreases with increasing exposure. Even when yellowness index is plotted as a function of the logarithm of the exposure time, a straight line function is not obtained. There are several factors which may be responsible for the decrease in rate of yellowing. One is the depletion of photosensitive entities. The assumption here is that the photosensitive species formed during processing, rather than the structural units of the polymer, initiate the degradation under the influence of absorbed radiation. Another factor is that the degradation products also absorb ultraviolet energy and as they accumulate on the surface the incident radiation is progressively screened from the photosensitive species. The degradation products evidently do not act as photosensitizers for the polymer nor do they produce additional yellow species. In fact, the less energetic, longer wavelength radiation absorbed by the yellow degradation products generally destroy the yellow species. Consequently, as the concentration of yellow-colored species increases, the rate of their destruction will also increase. The relative rates of formation and destruction depend on the relative spectral energy distribution of the light source, i.e., on the relative intensities of those wavelengths which produce and those which destroy the yellow species. The wavelength sensitivities for these two processes will be discussed in another section.

It should be noted that the extent of degradation is not necessarily reflected by the yellowness index. For example, yellowing will not occur if the degradation products do not absorb in the visible region. Also, if the yellow species initially formed is destroyed by further irradiation or by diffusion of oxygen into the sample, then color formation is an unreliable indication of the total extent of degradation. It has been shown (*65*) that in cellulose nitrate degradation is not associated with discoloration but occurs as a result of decomposition of the colored component. The rate of decrease in viscosity was shown to be greatest during the bleaching part of the sequence and negligible during the postirradiation period when the largest increase in yellow color was observed. Gray and Wright (*75*) found poor correlation between the rate of color formation and oxidation products formed in poly(vinyl chloride). Because of the influence of temperature on the rate of fading caused by penetration of oxygen, color formation was apparently slower in summer months although the rate of formation of all degradation products was greatest at this time. Frey (*76*) also reported no direct relationship between color and impact strength in the same polymer type.

d. Spectroscopic Analysis of Complexation Products. While photooxidative degradation products can be identified and quantitatively measured in hydrocarbon polymers by means of infrared spectroscopy, this technique is not adequate for all types of polymers. The presence of strongly absorbing oxygenated structures in such polymers as polyesters and acrylics interfere with the analysis of photooxidation products by infrared techniques. A sensitive method of analyzing for degradation products which is applicable to many polymers consists of reacting the oxidation products on the surface with complexing reagents and measuring the ultraviolet or visible absorption of the complex formed. The reaction and/or absorption spectra are generally specific for a particular chemical group and can thus be used to identify the structure.

The colorimetric method developed for polyesters by Gray and Wright (*77*) was based on the reaction between a benzene–methanol solution of *N*,*N*-dimethyl-*p*-phenylenediamine (DMPDA) and the photodegraded polymer. The deep-yellow-colored complex formed on the surface of the plastic was measured by differential reflectance spectroscopy, i.e., by comparison in a double beam spectrophotometer with an undegraded specimen treated similarly. The extent of degradation as measured by change in visible reflectance showed good correlation with the amount of DMPDA that was used up. The concentration of DMPDA in the solution before and after reaction with the polymer was determined by spectrophotometric analysis at 520 nm of peroxide-treated DMPDA before and after reaction. The method was later extended (*78*) to rigid poly(vinyl chloride), acrylics, and cellulose acetate–butyrate. It was modified as needed for each of the plastic types by changing the solvent and by use of *o*-, *m*-, or *p*-phenylenediamines as replacements for DMPDA in the color reaction. Also, the amount of diamine reacted was determined directly by the color developed in the diamine solution with which the exposed plastic was treated. In an attempt to identify the specific photooxidation products, over 40 organic compounds containing oxygen were reacted with the different diamines and the ultraviolet and visible spectra were compared with those of the reaction products of the irradiated plastics and diamines (*79*). No significant amounts of peroxide and hydroperoxide appeared to be present in poly(methyl methacrylate) and cellulose acetate–butyrate, but α-keto acids such as α-ketobutyric, α-ketocaprylic, and pyruvic acids as well as glyoxal and pyruvic aldehyde were assumed to cause the color reaction with the diamines in these plastics.

A number of other complexing agents and techniques have been used by various investigators for specifically identifying carboxyl, carbonyl, peroxide, hydroperoxide, and other oxidation groups. Palit and Ghosh (*80*) developed a very sensitive test specifically for carboxyl end groups in polymers. The reagent used is an acid-indicating dye, rhodamine 6Gx (Calcozine Color Index No. 45160) which is extracted from a buffered solution of pH 10 by benzene under specified conditions. Carboxylic compounds and polymers containing carboxyl groups change the yellow color of the rhodamine reagent to pink at very low carboxyl concentrations and to pink with green-yellow fluorescence at higher carboxyl concentrations. The color developed in the polymer solution is measured spectrophotometrically at 515 nm. Although Palit and Ghosh used the technique for the analysis of end groups, the application to carboxylic acid degradation products is obvious.

Another chemical method specific for carboxylic groups was first reported by Heacock (*81*). The technique involves the treatment of photo-oxidized polyethylene and polypropylene films with sulfur tetrafluoride (SF_4) gas at 80°C. Carboxylic acids are quantitatively converted into acid fluorides. The concentrations of the latter are determined by their Absorbance in the infrared region at 1835 cm^{-1}. Ketones and aldehydes do not react with SF_4 under these mild conditions. Since prior to treatment the Absorbance near 1725 cm^{-1} shows the presence of carboxyl as well as carbonyl groups, the difference in absorbance in this region before and after treatment can also be used to estimate the concentration of carboxyl groups which reacted. The remaining absorbance at 1725 cm^{-1} after treatment is attributed to the content of aldehydes plus ketones.

The carbonyl content of polymers has been determined by means of nitrophenylhydrazines. The use of an aryl hydrazine eliminates the need to introduce other reagents for color as is the case when hydrazine is used. A colored hydrazone having an absorption peak at about 410 nm is formed by the reaction of *p*-nitrophenylhydrazine with the carbonyl groups formed in photodegraded cellulose. A method has been developed (*82*) for microanalysis of the carbonyl groups by means of diffuse reflection spectra of the condensation product in the solid material. The formation of carbonyl groups in irradiated proteins has also been followed by the color reaction with 2,4-dinitrophenylhydrazine.

Campbell and Wright (*83*) have shown that peroxidic compounds as well as carbonyl compounds can be determined by the use of the diamine colorimetric technique. The differences in reactivity in acidic and basic solutions allow differentiation between the two types of structures. Eiss and Giesecke (*84*) showed that peroxide and hydroperoxide concentrations can be determined by reaction with benzoyl leuco methylene blue. In benzene-trichloroacetic acid solution the characteristic methylene blue color having an absorbance peak at 662 nm is produced. Beachell and Smiley (*27*) used this method for determining the peroxide concentration of photodegraded polystyrene. A method developed by Mitchell and Perkins (*34*) for detecting hydroperoxide groups formed during the early stages of oxidation of polyethylene depends on the reaction of these groups with sulfur dioxide vapors. When the oxidized polymer is exposed to sulfur dioxide vapors at room temperature, sulfate groups are formed by selective reaction with the hydroperoxide groups. Analysis is based on the strong infrared absorption of the sulfate groups at 1195 cm^{-1}.

Many of the spot tests (*85*) used for qualitative analysis of organic chemical groups can be adapted to analysis of degradation products in polymers. Oxidized polystyrene has been tested (*27*) for dioxo compounds, esters, and anhydrides with reagents used in the spot tests. The esters and anhydrides react with hydroxylamine hydrochloride and dioxo compounds react with thiophene and sulfuric acid.

e. LUMINESCENCE MEASUREMENTS. Many polymeric materials have been observed to fluoresce or phosphoresce when excited by appropriate ultraviolet radiation. This phenomenon has been ascribed variously to impurities, oxidative degradation products, and in some cases, to the polymer structure itself. In the case of low-density polyethylene, Boustead and Charlesby (*86*) identified the luminescence centers in the amorphous regions as free molecules of anthracene and a derivative of phenanthrene. The possibility of molecules of a naphthalene derivative attached to the polymer chain was also suggested. In the crystalline region the luminescence centers were ascribed almost entirely to benzoic acid and some alky–aryl ketones. In a study of the ultraviolet absorption spectrum of polyethylene, Partridge (*87*) also detected the presence of polycyclic aromatic molecules such as hydroxy derivatives of phenanthrene and *p*-terphenyl. These appeared to be formed in an oxidation reaction ascribed to simple benzene derivatives present as impurities in the ethylene gas used for manufacture. In addition to these hexane-soluble impurities, there also appeared to be isolated carbonyl groups attached to the polymer chains. Fluorescence measurements were utilized in this study to confirm the presence of the postulated impurities.

In an earlier study, Charlesby and Partridge (*88*) had assigned the phosphorescent centers in polyethylene as well as in polypropylene, polystyrene, and polytetrafluoroethylene to carbonyl groups formed as oxidation products probably attached to the ends of molecular chains or side chains. These groups were also associated with the visible components of the thermoluminescence. The origin of the fluorescence and ultraviolet component of the thermoluminescence of polyethylene and polystyrene was uncertain but carbonyl groups were also suspected as the centers. Zapolskii (*22*) has stated that the luminescence of unirradiated polystyrene in the 400–500-nm region is also due to ketone groups in the polymer.

The fluorescence of polystyrene has been reported by several investigators to be characteristic of the monomer or polymer structure rather than of the oxidation products. The strong fluorescence of polystyrene between 280 and 315 nm has been attributed to the presence of residual styrene monomer (*89*, *90*, *91*). It is a mirror image of the absorption band of the monomer. The exact position of the fluorescence peak and the shape of the curve appear to depend on the concentration of the monomer and whether the investigation is carried out in the solid state or in solution. The fluorescence at about 330 nm has been attributed to excimer formation. Yanari *et al.* (*90*) found this to be characteristic of the isotactic polymer while Vala *et al.* (*92*) reported excimer bands for both the atactic and isotactic polymer. The latter investigators also reported phosphorescence emission for polystyrene at about 380 nm extending into the blue region of the visible. In a study of cellulosic polymer films the wavelength and intensity of emission were also related to the chemical structure of the polymer itself (*93*). It was observed that saturated aliphatic esters and ethers of cellulose have a strong fluorescence maximum at about 350 nm while cellulose nitrate exhibits very weak fluorescence at 310 nm. A moderate fluorescence maximum appears at 440 nm for cellulose derivatives containing double bonds or carboxyl groups. Jortner (*94*) reported the fluorescence maximum for cellulose at 445 nm and for cellulose acetate at 480 nm.

Detection and analyses of fluorescence are very sensitive means of following structural changes which take place on exposure to ultraviolet radiation. Studies of wavelength sensitivity have shown (*1*) that in most polymers ultraviolet radiation of wavelengths longer than 350 nm destroys the fluorescent species present initially while shorter wavelength radiation produces different types of new fluorescent species, depending on the wavelength. This is discussed more fully in Sec. II, C. Under nondispersed radiation the fluorescence intensity generally increases and the character changes somewhat with continuation of exposure. In those polymers in which fluorescence is not observed initially it will generally appear with increasing intensity as the irradiation progresses. In the case of poly(ethylene terephthalate) it was noted by Pacifici and Straley (*95*) that exposure to ultraviolet radiation of 254 nm or longer resulted in the formation of a blue fluorescence having an emission peak at 459 nm. On prolonged exposure a second fluorescent product was observed having a weak emission band at 510 nm. Simultaneously with the accumulation of C=O groups in cellulose, the intensity of luminescence increases and its peak shifts toward longer wavelengths (*19*). The blue luminescence of polystyrene was also observed to increase in intensity and shift to yellow-green on irradiation with a mercury vapor lamp (*22*). Even slightly irradiated samples of poly(vinylidene chloride) exhibit a yellow-orange fluorescence and orange phosphorescence. The latter was observed at liquid-air temperature when excited by near ultraviolet or blue light (*57*).

In spite of the fact that luminescence in polymers is frequently associated with the formation of photodegradation species and that measurement of luminescence is a very sensitive analytical technique, the application of luminescence in evaluating the photochemical stability of polymers has received only limited attention. Several studies have been concerned with elucidating the mechanism of degradation through the use of fluorescent model compounds. In the work of Pacifici and Straley (*95*), for example, monohydroxylated and dihydroxylated photolysis products of polyester were identified by comparing the excitation and fluorescence spectra of the photolyzed sample with spectra of synthetic samples containing these compounds. Zapolskii (*22*) postulated the presence and mechanism of formation of α-diketone photooxidation products in polystyrene based on luminescence studies with biacetyl and benzil as model compounds. Studies on the luminescence of cellulose in conjunction with other evidence led to the assumption that the luminescence centers are closely related to C=O groups which form on rupture of the glycoside bonds (*19*). The weak luminescence of unirradiated cellulose was ascribed to the presence of a small number of C=O groups which serve as initiators of the photochemical reaction.

Recently, Pivovarov and co-workers (*96*, *97*) have claimed that it is possible to assess the number of molecules liable to decompose under the action of light from measurements of the intensity of luminescence of polypropylene and polystyrene. Such studies have been proposed for evaluating the effectiveness of a stabilizer through its ability to reduce the luminescence. It is suggested that both the luminescence and the primary act of photodegradation are governed by the carbonyl groups present in polypropylene. By nonradiative transfer of some of the energy absorbed by the carbonyl group to the stabilizer, the luminescence will be decreased and the stability will be improved. The same mechanism is in effect in polystyrene, although the luminescence here was not attributed to carbonyl groups.

The correlations noted by Gachkovskii (*98*) between the visible fluorescence of polymers and their degree of polymerization may be useful in studies of the effects of ultraviolet radiation. He investigated a number of different types of polymers and found that with increase in molecular weight the overall fluorescence intensity decreases while the emission band in the red region increases in intensity in relation to the intensity of the blue-violet band. In addition, the latter is displaced considerably into the red region. These spectral effects are similar in all polymers but are particularly strongly emphasized in polymers having a continuous system of conjugated bonds. In the latter types, the red bands become very intense in comparison with the blue-violet band at a molecular weight of only 1000. A deterrent to the use of this technique for detecting changes in molecular weight may be the observation by Silin *et al.* (*99*) that in a polydisperse system the position of the luminescence band will depend on the energy levels in the molecules with the longest effectively conjugated segment.

Qualitative correlations have been shown to exist between the luminous intensity produced when a polymer is heated in the presence of oxygen and the stability of the polymer to oxidation (*100–102*). The emission of light under these conditions is referred to as oxyluminescence. The intensity of light emission reflects the ease of oxygen absorption as measured by oxygen uptake (*103*). The emitted light is generally in the blue or blue-violet region and, in the case of some polymers, in the ultraviolet region. The measurement of oxyluminescence is applicable to many polymers and has been primarily used for rapid screening of chemicals for antioxidant activity. However, it may conceivably be applicable to studies of the photooxidative degradation of polymers in which the mechanisms of radiation and heat degradation are related. Oxidative degradation of polymers is believed to proceed via a series of consecutive radical reactions. Light is thought to be emitted when the peroxy radicals formed in the presence of oxygen recombine with the hydrocarbon radicals. This exothermic process results in a high concentration of energy in the new molecules. Most of the energy is lost by collisions, but a small fraction is lost in the form of light. A linear relation has been found between the integrated intensity of emission and the number of carbonyl groups formed (*104*).

2. Determination of Appearance and Mechanical Properties

While spectroscopic techniques aid in early prediction of the long-range stability of a polymer, the ultimate criteria of its usefulness are retention of appearance and mechanical properties. In a sample more than a millimeter thick, degradation is essentially a surface phenomenon. Therefore, surface features such as gloss, haze, and microscopic topography show the effects of irradiation before changes in mechanical properties can be detected. The particular test method used in evaluating the photochemical stability of a polymer is often determined by the material, its shape, and the property which makes the polymer useful. Retention of surface properties is of most interest when the polymer is used as a coating. The most frequently used criteria for assessing weathering resistance of polymers that will be subjected to mechanical stress are tensile strength and elongation. For applications where the rate of straining differs widely from the rate used in the stress–strain tests, then tests such as impact, creep, and fatigue are used. Impact tests are also generally used when the flexibility of structure deteriorates so that the material becomes sufficiently rigid to break under impact. When the polymer is used for its insulating properties, then change in electrical resistance is measured.

Test of all these properties have had widespread use in evaluating the effects of irradiation on polymers. They have been established for many years as standard tests of the physical properties of polymers. Most of the procedures have been incorporated into the test methods of the American Society for Testing Materials.

a. Gloss. Gloss is due chiefly to specular reflection of light from the surface. It may be defined as the degree to which a surface simulates a perfect mirror (*105*). Anything that changes the surface physically or chemically is likely to affect its gloss. Glossiness is considered to be a subjective phenomenon in which the observer is influenced by several factors. Consequently, it is not possible by means of a single physical measurement to make instrumental assessments which are strictly comparable to those made by visual observation. However, a number of different types of glossmeters have been developed (*105*, *106*) in an attempt to put the assessment of gloss on a more quantitative basis. Precise comparisons of gloss values are meaningful only when they refer to the same measurement procedure and the same general type of material.

Specular gloss is the ratio of the luminous flux reflected from a specimen to that incident on a specimen measured in the specular direction. Since it usually varies with surface smoothness and flatness, it is sometimes used for comparative measurements of these surface properties. Deviations of planarity of the surface as small as one-fourth the wavelength of light are large enough to affect the specular reflection. Standardized methods of test are reported in ASTM procedures (*107*) for 20°, 60°, 45°, and 85° geometries for both opaque and transparent samples. The geometry used is dictated by the degree of glossiness of the specimen. High-gloss specimens can be differentiated best when illuminated and viewed at angles near the perpendicular to the specimen. The Dori-Gon glossmeter Model 47 (*108*) is especially designed for high-gloss finishes. Since the latter are often judged by the distinctness of image reflected in them, the instrument measures distinctness of image as a percentage of the specular reflectance at 0.27° from the specular angle. A test method for gloss has also been described which represents in one expression the "effective" gloss over the entire range of illuminating angles (*109*).

The geometric manner in which incident light is reflected in different directions is measured with a goniophotometer, an instrument in which the direction of illuminating and viewing beams can be varied. Measurement of reflectance or transmittance as a function of the direction of incidence or view can be made manually with the Gardner goniophotometer (*110*) or automatically with the Hunterlab D 10 recording goniophotometer (*108*). Typical curves and descriptions of the characteristics and types of gloss measurements have been given by Hunter and Boor (*105*). Colling *et al.* (*111*) recently designed a spectrogoniophotometer to study the intensity distribution in a specularly reflected beam as a function of wavelength and angle of incidence. Since many of the instruments have optics with wide angles of acceptance and nonparallel or nonuniform beams, they are capable of detecting only gross distortions rather than specific details. Colling and co-workers state that the two most important factors in assessing the gloss of a film is the sharpness of the image and the total specular energy. Most conventional glossmeters measure only the latter.

b. Transparency and Haze. The specular transmittance or transparency of a plastic sheet is related to its clarity, i.e., to an observer's ability to distinguish clearly a relatively distant target viewed through it. Transparency is the transmittance obtained when the radiant flux measured is only that which is transmitted in essentially the same direction as the incident flux. The ASTM Method of Test, Designation D 1746-62T (*112*), is applicable to any translucent or transparent material having a range of transparency of about 1–90%. However, it is principally intended for use with nominally clear and colorless sheeting. The transmission of light at 560 nm has also been used to monitor photochemical degradation of plastics (*113*).

When a portion of the transmitted light passing through the specimen deviates from the incident collimated beam by forward scattering by more than 2.5°, it is considered to be haze. Specimens with haze values greater than 30% are considered to be diffusing or translucent. Measurements are generally made by means of a hazemeter. Specifications for an instrument and derivations of the formulas for haze are given in ASTM Standard Method of Test, Designation D 1003-61 (*114*). A description is given for the use of a recording spectrophotometer such as the G. E. Hardy for measuring haze as a function of wavelength. More frequently, instruments such as the Gardner hazemeter (*110*) and Hunterlab D25P tristimulus colorimeter (*108*) are used. Calculations of haze are based on measurements of transmission and scatter. Baum (*115*) has determined haze in translucent poly(vinyl chloride) by reflectance measurements on a Meeco Colormaster using both a white and black backing and calculating contrast ratio according to ASTM Tentative Method of Test, Designation D 1738-66 (*116*). However, the contrast ratio method does not give absolute values of percent light scatter.

c. Crazing and Microscopical Topography. Among the early signs of photochemical damage in polymers are the surface cracks and crazes which are readily detected by microscopy. Crazing is the appearance of a random network of small fissures or cracks found on the surface or interior of the material. It is produced by internal tensions or by the action of exterior stresses such as radiation. It starts in a direction parallel to the orientation, as, e.g., in polythenes. The fissures always lie perpendicular to the direction of the tension. Crazing results from the same phenomena that produce stiffening, but the latter is only evident after prolonged irradiation. For applications where stresses are present, crazing is studied as a function of weathering while the material is subjected to a constant load (*117*).

Photomicrographs of surface cracking are commonly used in weathering studies. Electron microscopy has also been found of value in observing surface roughness resulting from photochemical degradation. It is often found that the surface does not change progressively from a smooth to a rough one, but goes through cycles of roughness and smoothness. Presumably, the roughened surface soon blisters to uncover a pitted but smoother surface whereupon another cycle of roughening begins. Electron micrographs of polyamides exposed to ultraviolet radiation (*10*) showed that the crystallinity of some of the samples appeared to increase. It is assumed that exposure to ultraviolet energy may permit rotation of a segment of the polymer chain into a more highly ordered state. The methods for assessing the extent of these microscopic defects are generally of a qualitative nature and often based on visual observation. Newman and Wolock (*118*) have described in detail the light and electron microscopy and multiple-beam interferometry techniques for optical studies of crazed surfaces of polymers.

Since not only the optical but also the mechanical properties are affected by the minute cracks of crazing, one of the more popular methods for qualitatively evaluating the crazing resistance is based on the application of a bending stress that gives rise to an increased strain along the specimen (*119*). The amount of stress is then determined in the zone where the last signs of crazing appear. Leghissa and Salvatore (*120*) have described a photometric method which permits quantitative characterization of crazing in transparent plastic materials. The crazed sample is scanned with a light beam and the variations of the transmitted beam intensity are recorded.

d. STRESS–STRAIN PROPERTIES. The stress–strain properties of polymers are evaluated in terms of tensile (or flexural) strength, elongation, and elastic (or flexural) modulus. Stress is the intensity of the internal forces that resist a change or the load per unit area of minimum original cross section carried by the test specimen at any given moment. It is expressed in force per unit area, i.e., pounds per square inch or kilograms per square centimeter. Tensile strength is measured by the maximum stress under tension sustained by the specimen just prior to the moment of rupture. Flexural strength is the maximum stress sustained when the sample is flexed. More often, the strength at the breaking point is the value of interest. Instruments such as the Instron (*121*) or Scott L-6 Tensile Tester (*122*) put a gradually increasing load on the specimen at a standard speed such that rupture occurs in 3–6 min. The load is recorded at start and at moment of rupture to a precision of $\pm 1\%$.

Strain is the change in length per unit of original length in a linear dimension of the body accompanying a stress and is described as elongation or extension. If an extensometer is used to determine the distance between two fixed points at any time during a stress test, then a complete stress–strain curve of the stress versus elongation can be obtained. The elongation expressed as a percentage of the original gage length, i.e., percent elongation, is very sensitive to photochemical changes in the material. The elongation at the break is frequently used as a measure of photochemical stability.

The slope of the straight line portion of the stress–strain curve, i.e., where stress and strain are more or less proportional, is a measure of the stiffness or rigidity. The ratio within the elastic unit of stress to the corresponding strain in this region is known as "Young's modulus" or "flexural modulus." The "secant modulus of elasticity" is the ratio of stress to corresponding strain at any point on the stress–strain curve. Often creep and stress relaxation are used as criteria of retention of physical properties. Creep refers to the gradual deformation caused by a constant stress. The material is put under load and the increasing deformation is measured. In determinations of stress relaxation, the gradual decay of stress is measured while the material is maintained at a constant deformation, i.e., constant degree of elongation.

Specifications and procedures for testing the stress–strain properties of plastics are described for rigid materials and for thin plastic sheeting in several ASTM methods (*123–127*). Reviews of the basic principles and types of tests and equipment have also been published by Bergen (*128*), Loveless (*129*), and Harris (*130*). The particular stress–strain property used to evaluate photochemical stability is often determined by the material. In polyurethane, for example, weathering is characterized by some loss in tensile strength without appreciable changes in the modulus and ultimate elongation values (*11*). For polyethylene, the physical property that seems to be most sensitive to the photolytic processes of deterioration is the percent elongation at rupture (*131–133*). Polyethylene samples which showed almost complete loss of elongation on weathering still retained more than 80% of their original tensile strength and showed little or no surface cracking.

Caution should be observed in attempting to use a single physical property as a criterion for predicting the ultimate photochemical stability of a polymer, particularly if the judgment is based on a single period of exposure. Often there is an initial increase in flexural or tensile strength followed by a decrease after a longer exposure period. Tensile strength is closely connected with the extent of cross-linking (*134*), often being a maximum at a particular degree of cross-linking (about 1 % in many cases). A moderate degree of cross-linking enables a polymer to be stretched more easily so as to relieve the stress. As cross-linking increases, the fraction of all chains that can be so extended falls off. In addition, scission and volatilization contribute to the loss in strength. Elongation generally shows a continual decrease. The rigidity imparted by cross-linking shows itself in a decrease in elongation which is also brought about by scission processes. Degradation of physical properties is caused by both chain scission and cross-linking, which occur simultaneously but to different degrees at any one time. The processes are interrelated, one leading to the other. Depending on the polymer and conditions, scission may predominate in the early stages and later lead to cross-linking (*135*), or vice versa (*136*). Cross-linking is only significant in molecules with fairly reactive hydrogen atoms. The relative rates of these concomitant reactions will determine in most cases the ultimate effect on physical properties.

It has been shown (*11*, *19*, *136*, *137*) that the degradative process is determined by the energy of radiation as well as by the extent of exposure. In some polymers, short wavelength radiation may cause predominantly cross-linking and consequently increase in toughness. While under long wavelength radiation, cross-linking is accompanied by scission and volatilization resulting in brittleness. In polyethylene, the shorter wavelengths have been shown to have a greater effect on the tensile strength, while the longer wavelengths have a greater effect on elongation and brittleness. In other types of polymers, long-wavelength ultraviolet radiation is primarily responsible for cross-linking.

The initial loss of tensile strength in polyurethane is believed to be due to the embrittled surface resulting from a combination of scission, cross-linking, and volatilization. In polyethylene, surface cracking and embrittlement occur at about the same time as a significant decrease in tensile strength, elongation, and melt index. Tests which are particularly sensitive to embrittlement therefore provide an early indication of degradation. The cantilever beam test has been used for detecting embrittlement in a variety of thermoplastic and thermosetting materials (*132*, *138*). A Tour-Marshall Olsen (or Tinius Olsen) stiffness tester is used according to the ASTM Standard Method of Test D 747-63 (*124*). The sample can be bent at a maximum angle of 90° with the exposed side of the bar in tension. Exposure effects are judged by changes in stiffness and by the occurrence of breaks in the specimen during test. The angle at break provides an early indication of degradation since embrittled materials frequently break before significant changes in stiffness take place. Thin samples show the effect of aging more quickly. It should be noted that relative ratings of light stability in terms of stiffness are not necessarily the same as for tensile data since the two tests measure different properties at different rates.

A new technique was described recently for accelerating failure in high-density polyethylene by artificially inducing high stresses during irradiation (*139*). The technique utilizes the Bell Telephone Laboratories bent-strip-type specimen which is commonly used for determination of environmental stress crack resistance of ethylene plastics by ASTM Method of Test D 1693-66 (*140*). The specimen is bent into a U-shape after cutting a stress-inducing longitudinal slit (0.020-in. deep) into the center on the outer surface. Failure occurs when a crack, or a hole which later forms a crack, appears in the stressed area.

Martinovich and Hill (*141*) compared a number of test methods to determine their sensitivity and correlation in evaluating the effect of ultraviolet radiation on the physical properties of polyolefins. They included tensile properties, flex tests, brittleness temperature, aging of stressed specimens, and surface crazing. The flex-life test (*142*) in which tensile specimens are slowly bent 180° with the exposed surface on the outside, appeared to indicate some correlation with absolute elongation. Generally, only samples in which the elongation was reduced to less than 10% of their original value broke under these conditions. There was a reasonable correlation between flex tests, tensile properties, and aging of stressed specimens. The stressed specimen became brittle at almost the same point at which tensile strength was decreased by one-third. Failure could be detected by brittleness-temperature measurements in half of the exposure hours required to reduce elongation to an absolute value of 10%. Surface crazing also appeared to correlate with tensile properties, but the evaluation depended to a large extent on personal judgment. It was concluded that ultraviolet degradation of polyolefins could be satisfactorily followed by tensile strength, elongation, or embrittlement of a bent strip.

e. IMPACT STRENGTH. Qualitatively, the impact strength of a plastic is the amount of energy it can take up before some permanent damage is done. Approximately, impact strength is the product of tensile strength and elongation. The area under the stress–strain curve (carried out to break) is taken somewhat arbitrarily as the "impact strength." There are several empirical tests to measure impact strength which range from cracking the specimen to actually breaking it. They are designed to determine the resistance of a material to fracture by shock, i.e., to stress concentration. For many practical purposes, where the material is subject to shocks, its impact strength is more important than its tensile strength. Reduction in impact strength results from increase in brittleness. This can be partly ascribed to breakdown of the molecular chains and to oxidation reactions and cross-linking. Some of the material's strength is lost by partial deorientation which is accelerated by weathering. Embrittlement, and consequently loss of impact strength, is observed after very short periods of exposure of strongly oriented extruded sheets of poly(vinyl chloride).

Two kinds of pendulum-type impact machines are used to measure the energy expended in breaking, in one blow, a standard specimen (*143*). In the cantilever beam or Izod-type test, a test piece is rigidly supported at one end only and is struck at the other end by a weighted pendulum. The specimen is notched on the side parallel to the direction of molding pressure in order to further concentrate the stress and prevent plastic deformation. In the simple beam or Charpy-type test, the specimen is freely suspended at both ends (supported as a simple beam, usually horizontal) and is struck in the center by a weighted pendulum. Both the Izod and Charpy impact test can be conducted in a single machine. The impact value is measured by the difference between the energy in the pendulum blow and the energy remaining in the pendulum after breaking the specimen, with suitable correction for windage and friction.

Another type of impact test is the ball or dart (*144*) drop in which the force necessary to break a sample is measured in foot-pounds, by dropping a ball or dart of known weight from a known distance. The dart method is considered to be more accurate since regardless of the weight of the falling object the same size area makes contact with the plastic surface. This type of impact test has been used predominantly for polyethylenes but it is suitable for other polymers as well. In addition to the pendulum and ball-drop methods, there are flywheel-type machines that are designed to investigate the impact strength at varying speeds (*145*).

Impact tests are often carried out as a function of temperature. That temperature, estimated statistically, at which 50% of the specimens will fail in a specified impact test is used to predict the behavior of the polymer at low and high temperatures, i.e., the brittleness temperature (*146*, *147*).

The latter is a more sensitive test of photochemical degradation than are impact strength tests at ambient temperature. In polyethylene, low-temperature brittleness is so much more sensitive than stress-strain or electrical properties as a criterion of degradation that only half the aging time is required for detection.

f. Electrical Properties. Weathering of plastics is mainly a surface phenomenon resulting in accumulation of polar oxygenated groups on the surface with a consequent decrease in surface resistance. The resistance of a material is measured by the current or voltage drop under specified conditions. With the use of appropriate electrode systems, the electrical characteristics of the surface can be determined independently of the volume resistance. Due to some contribution from the latter, the surface resistance cannot be determined accurately, but changes in surface resistance can serve as a sensitive indicator for early detection of photochemical effects in polymeric materials. It has been used as a criterion of photooxidation of saturated polymers such as polyethylene in which physical property changes develop more slowly. Although surface property changes do not necessarily correspond to deterioration of physical properties, studies of the degradation of polythenes showed that surface resistance measurements could be related to stress–strain properties (*148*).

The dielectric constant and dissipation factor, which essentially determine bulk properties of a material, are dependent to some extent on the surface resistance. In thin films where the surface represents a large proportion of the total volume, dielectric property measurements can be satisfactorily used to determine the effects of radiation. In the case of polymeric materials used as insulating components, it is particularly important to know how well these properties are retained. Irradiation of poly(vinylidene chloride) at 254 nm has been reported (*57*) to result in a profound change in electrical characteristics. The highly insulating character (resistivity about 10^{16}) of the material is reduced by nearly a factor of 10^3 when irradiated for 5 hr. It was demonstrated that the photoconductivity produced is associated with the free radicals, being related to the electron spin resonance signal. It has also been shown (*149*) that on relatively short exposure the useful, low electrical power factor of polyethylene is increased manifold. Ultraviolet radiation has also been shown to reduce the dielectric constant of polyamide films (*10*), but the effects are often obscured by electrical nonuniformity of samples and change in dielectric constant under the influence of the electrical field during measurements. No trends were noted in the dissipation factor in the polyamide films.

Standard methods for testing the dielectric and surface electrical properties of polymers are described in detail in ASTM procedures (*150*) and have been reviewed by several authors (*151–155*).

3. Molecular Weight Determinations

In the previous section it was pointed out that many, if not all, of the physical properties of polymers are affected by the cross-linking and scission reactions which occur during photolytic aging of the polymer. Depending on the polymer type, wavelength of irradiation, and atmospheric conditions, either cross-link formation or scission, or both can occur. The physical properties after exposure depend on whether scission or cross-linking predominates. Cross-linking in moderate amounts is usually considered beneficial, whereas scission causes progressive degradation. Equations have been derived relating physical properties such as tensile strength to the amount of cross-linking in a plastic (*156*). However, the equations are based on the assumption of equilibrium conditions, which rarely exist during tensile testing.

The structural effects produced by cross-linking and chain scission are, respectively, increase and decrease in average molecular size of the polymer. Methods for detecting changes in molecular weight distributions can therefore be utilized in determining the amounts of scission and cross-linking which have taken place. Due to the abnormally large molecular size of polymers, cross-linking or scission involving only a few of the repeating units in the chain (an undetectable extent of reaction by chemical or spectroscopic means) will change the average molecular weight by a large amount. Thus, polymer properties such as intrinsic viscosity, melt index (melt flow viscosity), gel fraction (extent of insolubility), and volume of gel swollen in an organic liquid, which depend on average molecular weight, are particularly sensitive means of detecting cross-linking and scission. They are useful for providing a quick indication of deterioration and in elucidating the mechanism of degradation. They are less effective in revealing changes in mechanical properties. Simultaneous measurements of several of these property changes are often required.

In the initial stages of degradation, often the degree of polymerization and thus the molecular weight and viscosity increase as a result of cross-linking. With continued formation of intermolecular chemical bonds, a gel structure consisting of an insoluble network may be formed. Polymers which were previously completely soluble in hot solvents now imbibe some of the solvent and swell without entirely dissolving. A heterogeneous material is produced which is partly insoluble (gel) and partly soluble (sol).

Both swelling ratio and gel content measurements can be related to the extent of cross-linking. As the cross-linking increases, the amount of insoluble fraction increases, but becomes less swollen. In addition, on extended irradiation the intrinsic viscosity of the soluble fraction decreases due to bond breaking, i.e., scission reactions. The ratio of the rate of scission to that of cross-linking can be determined from solubility (gel content) data (*157*).

a. INTRINSIC VISCOSITY. The intrinsic viscosity of a polymer is its capacity to increase the viscosity of the liquid in which it is dissolved. It is the specific viscosity divided by the concentration extrapolated to infinite dilution. The higher the molecular weight in a given series of linear polymer homologs, the greater the increase in viscosity produced by a given weight concentration of polymer. Any reduction in intrinsic viscosity is consistent with a process that degrades the polymer, i.e., results in a permanent breaking of a chemical bond by irradiation so as to form molecules of smaller molecular weight than the parent molecule. In the case of polymers undergoing simultaneous cross-linking and degradation, intrinsic viscosity measurements alone may give erroneous results for chain scission. Corroborative evidence should be obtained from other methods such as gel measurements from which both cross-linking and chain scission can be computed provided certain conditions are met (*158*).

Although viscosity measurements do not afford an absolute value for the molecular weight, the technique is so much simpler than the methods used for determination of absolute molecular weight that it is extremely valuable for studies of molecular weight changes. Three general types of viscosity measurements are available: (a) the Gardner–Holdt bubble-tube method (*159*); (b) the torsion types exemplified by the MacMichael (*160*) and Brookfield (*161*) instruments; and (c) the use of various types of capillary viscometers (*162*). Marshall (*163*) recently described a method for measuring viscosity by parallel-plate plastometry and Roth and Rich (*164*) have described an ultrasonic method. The simplest and most widely used method for viscosities of dilute polymer solutions utilizes capillary viscometers in which gravity provides the driving pressure. Two types of capillary viscometers commonly used are the Cannon–Fenske modification of the Ostwald U-tube (*165*) and the Ubbelohde dilution viscometer (*166*). The most important variables that must be controlled to ensure adequate measurements of viscosity are temperature, flow time, and capillary alignment as well as cleanliness. All solutions should be either filtered or centrifuged. The viscosity is generally measured at two or more concentrations so chosen as to give relative viscosities in the range of about 1.10–1.50 or higher. The ratios of specific viscosity over concentration are then extrapolated graphically to zero concentration.

Empirical constants relating intrinsic viscosity to average molecular weight can be established through absolute determinations on a series of polymer samples of different molecular weight. Osmotic pressure measurements are generally used for number-average molecular weight. Light scattering and sedimentation equilibrium are used for weight-average molecular weight. Separate equations have been developed for different types of polymers (*167–172*). For example, the weight-average molecular weight of polyolefins can be obtained with the use of a version of the Staudinger equation (*173*). The same equation with different constants is used for other polymers.

Intrinsic viscosity measurements have been used as one of the main tools for studying photochemical effects in polymer solutions and have been used frequently for bulk polymer or film studies. The viscosity of the soluble fraction along with solubility and swelling ratio measurements give significant information about the type of degradation occurring. Jellinek and co-workers (*58*, *174*) showed by intrinsic viscosity measurements that photochemical degradation of polyacrylonitrile in solution follows a random chain scission process with very low quantum yield. They obtained rate constants from intrinsic viscosities measured as a function of time of irradiation for different concentrations, incident light intensities, and initial polymer chain lengths.

Fox *et al.* (*59*) investigated the differences in photochemical degradation mechanisms of poly(methyl methacrylate) and poly(ethyl acrylate) through effects on molecular weight changes along with other techniques. Since no cross-linking appeared to occur in poly(methyl methacrylate) irradiated in air, molecular weight changes and apparent quantum yields for random scission could be assessed quantitatively on the basis of viscosity measurements alone. The simultaneous scission and cross-linking reactions of poly(ethyl acrylate) were also investigated by viscosity changes along with solubility and swelling measurements by Jacobs and Steele (*175*).

Kato (*176*) followed the molecular weight changes produced by ultraviolet irradiation in the surface layers of polystyrene films by light scattering and intrinsic viscosity measurements, using a rate-of-shear viscometer for the latter. The amount of solvent-insoluble (gel) fraction was also determined. It was shown that chain scission and cross-linking occurred simultaneously by the fact that the molecular weights showed a tendency to increase while the intrinsic viscosity decreased. The falloff in degree of polymerization with ultraviolet irradiation of cellulose acetate was also determined by comparison of viscosimetric measurements and osmotic data (*94*). Based on a similar technique in conjunction with a variety of other techniques for studying the photochemical degradation of polyamides (*10*), the number of peptide linkages broken could be estimated. Kwei (*177*) calculated the weight-average molecular weight of poly(vinyl chloride) as a function of ultraviolet irradiation with the use of intrinsic viscosity measurements, and Cotten and Sacks (*15*) followed the changes in molecular weight produced by ultraviolet irradiation in polyethylene by this technique.

It was shown by Osborn (*178*) that the photolysis of poly(ethylene terephthalate) could be followed conveniently by its change in molecular weight based on solution viscosity measurements. An equation relating viscosity to osmotic pressure molecular weight was used. However, the quantum yield for fracture of poly(ethylene terephthalate) irradiated in air was found by Marcotte *et al.* (*44*) through sol–gel analysis to be about four times the value reported by Osborn.

b. MELT FLOW VISCOSITY (MELT INDEX). If the average molecular weight is not too great ($< 100{,}000$), bulk viscosity measurements on undiluted molten polymer may also be used as an empirical measure of the molecular weight. The melt flow viscosity of thermoplastics with relatively low melt viscosity is determined by means of an extrusion plastometer as described in the ASTM method (*179*). Measurements are made under prescribed conditions of temperature and pressure of the rate of extrusion of the molten resin through a cylinder having a specified length and orifice diameter. A weighted piston is operated within the cylinder which has thermostatically controlled heating. The value obtained by this technique is referred to as "melt index" and is reported as the number of grams of polymer extruded per 10 min. A rapid decrease in melt index, i.e., increase in viscosity, generally indicates that cross-linking is the predominant reaction.

Peticolas and Watkins (*180*) have demonstrated that the melt viscosity of linear polyethylene is proportional to weight-average molecular weight raised to the 3.4 power. The method is particularly applicable to a number of linear condensation polymers. However, it is not applicable to all types of polymers.

c. GEL FRACTION. Determination of percent insolubles is a convenient method of following decomposition when cross-linking is an important mechanism. For a polymer of arbitrary molecular weight distribution which is cross-linked at random, gel formation will first occur when there is one cross-linked monomer unit per weight-average molecule. Since beyond the gel point intrinsic viscosity measurements can only be made on the sol fraction, gel point measurements are most frequently used for determinations of cross-linking. The variation of gel or sol with radiation dose is readily determined in the region of cross-linking in which both gel and sol exist in appreciable quantities.

Measurements of gel fraction are made by extracting the soluble fraction in a Soxhlet apparatus or by the "bottle" extraction method (*181*). In the latter, tubular gauze bandage is used inside a glass-stoppered bottle to hold the gel fraction, which is swollen with the solvent. The soluble portion is in the solvent outside the bag. The swollen fraction is removed and weighed to determine the volume of gel swollen. The gel is then dried to constant weight either by evaporation under vacuum or by heating. The solvent is also evaporated from the sol fraction and the latter weighed. Polyethylene is extracted with boiling xylene which is evaporated under vacuum at 70°C or by prolonged heating at 140°C. Solvents used for acrylonitrile, polyester, and polyamide are, respectively, *N*,*N*-dimethylformamide, trifluoroacetic acid, and 90% formic acid (*182*). Cyclohexanone has been used for poly(vinyl chloride) (*183*).

Charlesby's technique (*184*) for evaluating the relative magnitudes of chain scission and cross-linking produced by high-energy radiation based on gel fraction measurements has been extended to ultraviolet irradiation studies. The gel content is measured as a function of time of exposure, and Charlesby's equation is fitted to the experimental results with the assumptions that both cross-linking and scission occur randomly and that each process is proportional to the radiant energy dissipation within the polymer. Stephenson *et al.* (*182*) showed that gel formation measurements of several different types of polymers irradiated at 254 nm gave evidence for scission-to-cross-linking ratios which were supported by the rates of change in tensile properties. Jacobs and Steele (*175*) found by this technique that two cross-linked units form in poly(ethyl acrylate) for each chain scission. Cotten and Sacks (*15*) determined that in polyethylene chain scission is favored in the thicker films and more highly crystalline regions, although it occurs simultaneously with cross-linking in all the films examined. They pointed out that the values obtained are an average of that at the film surface and at the center of the film so that the results are more representative of surface effects in the thinner films. Marcotte *et al.* (*44*) used extremely thin films of poly(ethylene terephthalate) for estimating the quantum yield for fracture and cross-links by sol–gel analysis. In this way they attempted to avoid the nonuniform energy dissipation produced in the thicker films by the limited penetration of short wavelength radiation. Kwei (*177*) used gel fraction analyses to investigate the effects of additives and substituents on cross-linking and main chain scission efficiencies in poly(vinyl chloride). Purification had a significant effect in reducing the extent of chain scission, but little effect on cross-linking. Both cross-linking and chain scission increased with the addition of chelate substituents. In another study, simultaneous gel fraction analyses and intrinsic viscosity measurements on the soluble phase of irradiated poly(vinyl chloride) films showed that both cross-linking and chain scission occur initially, but on continued irradiation cross-linking becomes the only important effect (*183*).

The solubility of a cross-linked polymer depends not only on the degree of cross-linking but on the average molecular weight and distribution function. A theoretical derivation for the effect of the molecular weight distribution of the original polymer on the relationship between the gel fraction and cross-linking index was given by Charlesby (*157*).

d. SWELLING PROPERTIES. The proportion of insoluble polymer in any polymer-solvent system is closely related to the swelling properties of the polymer. If there are no cross-links present in a linear high polymer, it will generally swell when placed in a solvent and then gradually pass into solution. Cross-links, however, place a limit to the extent of swelling. The exact relationship between the gel content and swelling ratio in any particular case depends on a large number of factors, including the mechanisms of scission and cross-linking.

The swelling ratio is determined from the weight and a measured dimension of the polymer before and after swelling. The swollen volume of a polymer film can be determined from measurements of the length of a strip of the cross-linked polymer on immersion in the solvent until the swollen length has become constant. Another, more accurate, method used by Schaefgen and Flory (*185*) is to remove the strip of cross-linked polymer from the solvent from time to time, blot off the excess solvent, and then weigh the polymer plus imbibed solvent in a capped bottle, repeating until constant weight is attained. Marcotte *et al.* (*44*) determined equilibrium swelling values of poly(ethylene terephthalate) by titrating the volume of trifluoroacetic acid taken up by a known weight of dried polymer. The special technique was necessary because the surface of this polymer cannot be dried of excess solvent satisfactorily without serious loss of volatile acid swollen in the polymer.

The extent of cross-linking can be determined from the ratio of the swollen volume to the dry, unswollen volume of the cross-linked material. Jacobs and Steele (*175*) used swelling measurements on ultraviolet irradiated poly(ethyl acrylate) to estimate the degree of cross-linking based on the Flory–Rehner theory (*186*). The value was in good agreement with that from solubility data when the cross-linking was less than 10^{-5} mole of cross-linked units per cubic centimeter. The determination of cross-linking from swelling data has the advantage that it is unnecessary to know the initial molecular weight or molecular weight distribution before cross-linking, at least in the use of the approximate equation (*187*). For very high degree of cross-linking, when the sol fraction becomes negligible, one can no longer obtain useful data on the molecular weight distribution. However, it is still possible to trace the increase in cross-linking either from swelling properties or from its elastic behavior.

e. GEL PERMEATION CHROMATOGRAPHY. Gel permeation chromatography (GPC) is a relatively new technique (*188*) for determining molecular weight distribution of polymers. It is based on the separation of molecular sizes by means of selective permeation on a microporous support such as a gel of polystyrene cross-linked with divinylbenzene. The larger molecules pass through the column more rapidly since they cannot penetrate the pores of the gel as readily as the smaller molecules. The components of the solute mixture thus leave the column in the order of decreasing molecular weight corresponding to the holdup time in the stationary gel phase which in turn is dependent on diffusion equilibrium. M. F. Vaughan (*189*) was the first to propose this as a simple, rapid technique, and J. C. Moore (*190*) first prepared a hydrophobic gel with micropores covering the size range of most polymer molecules.

The composition of the column effluent may be continuously monitored by a differential refractometer or other means. A relationship is provided between elution volume and weight percent polymer in the effluent. With proper calibration techniques, this can be used to obtain molecular weights and distributions. The method was used initially primarily for proteins but has more recently been applied to studies of synthetic polymers (*191*). Although it has not yet been used extensively in studies of photodegradation of polymers, it should prove to be a valuable tool in such studies. Shultz (*192*) derived an expression for the weight distribution function of polymer molecules in a photodegraded film from which he computed theoretical gel permeation curves. Kwei and Kwei (*193*), in studies of poly(vinyl chloride) and its derivatives irradiated by 254 nm, showed that the ratio of the weight-average over number-average molecular weights could be estimated by gel permeation chromatography. The value obtained agreed with the results based on intrinsic viscosity and osmometry (*177*).

C. *Wavelength Sensitivity*

An understanding of the composite effects of ultraviolet radiation on polymers can be achieved by isolating narrow bands of ultraviolet radiation and determining the effects of the individual regions. Such studies also enable one to determine the wavelength region principally responsible for the degradation of a particular polymer. This radiation can then be effectively screened by a protective ultraviolet absorber. The relative extent of degradation produced by the different wavelength regions is termed an *activation spectrum*.

The extent of degradation caused by a specific wavelength region is a function of the intensity of the radiation in that region and the absorption of the radiation by the photosensitive entity of the polymer. Since the relative intensity of radiation of different wavelengths varies from source to source, the activation spectrum will be different for each source. Generally, no attempt is made to correct for the energy distribution of the light source and thus put the wavelength sensitivity data on an equal intensity basis. Such information would allow the data obtained with one source to be translated to wavelength sensitivities using any other source of known spectral distribution, assuming constant quantum yield. However, it would also be necessary to know the dependence of degradation on intensity, which is not a simple function.

1. Irradiation Techniques

A number of techniques can be used to isolate specific wavelength regions for irradiation. Some of these have been discussed in the first chapters of this book. The use of a monochromator or interference filters to isolate narrow bands of energy is common to all types of photochemical studies, including wavelength sensitivity studies of polymers. Broad-band transmission and short-wavelength cutoff glass filters have also been used in attempts to isolate bands of radiation. Generally, several are used in combination or they are used in conjunction with solution filters to further reduce the bandpass (*194*). However, in polymer degradation studies single glass filters have been used to study the effects of broad regions of the spectrum. The use of a spectrograph for irradiation of polymers in the form of films or fibers is a very specific technique for determining the wavelength sensitivity of these materials. A self-filtering technique has also been used specifically for polymers.

a. ISOLATION WITH A MONOCHROMATOR OR INTERFERENCE FILTERS. A monochromator can conveniently be used to isolate any region of the spectrum of interest emitted by the source. Although the bandpass can be selected at will, it is limited at the narrow end by low light intensities and at the wide end by scattered light of wavelengths other than those nominally selected. Because of the low light-gathering power and excessive light scattering of most monochromators, they are most efficient for radiation studies when used with a source such as a medium-pressure mercury arc that has most of its radiation concentrated in a few emission lines.

Interference filters are designed to isolate relatively narrow bands of radiation. They are somewhat better than a monochromator in terms of scattered light, provided the higher orders of transmission are filtered out. However, the transmission factors are very poor and for this reason interference filters are also most efficient for sources with line emission. Their use has been limited in radiation studies because they are not readily available for all wavelengths in the ultraviolet region, particularly in the short wavelength region. The photochemical effect of ultraviolet radiation on the filter itself is a factor which must be taken into consideration.

Wavelength sensitivity studies by means of a monochromator or filters require exposure of a series of identical samples, each under a different wavelength region. Physical as well as optical measurements can be made if a large area of the sample can be uniformly illuminated. Filters are better than monochromators for irradiating a large area.

Several examples can be cited of wavelength sensitivity studies of polymers utilizing either interference filters or a monochromator, or both. In a study of the sunlight aging of a polyblend film containing poly(vinyl chloride) (*195*) five interference filters peaking between 330 and 410 nm were used. A Corning glass filter was used in conjunction with each to eliminate higher order visible and near infrared radiation transmitted by the interference filters. The 2-in.^{2} irradiated area of each film allowed measurement of stiffness to be made as a criterion of degradation. In a study of several types of polymers, Stephenson and co-workers (*137*) used either interference filters or a grating monochromator to study the relative effect of different ultraviolet wavelength regions. They isolated bands from a high-pressure AH6 mercury arc centered at 244, 314, and 369 nm. A low-pressure mercury G30T8 germicidal arc was used for 254-nm radiation. The material was irradiated inside a quartz cell which could be evacuated or flushed with an inert gas. Changes in tensile strength, ultimate elongation, scission, cross-linking, and molecular weight were determined. Selivanov *et al.* (*55*) determined the effect of different wavelengths on the photooxidation of polystyrene by means of a grating monochromator and medium-pressure mercury lamp. The accumulation of carbonyl groups and formation and decay of free radicals were followed.

b. BROAD-BAND TRANSMISSION AND SHORT-WAVELENGTH CUTOFF FILTERS. Broad-band ultraviolet transmission and short-wavelength cutoff filters are frequently used for wavelength sensitivity studies because they are relatively inexpensive. However, it is not possible to isolate narrow regions of the ultraviolet with such filters, and therefore the effects of individual wavelength regions cannot be accurately defined in this way. The broad-band ultraviolet filters include not only a large portion of the ultraviolet but some of the red region of the visible as well. The short-wavelength cutoff filters transmit all of the visible and some of the near infrared radiation along with portions of the ultraviolet. In contrast to the studies using interference filters or a monochromator, the composite effects of radiation in these regions are thus investigated. The effect of increasing amounts of short-wavelength ultraviolet energy is determined by comparisons of exposures behind filters with progressively shorter wavelength cutoff points in the ultraviolet. An advantage to this method of irradiation is that composite effects are a more realistic simulation of the normal exposure conditions. As with interference filters, the transmission characteristics of the filters are affected by the radiation they absorb. It should be pointed out that depending on the polymer absorption characteristics, nearly monochromatic radiation can sometimes be obtained with the proper combination of light source and short-wavelength cutoff filter. For example, Marcotte *et al.* (*44*) were able to study the effects of 313-nm radiation on a thin film of poly(ethylene terephthalate) using a mercury source and a solution filter of lead chloride in concentrated aqueous sodium chloride which screened out wavelengths shorter than 303 nm. The wavelengths longer than 313 nm were only absorbed to a minor extent by the film. Most solution filters, including the one commonly used to absorb 303 nm and pass 313 nm, i.e., aqueous potassium biphthalate, are much too unstable toward photodegradation to be of use in extended exposures. However, the lead chloride solution was found to be relatively stable.

Studies with glass filters have been carried out with sources such as the Fade-Ometer (enclosed carbon arc), mercury arc, sunlight, and fluorescent sunlamp alone and in combination with the fluorescent blacklight. Occasionally two different types of sources are used in one study in conjunction with glass filters as a means of obtaining separate regions of the spectrum. Analyses of degradation following irradiation behind filters have included measurements of physical properties: i.e., elongation, tensile strength, and brittleness in polyolefins (*137*, *196*) and poly(ester-urethane) (*11*); spectral properties in polyolefins (*14*), polyester (*197*), poly(ester-urethane) (*11*), and poly(vinylidene chloride) (*53*); and evolution of gases in poly(methyl methacrylate) (*198*) and poly(vinyl chloride) (*199*). The results are generally in terms of the wavelength limit beyond which little additional destruction occurs. For example, radiation shorter than 300 nm was judged to be primarily responsible for physical changes in polypropylene and for formation of trienes in polyethylene. Radiation shorter than 350 nm was found to be responsible for photochemical effects in poly(methyl methacrylate). In the latter study a mercury arc source was used unfiltered and with either of two filters, one transmitting ultraviolet longer than 300 nm and the other transmitting ultraviolet longer than 350 nm. In polyesters and poly(ester-urethane) color change appeared to be caused by radiation shorter than about 410 nm, but the rate increased with decreasing wavelength. The rate of liberation of hydrogen chloride from poly(vinyl chloride) was shown to be largest when wavelengths as short as 235 nm, having energy capable of breaking the C—Cl bond, were included. The effect of longer wavelengths appeared to be related to the absorption of radiation by carbonyl groups present as trace impurities in the polymer.

c. RADIATION DISPERSED BY A SPECTROGRAPH. A technique in which the sample is irradiated in a spectrograph allows a detailed evaluation of the relative effect of different wavelengths to be obtained from a single exposure. The sample is substituted for the photographic plate. Separate areas of the sample are thereby irradiated simultaneously by different wavelengths which are dispersed by the prism or grating in the spectrograph. The radiation source generally used is a high-pressure xenon arc, often filtered with Pyrex glass to simulate sunlight. Some studies have also been carried out with direct sunlight (*63*, *200*). The wavelength positioning on the sample is determined through exposure of a standard photographic plate to a low-pressure mercury arc line source.

A quartz prism spectrograph used for wavelength sensitivity studies of plastics has been described in detail (*200*). Energy between 200 and 600 nm is spread across a 2-cm-wide area of the sample. The dispersion in the ultraviolet region ranges from about 5 nm/mm in the short wavelength region to greater than 30 nm/mm in the long wavelength region. With an effective speed of *f*/2.0 the light-gathering power is high enough so that degradation can be detected after exposure times ranging from only a few hours to a few days, depending on the polymer. However, measurements are essentially restricted to optical changes since the dispersion is not large enough to allow comparisons to be made of physical changes in the different wavelength regions. A glass prism spectrograph used for wavelength sensitivity studies of dyes was reported to have a dispersion such that radiation from 300 to 800 nm is spread across a 6-cm-wide area (*201*). It also would probably be unsatisfactory for determination of the wavelength effect on physical properties. It has been shown by Tincher and co-workers (*202*, *203*) that a Jarrell-Ash Model 7500 *f*/6.3 plane grating 0.75 meter spectrograph having a dispersion of 20 Å (2 nm) per millimeter in the first order at the plate holder (5 in. × 7 in.) can be used for tensile strength measurements on fibers as well as for optical studies on films. Fibers are wrapped in a transverse direction around the exposure plate thus exposing the center of each fiber to essentially monochromatic light. Breaking strength measurements are subsequently made on individual fibers. By studying both films and fibers of the same material, it is possible to determine if the same wavelengths are responsible for both discoloration and strength loss. The high dispersion of the spectrograph gave such low intensities that relatively long exposures, on the order of hundreds of hours, were required for meaningful changes of physical properties in the nylon samples investigated.

The nonlinear dispersion of a prism produces a higher concentration of wavelengths per unit area in the longer wavelength regions where the dispersion is poorer. Correction can be made to put the activation spectrum on an equal-wavelength-interval basis if the dispersion and the wavelength intensity distribution of the light source at the sample plane are known. However, the argument against attempting such a correction is the same as that given by Johnson *et al.* (*202*) against putting the activation spectrum on an equal intensity or equal energy basis. That is, the results would be badly distorted since the degradation is neither a linear nor other well-defined function of intensity or time.

d. INDIVIDUAL LIGHT SOURCES. As an aid in studying the effects of different wavelength regions, occasionally more than one light source is used in a study. A low-pressure mercury arc can be used unfiltered for predominantly 254-nm radiation. This is frequently compared with the radiation of a filtered medium-pressure mercury arc, from which the 313- and 365-nm radiation can be isolated by solution and glass filters (*35*, *37*). In a study of the relative effects on wool of two of the ultraviolet regions known to be present in sunlight, a Philips TLU 20W/12 ultraviolet sunlamp having a peak output at 305 nm and a half-bandwidth of approximately 30 nm was used for the short wavelength radiation. For 365-nm radiation an air-cooled Philips HOKI 2000-W mercury arc was used with a Chance OV 1 filter which excluded radiation of wavelengths shorter than 340 nm and longer than 420 nm except for a small amount of red (*204*).

The wavelength effect of irradiation on photooxidation of polystyrene was investigated with the use of three ultraviolet light sources having different emission characteristics. A carbon arc Fade-Ometer emitting radiation longer than 300 nm and a fluorescent blacklamp emitting in the region of 350 nm were used for the longer wavelength radiation which was compared with 254 nm obtained from a low-pressure mercury arc (*52*). It was shown that the relationship between discoloration, as measured by change in ultraviolet Absorbance at 365 nm, and oxidation, as measured by the change in carbonyl Absorbance at about 1710 cm^{-1} in the infrared region, depended on the wavelength of irradiation. The longer wavelengths produced colored products which were mainly carbonyl-containing compounds, while the discoloration produced by the shorter wavelengths appeared to be associated predominantly with conjugated species which did not contain carbonyl groups.

e. SELF-FILTERING. A completely different technique was used by Shultz and Leahy (*205*) to determine the active wavelength causing chain scission of poly(ethylene terephthalate). The polymer films were arranged in a stack to provide an increasing amount of filtering with increasing layers of the film. From measurements of the ultraviolet absorption characteristics of the nonirradiated film and the viscosities of the films after irradiation with a high-pressure mercury vapor lamp, they calculate that the wavelength region of 314 nm is principally responsible for molecular weight changes. It appears to be adequate as a preliminary screening test for those films having reasonably small extinction coefficients that increase fairly rapidly in the wavelength region causing maximum degradation.

5
4
2
3
1
(a)

(b)

2. Optical Detection of Degradation Produced by Dispersed Radiation

Following irradiation of a sample in a spectrograph, the photodegradation produced in the different areas of the sample can be measured by one or more optical methods. Ultraviolet, visible, infrared, and fluorescence techniques have been utilized (*206*, *207*). A microdensitometer attachment designed to be used with the Cary 14 spectrophotometer has been employed to determine the wavelength sensitivity of yellowing and change in ultraviolet absorption. It is shown in Fig. 9-5a with the sample holder (5) removed from its normal position, which is opposite the plate containing the 0.5-mm slit. The 4-rpm synchronous motor moves the sample across the slit at a rate of 2 mm/min while the Absorbance at a fixed wavelength is recorded. The sample is scanned before and after irradiation. The change in Absorbance at a specified wavelength, caused by different wavelengths of irradiation, produces an activation spectrum. Figure 9-5b shows the synchronous motor on the outside of the cell compartment along with a typical activation spectrum trace.

Calibration of the chart trace in terms of sample position in relation to the wavelength of irradiation is obtained by scanning a small photographic plate which has been exposed, in the sample position of the spectrograph, to the emission lines of a low-pressure mercury arc. Through the readings on the revolution counter attached to the synchronous motor and the known wavelengths of the mercury emission lines, the chart trace can be related to the wavelength of irradiation. In studies in which the sample is particularly sensitive to radiation, the wavelength calibration can be obtained directly by recording the mercury lines as darkened images on the sample itself. The relative extent of degradation has been determined in this way by optical density measurements, using a Hilger densitometer, of the blackened areas at positions corresponding to the mercury emission lines.

FIG. 9-5a. Microdensitometer attachment to Cary 14 spectrophotometer. The switch (1), motor (2) and gear coupling (3) are placed outsied the cell compartment. The platform containing the slit (4) on its carriage, and sample holder (5) are placed inside the cell compartment.

FIG. 9-5b. The motor, coupling, and gear of microdensitometer attachment are shown engaged to revolution counter on outside wall of Cary 14 cell compartment. A trace of the density across the sample before and after irradiation (activation spectrum) is shown on the chart.

Typical activation spectra are shown in Fig. 9-6. The relative extent of yellowing in polyester produced by the wavelengths recorded on the abscissa is shown by the curve representing the 400-nm scan of the sample. Although yellowing is ideally measured by blue light (430 nm), sensitivity for detecting degradation is enhanced as the analytical wavelength is decreased. The 400-nm choice is a compromise since a shorter wavelength than this would not be representative of visible changes in some materials. It is seen that radiation below 350 nm causes yellowing, i.e., an increase in absorption, while the longer wavelength radiation bleaches the polymer, i.e., decreases the original absorption of 400 nm. Investigation has shown that the longer wavelength radiation will also destroy the yellow species produced by the shorter wavelengths. Since yellowing and bleaching are not simple reversible reactions, yellowing may not be representative of the total extent of degradation.

The diverse effects of different wavelength regions are shown more strikingly when the sample is scanned with a wavelength in the ultraviolet region. Thus, the curve in Fig. 9-6 representing the 325-nm scan of the sample shows that the long wavelength radiation greatly reduces the

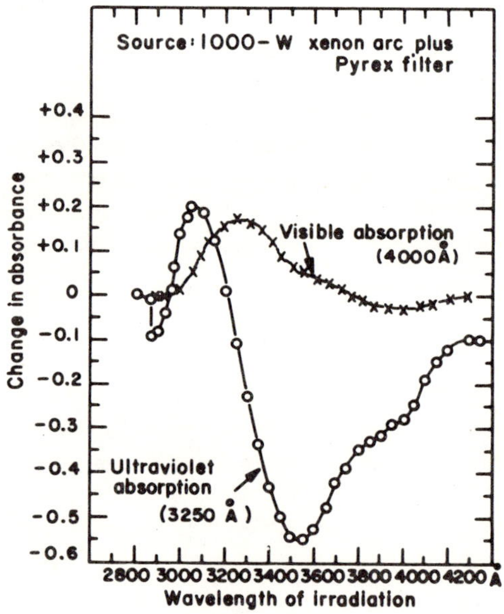

FIG. 9-6. Visible and ultraviolet activation spectra of a polyester casting. Reprinted from Hirt and Searle (*1*, p. 63), by courtesy of John Wiley & Sons.

absorption of polyester for 325 nm and thus makes the polymer more transparent to ultraviolet and less susceptible to yellowing. The opposing effects of different wavelength regions are exhibited by most polymers. In cellulose nitrate, competitive reactions by different wavelength regions were also observed in relation to yellowing (*65*). In contrast to the studies on polyester, it was concluded that the longer wavelength radiation produces a faster rate of yellowing, while shorter wavelengths destroy the yellow species, i.e., cause bleaching. However, the interpretation of the spectral characteristics of the light sources used appears to be in error. The mercury arc radiation, the source which caused bleaching predominantly, probably subjected the samples to longer wavelength radiation than the sunlamp, which produced a greater rate of yellowing. In another study, the ultraviolet- and visible-absorbing species produced by 254-nm radiation in poly(vinylidene chloride) was destroyed by 310- and 365-nm radiation (*57*). The products were different in the two cases and depended on the presence of oxygen. Neither of the longer wavelengths had an effect on the unirradiated sample. The dependence of the type of reaction on wavelength of irradiation is not limited to optical effects. In some materials the long wavelength radiation fosters cross-linking while the shorter wavelength radiation produces predominantly scission reactions. In cellulose nitrate the wavelengths which produced bleaching also caused chain scissions. The net effect of ultraviolet radiation on a polymer is thus a composite of the effects of the different wavelengths present in the light source and is dependent on the spectral characteristics of the source.

Changes in infrared absorption characteristics were measured with a microdensitometer attachment designed for the Beckman IR-4 infrared spectrophotometer (*206*). Microdensitometry in the infrared region was accomplished manually. Successive 0.5-mm-wide areas of the sample were positioned opposite the slit in the beam of the instrument and a spectrum was obtained in the infrared region of interest (usually the carbonyl region). The calibration of sample position in the holder with respect to wavelength of incident radiation on the sample in the spectrograph was made by means of the same spectrographic plate described previously.

A special precaution is needed for infrared work. In any instrument in which the total source radiation is incident on the sample prior to dispersion by the prism, it is advisable to insert a heat filter such as carbonized ZnS in the measuring beam. This prevents the sample from being overheated and thus decomposed by the visible and infrared radiation outside the portion of the beam used to detect the degradation. Polymers which

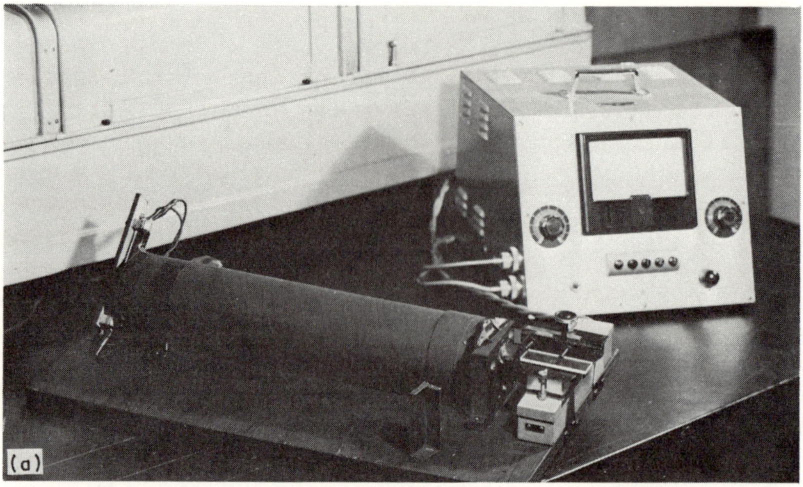

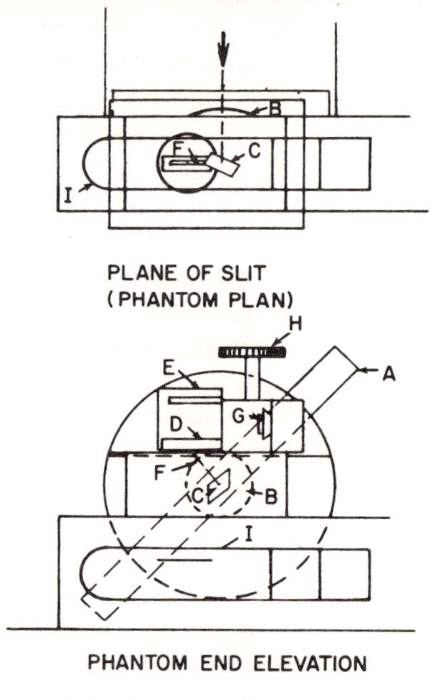

(b)

FIG. 9-7a. Photograph of fluorintensometer showing fluorescent blacklight, tubular light baffle, and sample in position on microscope stage over photomultiplier.
FIG. 9-7b. Schematic diagram of fluorintensometer.
A, 4-W blacklight tubular lamp, Sylvania 4T5/BLB; B, image-forming double convex lens, $F_{\infty} = 54$ mm, diameter 23.5 mm; C, rectangular aluminized front surface diagonal; D, sample having an area of 1 × 3 in. maximum; E, traversing light-tight housing of sample; F, slit 0.5 × 10.0 mm; G, rack and ways of converted microscope stage; H, knurled knob on shaft of driving pinion; I, envelope of 1P21 photomultiplier tube.

have been exposed to ultraviolet radiation are especially susceptible to heat degradation.

Another optical method used to detect the relative extent of degradation produced by different wavelengths utilizes measurement of the intensity of fluorescence (*206*, *208*). The blue-white fluorescence characteristic of many polymers is destroyed or reduced in intensity by long-wavelength ultraviolet radiation while shorter wavelength radiation produces a different type of fluorescent species. Although the latter appears to vary with the wavelength of incident ultraviolet radiation no attempt has been made to identify the species. Only the change in intensity of fluorescence as a function of wavelength of incident ultraviolet radiation to which the sample has been exposed is measured. This is termed a "fluorescence activation spectrum." It should not be confused with an excitation spectrum which represents the fluorescence intensity produced by different wavelengths of excitation of the fluorescence; the excitation spectrum is usually similar to the absorption spectrum of the material. Measurement of the change in fluorescence intensity is an extremely sensitive criterion of photochemical damage. It can often be detected prior to changes in ultraviolet absorption.

A fluorintensometer specifically designed to measure the relative intensity of fluorescence across the irradiated sample is shown in Figs. 9-7a, b (*208*). Fluorescence is excited by a 4-W Sylvania tubular fluorescent blacklight in a Harco Black Light Kit (*209*). The exciting energy is composed of a relatively broad band extending from 320 to 392 nm and peaking at 353 nm. A reduced image of the source is focused on the slit at a 45° angle so that any specular reflection can be caught by the baffles. A microscope stage with a scale and vernier holds the sample and allows it to be moved across the slit. The fluorescence intensity of 0.5-mm-wide sections is measured by a 1P21 photomultiplier 50 mm directly below the slit on the illuminated side of the sample. The fluorescence emission is collimated by baffles before reaching the photomultiplier.

III. Mechanistic Studies of Photodegradation of Polymers

It is difficult to draw a sharp line of distinction between the techniques used for mechanistic studies and those used for evaluating the stability of polymers. Many of the techniques for determining extent of degradation have proven of value in investigating mechanisms of degradation. For example, most of the spectroscopic techniques aid in identifying the chemical structures of the degradation products. Molecular weight determinations as well as some of the physical property changes also help to establish the mechanisms of degradation. Likewise, quantitative studies on gas evolution and oxygen uptake as well as investigations of free-radical processes yield information not only on the mechanism but also on the extent of degradation. However, mass spectrometry, gas chromatography, and electron spin resonance techniques are used primarily for mechanistic studies because of their requirement for expensive instrumentation. Model compounds are used exclusively as a means of investigating degradation mechanisms.

A. *Kinetic Studies of Volatile Decomposition Products and Analyses by Mass Spectrometry and Gas Chromatography*

Carbon dioxide, carbon monoxide, and water vapor evolve early in the photodegradation of many polymers. Frequently monomers are also among the low-molecular-weight products produced. Measurements of the rate of evolution of gaseous products and analyses of the components provide direct means of studying the primary process in the photochemical degradation reaction. The techniques involved are therefore among the most frequently used methods for such studies. Determination of the total pressure of evolved gases has been a common technique for evaluating the extent of photochemical decomposition. The development of mass spectrometry and gas chromatography as analytical tools has made the identification and quantitative analysis of small concentrations of volatile products feasible. The major problem in the past in tracing the mechanism of photochemical degradation of polymers from the gaseous products given off has been the collection of sufficient material to enable the products to be identified. Having the capability of identifying individual compounds in small concentrations allows investigation of the degradation reactions under conditions more comparable to service conditions.

1. Kinetic Studies

The kinetics of gas evolution are frequently determined by pressure-measuring instruments such as a thermocouple or McLeod gage. The polymers are prepared either in the form of dried films or powders which are sealed inside quartz irradiation tubes under vacuum or nitrogen and occasionally in an oxygen atmosphere. The tubes are connected to a vacuum manifold through cold traps to eliminate mercury vapor. The primary volatile degradation products are removed by condensation as they are formed, to prevent secondary reactions from becoming appreciable. At intervals, the pressure of those gases that are not condensable at liquid-nitrogen temperature is measured. Following irradiation, the condensed gas is fractionated and the pressure of the volatiles is determined at different temperatures.

Measurement of the pressure and thus amount of gases evolved as a function of the amount of energy absorbed can be used to determine the quantum yield and often the mode of degradation. For example, heat or light decomposes poly(vinyl chloride)-type polymers predominantly into HCl, H_2O, and a colored residue. Wilson (*210*) found that the amount of HCl evolved on irradiation of a poly(vinylidene chloride) film was only about one molecule per molecule of polymer. Since the appreciable color developed could not be accounted for by conjugation of double bonds on the basis of this small amount of HCl evolved, it was concluded that the color probably arose from oxidation products. Dehydrochlorination presumably leaves a polyene "backbone" of free radicals which can react with oxygen to form carbonyl groups. The question of whether oxygen plays a direct role in causing dehydrochlorination or whether its action is subsequent to the removal of HCl has been the subject of many investigations utilizing the measurements of HCl evolution, often in conjunction with infrared measurements, color formation, and oxygen uptake (*53*, *210–212*). The HCl can be measured by carrying it over in an airstream into a solution of sodium hydroxide or silver nitrate for analysis by alkalimetry or argentometry, respectively (*213*).

The kinetics of gas evolution have also been investigated as a function of wavelength of irradiation. Frolova and Ryabov (*198*) determined the quantum yield of photodegradation of poly(methyl methacrylate) films in the absence of oxygen. They compared the effects of radiation of wavelengths longer than 350 nm with radiation which included the 300–350-nm region. The gas pressure was determined by means of a McLeod gage and the quantum yield was calculated from the number of moles of gas formed per basic mole of polymer for each quantum absorbed by the polymer. No appreciable amount of gaseous products was formed by wavelengths longer than 350 nm. Marcotte *et al.* (*44*) studied the gas evolution from poly(ethylene terephthalate) film irradiated in vacuum with monochromatic light of wavelengths of 254 and 313 nm. The rate of formation of the gaseous fractions condensable at -196 and -78°C was measured as a function of time. The gas fractions volatile at these temperatures were identified by mass spectrometry as CO_2 and CO, and it was determined that the quantum yields for their formation were the same for both wavelengths. The pressure–temperature relationship for the −78°C fraction agreed with standard data for CO_2, thus confirming the formation of CO_2 which had not been detected in previous investigations. A mechanism based on primary reactions leading to formation of CO and CO_2 was proposed consistent with other experimental observations of the photolysis products.

2. Analyses by Mass Spectrometry

A detailed analysis of the volatile products of degradation combined with rate measurements can yield a great deal of fundamental information on the kind of degradation reaction that occurs. Since mass spectrometry is capable of analyzing very small amounts of material, the products of photolysis can be detected early in the degradation process without the necessity of concentration by high-intensity or extended exposures. The chief limitation of measurements at low levels of reaction is that the volatile components are invariably contaminated with trace impurities that may interfere with the analyses. Mass-spectrometry studies are generally restricted to decomposition products with a vapor pressure at room temperature of 1-mm mercury or more, particularly when small quantities are involved.

The experimental technique for obtaining mass spectra of volatile decomposition products was described in detail by Achhammer *et al.* (*214*). The exposure tube usually has a cooling arm and adaptor which can be fitted directly into the inlet system of the mass spectrometer. To simplify interpretation of the mass-spectral data, the gaseous products are fractionated by first analyzing the products that are volatile at −196°C (liquid-nitrogen temperature), then those volatile at −78°C (dry ice–acteone temperature), and finally those volatile at room temperature. Samples are often withdrawn and analyzed during the course of the exposures.

Qualitative estimates can be made rapidly, but precise quantitative analyses require the use of standards. The partial pressure of each component is determined by means of the known pattern sensitivities. The sensitivity of a compound is defined as the height of the characteristic peak produced when the pure substance is in the inlet system at 1-μm pressure. Knowing the total volume of the inlet system and sample tube, the gas law is then used to compute the weight of the component produced. Because of the presence of water and carbon dioxide in the background, small concentrations of these cannot be measured accurately.

Substitution of radioactive carbon (^{14}C) for stable carbon or of the stable isotopes of carbon, hydrogen, or oxygen (i.e., carbon-13, deuterium, or oxygen-18, respectively) for the more abundant forms in certain structural positions in the polymers is a valuable adjunct to mass spectrometry in studies of the degradation mechanisms. By determining the structure of the volatile fragment that contains the isotope or by merely determining the concentration of radioactive species in the volatile products previously identified by mass spectrometry, the mode of breakdown of the polymer can be deduced (*64*, *215*, *216*). Since deuterium is considerably heavier than hydrogen, it is abstracted at a much slower rate, and the reactivity of deuterium-containing radicals is also slower. Information on the origin of the radical species can thus be obtained from comparisons of reaction rates of deuterium- and hydrogen-containing structures (*54*).

Achammer, Wall, and Madorsky and co-workers have been instrumental in demonstrating the usefulness of mass spectrometry for the analysis of degradation products in a variety of polymers. In a study of the degradation of polystyrene, Achhammer *et al.* (*214*) showed that the mechanism involves two different processes. When the polymer is exposed to ultraviolet radiation in the presence of oxygen, in addition to oxidation there is a breakdown of the thermolabile groups in the polymer which are presumably formed by incorporation of oxygen during processing. Discoloration was associated only with photooxidation of the polymer. Insolubility occurred simultaneously with decomposition of oxygenated groups and hence is not considered to be a consequence of the primary reaction with oxygen.

Achhammer *et al.* (*10*) also proposed a photochemical degradation mechanism for polyamides based on the products detected by mass spectrometry. These products included H_2O, CO_2, CO, C_2H_5OH, and over 15 individual hydrocarbons. The latter were found in all experiments in which CO was detected. When the proportion of CO_2 was high, little CO was found, and vice versa. The principal component in the volatile products of most of the polyamide samples was water, and it was shown that the reaction producing water may be the one responsible for the cross-linking of the polyamide chains.

Two basic mechanisms proposed for the photolysis of poly(vinyl acetate) (*60*) were based on mass-spectrometry analysis of the gases evolved on exposure of the polymer in vacuum in conjunction with other studies of alkyl esters. The mechanism of color formation in several types of poly(vinyl chloride) was also studied by mass spectrometry (*33*). The effect of ultraviolet radiation on the intensity of color was shown to be related to the total amounts of products evolved, but not related to the amount of HCl. This confirmed the conclusion reached from kinetic studies showing that the color formed was the result of oxidation rather than from the polyene structure itself. The presence of benzene was attributed to the breakdown of the catalyst used.

In a study designed to provide basic information on the photochemical degradation of paper by ultraviolet radiation, Flynn and co-workers (*217*, *218*) determined the kinetics and types of gases evolved when cellulose is irradiated by 254 nm. The pressure of the evolved gases was measured by different types of pressure gages and the composition of the gases was determined with a mass spectrometer. The results indicated that the primary process using 254-nm radiation is the decomposition of the alcohol groups, resulting in the liberation of hydrogen and the formation of aldehyde groups. Although the gaseous products were composed of 90% hydrogen (the remainder being carbon monoxide and carbon dioxide), it was the first time that hydrogen had been reported as a product. By studying the rates of evolution of gaseous products as a function of absorbed radiation for starting materials containing different functional groups, it was concluded that photolysis of the alcohol groups plays a dominant role in both primary and secondary reactions. Flynn and Morrow (*216*) replaced the hydrogen in the hydroxyl groups with deuterium by exchange with heavy water and determined the composition of the irradiation products. The total deuterium-to-hydrogen ratio in the gaseous products, H_2, HD, and D_2, was obtained by mass spectrometry. From the relative amounts of these products, inferences could be drawn as to the type of reactions taking place.

Nevskii *et al.* (*64*) studied the photodegradation mechanism of polyurethanes through mass-spectrometric analysis of the volatile degradation products of two formulations containing tolylene diisocyanate. One product contained ethylene glycol and the other contained polyoxyethylene glycol. The similarity in the gaseous decomposition products indicated that most of the gas formation is due to the decomposition of the urethane groups. The gaseous products found could be rationalized by assuming breakage of the N—C and C—O bonds. Breakage of the N—H bonds was established by substituting deuterium for hydrogen in the urethane group. Heavy hydrogen was found in the gaseous phase. Mass spectroscopy was also one of the techniques used by Schollenberger and Dinsbergs

(*11*) to investigate the mechanism of ultraviolet degradation of poly(ester-urethane). They similarly concluded that the bonds of the urethane bridge groupings break on exposure to ultraviolet radiation. The presence of methane following irradiation through a Pyrex glass filter which transmits lower energy than that required to break a C—C bond showed that methane is formed as a result of secondary radical reactions occurring after decomposition of the ether chain of polyurethane (*64*).

Fox and co-workers (*59*) used mass spectrometry to identify the volatile degradation products produced in poly(methyl acrylate) and to determine the quantum yields of the main products. A mechanism of the photodegradation was proposed which was compatible with the findings that most of the volatile products appear to originate from the ester groups in the polymer. Depolymerization appeared unlikely since no methyl acrylate was detected. By comparison with studies on poly(methyl methacrylate), which gives rise to similar volatile products but does not cross-link, it was inferred that cross-linking in the methyl acrylate polymer occurs through radicals formed by abstraction of the tertiary hydrogen atoms along the polymer chain. Frolova *et al.* (*215*) used poly(methyl methacrylate) labeled with radioactive carbon in different positions to confirm mass-spectrographic data. The latter indicated that methyl formate is the main volatile product of degradation formed as a result of splitting off the pendant ester group. However, under ultraviolet irradiation methyl formate is degraded, with the formation of a complex mixture of gaseous products, making the assignment of the main reaction difficult by mass-spectrometric studies alone. It was shown through studies with radioactive carbon that monomeric molecules of methyl methacrylate do not split off.

3. Analyses by Gas Chromatography

Gas or vapor-phase chromatography (VPC) lends itself particularly well to studies involving small concentrations of volatile decomposition products since it can be used for separation as well as for identification. Separation of the interfering impurities makes the analysis somewhat less complicated and improves the quantitative accuracy. It is also somewhat more sensitive and more adaptable to quantitative analysis than mass spectrometry, particularly for such products as H_2O and CO_2.

Desai and Shields (*219*) have described various columns used for the separation and identification of the components of the gaseous products.

Nitrogen is generally used as the carrier gas for determination of hydrogen. Helium is used as the carrier gas for hydrocarbons, CO, and other volatiles. Volatile products formed on irradiation can be withdrawn from the exposure tube and transferred to the gas chromatograph with a gas-tight syringe. Hartley and Guillet (*220*) describe a technique in which the gases are collected on activated charcoal which is cooled in liquid nitrogen. The hydrogen used for flushing the sample is then pumped off and the charcoal-containing tube is connected to the gas chromatograph. After allowing the tube to warm to room temperature, any desorbed gases are swept into the gas chromatograph by a stream of hydrogen.

The retention times of the various peaks of the gaseous products are calculated relative to the air peak and compared to the retention times of the reference compounds. The peak areas are compared to the peak areas of known volumes of the reference compounds. An integrating unit attached to the chromatograph recorder can be used to measure automatically the area of the peaks.

Since mass spectrometry is often used for identification in the absence of reference compounds, the combination of VPC and mass spectrometry can be efficiently utilized. The former can be used for separation and quantitative analysis while the latter is used for identification. The combined technique is a relatively recent innovation and its use in photodegradation studies of polymers has not yet been reported.

Gas chromatography has been utilized in studies of the degradation mechanisms of acrylates and methacrylates subjected to ultraviolet radiation. Grassie and MacCallum (*221*) showed that poly(alkyl methacrylates) can depolymerize quantitatively to the monomer. Ester decomposition in some types has been interpreted as a radical process occurring in direct competition with depolymerization. A good deal of information was gained about the type of chain end through which initiation occurs by studying the rate changes during the course of reaction and by comparison of different polymer preparations. In contrast to the methacrylates photochemical degradation of the acrylates produces mainly carbon dioxide and the corresponding alcohol together with a variety of volatile materials in smaller yield. Precise analysis of these products using gas chromatography is enabling the acrylates to be fitted into the general picture of the decomposition of acrylic polymers (*222*).

Rode *et al.* (*62*) showed that ultraviolet decomposition of the polyarylates is accompanied by the loss of weight associated with evolution of gaseous products. It was determined by chromatographic analysis that only CO and CO_2 are present in the evolved gas and that the latter is produced in higher concentration only during the early part of the reaction; on prolonged irradiation, only CO is evolved. In order to study the kinetics of gas evolution, the pressure of gas in the system was measured manometrically and at certain intervals the CO_2 was frozen out in liquid nitrogen. From the fall in pressure in the system, the quantities of CO and CO_2 were calculated. In addition, samples of gas were withdrawn and analyzed by gas chromatography at intervals during the course of reaction. There are two main competing processes in the degradation, both involving the carboxylic groups, which are the weakest sites. One process is the rupture of the ester group in the main chain, and the other is the opening of the lactone ring in the alcoholic part of the molecule. Rupture of the ester group will result mainly in loss of molecular weight, while opening of the lactone ring will lead to cross-linking. If it is assumed that each rupture of an ester group is accompanied by evolution of one gas molecule, then from determination of changes in molecular weight, in quantities of cross-linked material, and in gaseous products, it is possible to determine the number of ruptures of ester groups in both the main chain and alcoholic parts of the polymer.

Indications of the primary processs of photodegradation of cellulose material can also be obtained from a study of the nature of the volatile degradation products and their sequence of evolution along with changes in molecular weight and formation of free radicals in the substrate. Desai and Shields (*219*) recently presented the results of the first gas-chromatographic study of the volatile products of photodegradation of cellulose. The cellulose material was subjected to the unfiltered radiation of a high-pressure mercury vapor lamp at ambient conditions of temperature and pressure and a relative humidity of 50%. The results showed that with exposures of about 1 hr the main volatile products formed are carbon monoxide, carbon dioxide, and hydrogen. On longer exposure, the presence of a variety of low-molecular-weight organic compounds were observed for the first time. These observations along with studies of model compounds suggest that the initial reaction is mainly confined to chain cleavage, yielding glucose and oligosaccharides. These degrade on further exposure, yielding low-molecular-weight aldehydes, ketones, esters, alcohols, and hydrocarbons.

Photochemical degradation studies of a number of other types of polymers have also been carried out by means of gas chromatography. Stephenson *et al.* (*223*) analyzed the volatile decomposition products of polyethylene, polyimide, poly(ethylene terephthalate), and polyacrylonitrile. Hydrogen and methane were the principal decomposition products of polyethylene, and occur along with carbon monoxide in the decomposition of other polymers. There was a considerable difference in the relative abundances of methane and hydrogen, depending on whether the irradiation was performed in vacuum or in a nitrogen atmosphere. Mechanisms were proposed which were consistent with the observed results of photolysis. Several studies of the photochemistry of ketone polymers have also utilized gas-chromatography techniques. Harley and Guillet (*224*) used model compounds to show that the quantum yields of the Norrish types-I and -II free-radical processes are functions of chain length, as would be expected. David *et al.* (*225*) determined through analyses of volatile products, along with other techniques, that the principal mechanism in the photolysis of polyphenylvinylketone with 365-nm radiation is the abstraction of a hydrogen in the gamma position with respect to the carbonyl group, followed by breaking of the main polymer chain (Norrish type-II mechanism). The quantum yield of carbon monoxide resulting from a Norrish type-I radical split is less than 10^{-3}, as determined by gas chromatography.

B. *Electron Spin Resonance*

Electron spin resonance (esr) is a very useful and sensitive method for detecting, identifying, and quantitatively measuring free radicals. The esr spectra are associated with the unpaired electrons that exist in free radicals. The spectra exhibit hyperfine splitting due to the effect of neighboring hydrogen atoms on the unpaired electron and consequent coupling of the electron and proton magnetic moments. Thus, useful information about the immediate environment of the unpaired electron, and consequently the nature of the free radical, can be obtained from the fine structure. The concentrations of the radicals can be determined from the intensities of the spectra. Since so many polymer degradation reactions proceed via free radical mechanisms, esr has become a valuable tool for investigating these processes. Until its development, the presence of free radicals in a reacting system could only be postulated indirectly through observation of the intermediate and end products. Wall and Florin (*226*) have given an excellent review of the principles and experimental techniques of esr as they relate to the studies of polymers.

The major portion of esr studies of polymer degradation have been concerned with the effects of ionizing radiation. The difficulty in detecting the very low concentrations of radicals produced by ultraviolet radiation has limited the use of the technique in this application. However, it has been able to reveal the types of radicals produced by ultraviolet radiation in a number of polymers, thus giving information on the mechanisms of the photochemical reactions. Although these are frequently shown to be similar in nature to the reactions produced by ionizing radiation, important differences in mechanism do exist in some polymers. Therefore, an understanding of the radical mechanisms in photochemical degradations requires esr studies of the radicals produced by photolytic processes. A review of the status of this field was given by Grassie (*227*) in 1968.

Detection of the primary radicals formed on irradiation is particularly difficult in studies carried out under ambient conditions because of the rapid reactivity of the radicals. If oxygen is present, usually only a low, steady-state concentration of radicals exists even under relatively intense irradiation. Often the concentration at ambient temperature approaches the limit of detection of esr measurements which is of the order of 10^{-8} mole/liter. In very highly cross-linked systems the concentration of occluded radicals can be as high as 10^{-3} mole/liter, but generally the concentration is considerably lower. A higher concentration can be obtained under vacuum conditions. A few large polymer radicals can be trapped easily and may have lifetimes of months or even years at room temperature. However, most polymers yield esr spectra when irradiated and observed at liquid-nitrogen temperature, even in the presence of oxygen. The low-temperature spectra are frequently different from those observed at ambient temperatures, and are more likely to be representative of the primary radicals formed. Nevertheless, investigation of the entire reaction path requires measurements made at ambient as well as at lower temperatures. If the sensitivity of esr measurements were improved sufficiently to allow detection of all types of polymer radicals at ambient conditions, esr would become an even more valuable technique for investigating photochemical degradation mechanisms of polymers. Some of the studies that have already yielded valuable information are described in the following pages.

1. Poly(methyl Methacrylates)

One of the first irradiated polymers to be studied by esr was poly(methyl methacrylate) (*228*). A radical species which is stable at room temperature for long periods of time is produced under a variety of conditions. Radicals formed by ultraviolet or high-energy irradiation from several types of methacrylate monomers or polymers either at ambient or low temperature, as well as radicals occluded during polymerization or formed by mechanical destruction, give identical esr spectra under ambient conditions of detection (*229–232*). In spite of the numerous investigations in an attempt to understand the mechanisms of polymerization and degradation of methacrylates, the esr spectrum has not yet been identified unambiguously. Considerable controversy has existed over whether the well-resolved spectrum composed of a quintet with a quartet in between, each having a binomial distribution of intensities, belongs to a single radical or to two different radicals. A number of investigators (*233–238*) have favored the two-radical theory based mainly on the different rates of decay of the two sets of lines with change in temperature. The presence of both an alkyl and allyl radical or of two allyl radicals has been suggested. However, an equally large number of investigators explain the observed 9-line spectrum in terms of a single radical. Abraham *et al.* (*229*) and Symons (*239*) point out that the presence of two distinct radical species is not a reasonable assumption since it is highly unlikely that they would be trapped in the same relative concentrations irrespective of the sample, the type of radiation, or the ester group. Further, the allyl radical would be expected to react very rapidly in the presence of excess monomer. Comparisons with the corresponding esr spectra of other methacrylates (*229*, *240–242*) and of various deuterated poly(methyl methacrylates) (*242*, *244*) have been employed to show that the spectrum in question is due to a single radical species. Michel *et al.* (*245*) have attributed the different decay rates to the superposition of a weak broad spectrum of a minor component more stable than the species giving the 9-line spectrum and having a greater contribution to the 4-line than to the 5-line component. There have been a considerable number of studies which support the theory that the 4- and 5-line spectra belong to different conformations of the same radical which are present in statistically fixed proportions (*230*, *246*, *247*). It is plausible that the more stable conformation would decay at a slower rate as the temperature is raised. Iwasaki and Sakai (*248*) pointed out that the decay is determined not only by the matrix in which the radicals are trapped but also by the conformation angle of the radical species.

It is generally agreed that the alkyl radical has the structure $RCH_2—\dot{C}—(CH_3)COOCH_3$, and that the presence of a small amount of monomer and some degree of polymerization with partial trapping of the radical in the resulting matrix is required to obtain the 9-line spectrum (*237*, *249–251*). Atwater and McCue (*252*) suggest that the radical comes from a molecular species intermediate between the monomer and polymer, while Bresler *et al.* (*234*) and Sohmer and co-workers (*251*, *253*) propose that the room-temperature esr spectrum is due to radicals of highly polymerized molecules. Michael and co-workers (*254*) came to the conclusion that the esr signal arises from radicals directly associated with the residual monomer molecules rather than from breaks in the polymer chain. They suggest that the monomer molecule may combine with other radicals produced by the radiation on the main polymer chain, thus forming a radical identical to that which has been postulated as arising from a break in the polymer chain. The radical would thus be trapped in the polymer matrix. This mechanism could account for the stability of the esr signal and explain the fact that the counterpart radical which would be formed if chain scission occurred is not observed. Other investigations (*248*) have suggested that the presence of monomer is required to obtain the 9-line spectrum because it may serve to swell the polymer slightly, thus allowing greater freedom of motion to the polymer segments and hence better resolution of the spectrum.

The radical derived from the pure monomer has been shown to have a 7-line pattern (*253*). Although the occluded polymer radical might be expected to have an esr spectrum similar to that of the monomer radical, it is claimed (*229*, *255*) that the precise steric requirements of the polymer chain may permit only two conformations of the polymer radical which result in superposition of a 5- and 4-line spectrum. Inhibition of the rotation of the methylene or methyl group can result in greatly reduced interaction of either one or two of the five β protons with the unpaired electron. Symons (*239*) believes that the inhibition involves only the methylene hydrogens while Abraham *et al.* (*229*) claim that one of the hydrogens of the methyl group is inhibited in both conformations and that a methylene hydrogen is additionally inhibited in the conformation giving the 4-line spectrum.

The low-temperature esr spectra produced by ultraviolet-initiated polymerization of methyl methacrylate (*253*) or high-energy irradiation of the polymer (*226*, *256*) are different than the spectra observed at room temperature, but have given some indication of the mechanism by which the room-temperature species are produced. The mechanism proposed by Ormerod and Charlesby (*237*) provides explanations for most of the esr observations, including the formation of only one radical at room temperature. It also offers a possible reason why degradation resulting from chain scission is the main radiation-induced process in methacrylate polymers, while in acrylate polymers elimination of side chains followed by cross-linking predominates. Ormerod and Charlesby suggest that the radical formed by splitting off the ester group from the main chain rearranges to yield an olefin plus the so-called propagating radical that gives the 9-line spectrum.

2. Polyurethanes

Free radicals are rapidly formed in polyurethane and are extremely stable, existing at temperatures up to 180°C. In a study of the mechanism of photochemical degradation of polyurethanes, Tarakanov and co-workers (*257–259*) showed by esr measurements that there was no formation of peroxide-type radicals. The esr spectrum is apparently the result of secondary reactions originating from radicals of the type —RHN· and $\cdot CH_2$—CH_2—. These are known to be formed by cleavage of N—C and C—O bonds in the urethane group under the action of ultraviolet radiation. It was concluded that the paramagnetic products formed are connected with the development of a system of polyconjugated bonds in which strong delocalization of unpaired electrons takes place along the conjugated system.

The free radicals become more stable with increasing time of irradiation. This is explained by postulating that if the length of the conjugation chain increases with time, then the area of delocalization of the unpaired electron also increases, and thus the free radicals are increasingly stabilized. The progressive red shift in absorption accompanied by the appearance of color which occurs along with the accumulation of paramagnetic products is further evidence for the formation of a system of conjugated bonds. The yellowing resulting from ultraviolet irradiation at room temperature is thought to be due in large part to the formation of conjugated —N=N— groups. A possible mechanism for formation of polyconjugated systems during ultraviolet irradiation was suggested and it was shown that accumulation of these photolysis products eventually inhibits further photochemical degradation of polyurethanes.

3. Polystyrene

The majority of esr studies on polystyrene have been carried out on samples exposed to gamma radiation. Ultraviolet irradiation produces only a small concentration of radicals detectable by esr even at liquid-nitrogen temperature in the absence of oxygen (*260*). In well-evacuated samples a small concentration of radicals persists for several years at room temperature, but requires high sensitivity for detection (*261*). Cozzens *et al.* (*262*) showed that ultraviolet and gamma radiation produce the same esr spectrum in polystyrene and poly-α-methylstyrene. They obtained a triplet peak with hyperfine splitting by irradiating highly concentrated glassy solutions near liquid-nitrogen temperature. Although the samples were not evacuated, they were observed with simultaneous irradiation, thus maximizing the possibility of detecting short-lived primary species.

By comparison with esr spectra of poly(1-vinyl naphthalene) and deuterated polystyrene, Cozzens *et al.* verified the radical structure proposed by previous investigators (*260*, *263*). The radical appears to be formed by removal of a hydrogen on the tertiary carbon atom of the polymer chain attached to the ring and not by removal of a hydrogen on the ring itself, as suggested by Tsvetkov *et al.* (*264*). The rapid reactivity of the species is evidenced by the disappearance of the esr signal as the temperature is raised from −185 to −155°C. One of the reactions that it undergoes appears to result in the formation of a new radical having a structure closely related to a cyclohexadienyl radical (*261*, *262*, *265–267*).

Selivanov *et al.* (*51*) employed the esr technique to investigate the mechanism of formation of carbonyl groups during the photooxidation of polystyrene. They observed a symmetrical singlet which formed in vacuum at liquid-nitrogen temperature and changed in air to an asymmetrical singlet with rise in temperature. Based on this information and the results of kinetic studies of radical formation and destruction, it was concluded that the intermediate compound is not a peroxide but a peroxide radical formed by reaction of oxygen with the hydrocarbon radical at the tertiary carbon site.

4. Polyolefins

The esr spectra of irradiated polyolefins are generally weaker than those of poly(methyl methacrylate) and polystyrene. No signal is obtained at room temperature in air following ultraviolet irradiation. However, by carrying out the esr studies at liquid-nitrogen temperature under vacuum a considerable amount of information has been obtained with regard to the mechanism of photochemical degradation. Rånby and Carstensen (*268*) reviewed the esr studies of polyolefins carried out with both ionizing and ultraviolet radiation. Differences in observed radical structures produced were interpreted as evidence for differences in the mechanism of free radical formation. The radicals produced in polyethylene by ionizing radiation are attributed to alkyl, allyl, and polyenic structures in amounts depending on the dose rate. Radicals positively attributed to C—C bond scission either in main chain or side groups have never been observed with ionizing radiation. In contrast, the nature of the free radicals formed in polyethylene and polypropylene by ultraviolet light and the efficiency of radical production indicate that breakage of both C—C and C—H bonds occurs as primary reaction processes (*269*, *270*).

It was pointed out that under ultraviolet radiation the weaker C—C bonds are broken more easily than the stronger C—H bonds since the radiation energies are not much higher than the strength of the stronger bond. It therefore follows that ultraviolet radiation causes mainly degradation of the polyethylene chain rather than cross-linking (*271*). Further, since the small alkyl radicals formed under ultraviolet irradiation have only slightly more energy than is required to break the bond, they do not react readily and can therefore only be detected following ultraviolet irradiation.

The esr spectrum produced in polyethylene by ultraviolet radiation in vacuum at liquid-nitrogen temperature has been interpreted as composed of two components, a 5-line spectrum belonging to radicals resulting from main-chain scission and a sextet associated with breakage of a C—H bond (*230*). However, Tsuji and Seiki (*272*) have recently reported that different spectra are obtained depending on the pressure in the sample tube. The multiline spectrum is observed only when the gas pressure in the sample tube is about 10^{-3}-mm mercury. At lower pressures a broad low-intensity esr spectrum is obtained. It is suggested that the gas molecules around the polymer have an important role in the energy transfer or hydrogen abstraction processes resulting in C—H bond rupture. At low pressures these processes are not efficient and the spectrum is attributed to the free radicals produced at the energy-absorbing sites such as ketone groups and other impurities in the polymer.

Polypropylene produces a spectrum which is the superposition of a sharp 4-line multiplet that decays rapidly and a stable diffuse component. The most significant difference in the esr spectra of the two polymers is the presence of methyl radicals in polypropylene. It is thought that the 4-line spectrum is due to the methyl radicals resulting from side-chain scission and the diffuse spectrum is due mainly to the polymer radicals which are counterparts of the methyl radicals. The large radicals could also be interpreted as arising from main chain scission and hydrogen abstraction as in polyethylene. The free radicals are produced with seven times more efficiency in polypropylene than in polyethylene, probably due to the side chains in polypropylene enhancing free radical production.

Browning *et al.* (*273*), using monochromatic irradiation of 254 nm and liquid-nitrogen temperature, but without vacuum conditions, observed the same type of esr spectra for polyethylene and polypropylene reported by Rånby and Yoshida (*270*). They also examined other types of polyolefins and detected radicals corresponding to the side chain groups in all of them. However, they only observed the counterparts of the small alkyl radicals in some cases. Polymer radicals such as those resulting from hydrogen abstraction from the side chain were detected in many cases.

5. Poly(oxymethylene)

The copolymer of oxymethylene with oxyethylene is degraded in air by sunlight even though the polyacetal structure itself should not absorb the radiation of sunlight. In order to better understand the nature of the photolysis of poly(oxymethylene) (POM) and its copolymer, an esr study was undertaken by Hughes and Coard (*274*) to detect and identify the dominant radical intermediates. Photolysis studies of pure samples in powder form were carried out at liquid-nitrogen temperature under 254- and 313-nm radiation, both in vacuum and in the presence of oxygen.

Under vacuum conditions a small number of formyl and methyl radicals are formed, apparently resulting from traces of chromophores such as formaldehyde or formic acid present as impurities. This is evidenced by the fact that continued photolysis in vacuum does not produce increased radical concentration. Photolysis in oxygen, however, produces a continuously increasing concentration of radicals which were identified as the products of photooxidation.

The photolytic generation and subsequent decomposition of formyl radicals is considered to be an important process leading to the photolysis of poly(oxymethylene) materials because of the hydrogen atoms which are produced in the photodecomposition of the formyl radicals. The hydrogen abstraction reactions by the hydrogen atoms assume a predominant importance at ambient temperatures. In addition, the hydrogen atoms form hydroperoxy radicals on reaction with oxygen. Formyl radicals are produced under both 254- and 313-nm radiation although the quantum yield is considerably lower under the longer wavelength radiation. However, radiation of 313 nm and longer has a greater effect in decomposing the formyl radical. It was also suggested that formyl radicals may be produced through absorption of ultraviolet light by an acetal–oxygen charge-transfer complex which would absorb at longer wavelength than the polyoxymethylene. Sunlight would therefore have a greater effect on the complex than on the poly(oxymethylene) itself.

6. Poly(ethylene Terephthalate)

Although it may be expected that some of the more stable free radicals would be trapped at room temperature in a rigid medium such as poly(ethylene terephthalate), only very weak, poorly resolved esr spectra were obtained of this polymer following ultraviolet irradiation at room temperature under vacuum conditions (*44*). However, by comparison with esr spectra obtained by gamma irradiation, a tentative interpretation of the spectra produced by ultraviolet irradiation could be made. The strong singlet line was assigned to the aliphatic radical, $\cdot O—CH_2—CH_2—$, and a weak pair of triplets was assigned to the aromatic radical, $p\text{-}C_6H_3$. The type of cross-linking reaction postulated was considered to originate from the aromatic radicals.

C. *Model Compounds*

The complexity of the polymer degradation reactions and the attendant difficulties in isolating and identifying the degradation products and thus in interpreting the reactions have made the use of model compounds a valuable technique for obtaining information on the sort of reactions that occur in polymers. Simple compounds which are models of segments of the macromolecular chain or of an intermediate structure in the decomposition mechanism can be prepared in a fairly pure state so that the decomposition reaction can be investigated in the absence of complicating factors. Further, the primary step in photodegradation, a vital factor in interpreting the overall reaction, can be studied more readily.

The value of model compounds in polymer degradation studies was realized many years ago. Grassie (*275*) reviewed in detail investigations on the oxidation mechanism of olefinic model compounds and discussed the extent to which these can be related to polymer degradation mechanisms. There is a considerable background of knowledge available on the oxidation of small molecules representative of saturated and unsaturated polymers. Experimental evidence has shown that the oxidation characteristics of most polymers are qualitatively just about what one would expect by comparison with the small model compounds. The primary reaction is a hydroperoxidation and the degradation products, which can be accounted for by a mechanism involving hydroperoxide decomposition, are similar to those obtained from simple molecules. The lower reactivity of saturated compared with olefinic molecules is also reflected in the polymers. However, scission and cross-linking reactions which accompany oxidation are more important in polymers than in small compounds. In addition, it must be kept in mind in relating information obtained from model compounds to polymers that structural abnormalities present in small concentration in the polymers may have a very significant effect on the course of reaction since these can initiate chain reactions. When different types of reactions are found in the polymer than are expected based on model compounds, this suggests that the degradation reaction is initiated at abnormalities in the polymer structure.

Polyethylene is an example of a polymer which proved to be much more reactive toward oxygen than could be predicted from studies of its model compounds, i.e., short-chain normal paraffins. Small concentrations of carbonyl groups, carbon–carbon double bonds, and branched methyl groups are known to be present in the linear chain of methylene groups in the polymer. The carbonyl groups are the structural abnormalities which are thought to be responsible for the more rapid rate of photo-oxidation of polyethylene than was predicted (*276*).

Structural abnormalities were also found to be an important factor in the photochemical degradation of poly(vinyl chloride). *Sec*-butyl chloride was used in one study as a prototype of the polymer to obtain information on its mechanism of degradation (*199*). The effects of various conditions on the degradation and products of reaction of this model compound were determined. Radiation of wavelengths longer than 270 nm caused evolution of HCl only when a carbonyl-containing impurity was present, and the rate and amount of evolution was proportional to the concentration of the carbonyl compound. The wavelength dependence and effect of ketonic impurities on the liberation of HCl from poly(vinyl chloride) can be related to these observations. The effect of oxygen on the evolution of HCl and on the molecular weight of the degraded polymer could also be explained through investigation of its effect on *sec*-butyl chloride. The liberation of HCl appears to result from thermal decomposition of unstable compounds produced in the photochemical reaction. This was confirmed by a recent study in which photodegradation products of the model compounds, 2,4-dichloropentane and 4-chloro-2-butanone, were compared with those of poly(vinyl chloride) (*18*). No HCl was detected from either of the model compounds when photooxidations were carried out at low temperatures, but was detected from the ketone at the highest temperature used, i.e., 25°C. This compound was used as a model for oxidized poly(vinyl chloride) and is known to be thermally unstable at room temperature and to decompose partially to HCl and methylvinyl ketone. Thus, it was concluded that the evolution of HCl occurs subsequent to and not simultaneous with the photooxidation reaction. This explains the greater thermal stability of the unirradiated sample.

Although the presence of impurities can influence the reactions of the polymer to ultraviolet radiation, generally the type of reactions can be predicted from studies of model compounds. For example, by analogy with photooxidation reactions of ethyl benzene the most probable sites for oxidative attack on polystyrene are predicted to be the tertiary hydrogen atoms with formation of hydroperoxide as the primary product decomposing to ketone and hydroxide groups. This was borne out by the identifiable products of oxidation which include benzaldehyde (*214*) and ketones of the type expected (*277*).

Polyester resins are complex cross-linked structures of linear unsaturated polyesters. An understanding of the degradation mechanism through analysis of the degradation products is a difficult task. By studying the effects of irradiation on model compounds representing a number of types of linear unsaturated polyesters prepared from the purest starting materials, and by comparing them with their cross-linked counterparts, Voigt (*61*) was able to obtain some understanding of the degradation process of the resins. He assumed that photooxidations could not play a large part in the formation of degradation products because of the absence of oxygen during irradiation. On this basis, the simplest and most comprehensive explanation of the experimental results involved absorption of radiation by the maleic acid chromophore as the primary photochemical step. Two different secondary reactions were postulated following activation of the chromophore; one leads to cross-linking and the other to chain ends containing carbonyl groups. Buxbaum (*278*) showed, by using suitable esters as model compounds for poly(ethylene terephthalate), that interpretation of the kinetics and identification of the reaction products are facilitated. Segments with terminal hydroxyethyl groups were used in the model compounds.

Another example of the information which can be gained from studies of a small model compound of a polymer is the use of glutaronitrile for investigation of photochemical degradation of polyacrylonitrile (*58*). Ultraviolet absorption spectra of solutions of the model compound before and after irradiation showed characteristics very similar to those of polyacrylonitrile in the same solvent. This gave evidence that spectral changes in polyacrylonitrile were not related to the chain-scission reaction in the polymer. Rather, reactions at the nitrile groups leading to formation of small molecules with conjugated systems or possibly to substituted pyridines appeared to be responsible for the changes in spectra.

In another type of study, Hartley and Guillet (*224*) used model compounds to establish the molecular weight dependence of the photochemical processes in polymeric ketones. Variations in quantum yield with change in length of the alkyl radical of symmetrical aliphatic ketones were investigated. Carbon monoxide evolution, ultraviolet absorption spectra, and solvent viscosities were measured. The results were interpreted in terms of competition between the Norrish type-I free-radical process, the Norrish type-II photoelimination process to yield olefins and methyl ketones, and radiationless deactivation processes. Only the Norrish type-II process appeared to be a function of chain length when the number of carbon atoms exceeded three.

Finally, model compounds of the degradation products have also been used as an aid in their identification. Examples of such studies involve fluorescent-type compounds. Pacifici and Straley (*95*) identified mono- and dihydroxylated polyester as products of the photolysis of poly(ethylene terephthalate) by comparing the fluorescence spectra of the photolyzed sample with the spectra of synthetic samples containing the suspected products. Hydroxylation of the aromatic nuclei is believed to occur through reaction of the polyester with hydroxyl or hydroperoxy radicals produced by ultraviolet irradiation under atmospheric conditions. Model fluorescent compounds have also been used in analysis of the mechanism and products of degradation of polystyrene. Zapolskii (*22*) showed that diacetyl and benzil have luminescent properties similar to those of irradiated polystyrene. These structures were also confirmed by infrared spectroscopy. Quinoid structures were thus ruled out as the cause of yellowing, and a mechanism for the formation of α-diketones as a secondary process under the action of ultraviolet radiation was suggested.

Thus, the use of model compounds has been shown to be a valuable technique for investigating the photochemical degradation mechanisms of a variety of polymers. The types of model compounds have included those which are chemically similar to the structural units of the polymer as well as those which represent the suspected degradation products. The type and importance of structural abnormalities in the photodegradation process can frequently be established by such studies.

D. *Oxygen Uptake*

The appearance of oxygenated degradation products in polymers formed as a result of ultraviolet irradiation shows that deterioration is associated with absorption of oxygen from the air. Scission and cross-linking reactions probably occur as a result of the mechanism incorporating oxygen since deterioration of the useful physical properties frequently runs parallel with absorption of oxygen (*275*). Atmospheric oxygen freely penetrates disordered regions in polymers where it reacts with chain segments activated by ultraviolet radiation. Olefinic groups, tertiary hydrogen atoms, and other structural abnormalities are particularly vulnerable to oxygen attack. Therefore, measurement of the amount of oxygen taken up by a polymer under the influence of ultraviolet radiation provides a quantitative evaluation of the photooxidative instability of the polymer. Since the absorption of a small amount of oxygen can be measured with high precision, the degradation can be detected during the early stages of failure. Oxygen uptake measurements have been used in photochemical degradation studies of a variety of polymers.

Kelleher *et al.* (*279*) recently used this method for assessing the photochemical stability of films of acrylonitrile-butadiene-styrene (ABS) resins and for determining the protective effect of carbon black. It was shown that oxygen uptake measurements correlated well with stability ratings of the samples based on mechanical property changes produced by prolonged outdoor aging. The measurements were made by a modified method based on one previously used by Hawkins *et al.* (*280*) and Winslow *et al.* (*281*). The film was exposed at 50°C to 365-nm radiation in the presence of oxygen and the consumption of oxygen was measured volumetrically.

In a study of the ultraviolet-initiated autooxidation of poly(ester-urethane), Schollenberger and Dinsbergs (*11*) determined the absorption of oxygen during exposure of the sample in a carbon arc radiation source. The chemical changes attending the combined exposure to oxygen and ultraviolet radiation were followed by infrared, ultraviolet, and mass-spectrometry measurements as well as by elemental analysis. Mechanical property tests showed that in the presence of oxygen the tensile strength was reduced manifold compared with exposure under vacuum conditions.

Oxygen uptake studies were also carried out by Heacock *et al.* (*14*) during photochemical degradation of polyethylene films. The percentage oxygen in the films measured by the Unterzaucher (*282*) method as a function of radiation at different temperatures gave a family of curves which were similar to the curves for formation of carbonyl and hydroxyl groups and also paralleled the increase in density and crystallinity. It has been shown (*15*, *283*) that the relative magnitudes of scission and cross-linking in polyethylene can be correlated with the rates of oxygen uptake. Cotten and Sacks (*15*) based their conclusion on measurements of oxygen transmission through the films by the standard ASTM method (*284*). The assumption is that the oxygen transmission rate as measured by this technique bears a direct relationship to the rate of oxygen consumption during photooxidation reaction. It was suggested that cross-linking takes place through an oxygen linkage since carbonyl group formation was found to be appreciably higher in the thinner, more cross-linked films.

Based on the kinetics of initial rate of oxygen absorption, the photochemical stability of polyurethanes was shown by Tarakanov *et al.* (*258*) to be dependent on the chemical constitution of the diisocyanates and hydroxyl-containing compounds used. The order of stabilities was somewhat different than that based on initial rates of gas evolution.

Studies on poly(vinyl chloride) frequently involve measurement of the amount of oxygen taken up as well as measurement of the evolution of HCl during ultraviolet irradiation (*199*, *285*). Druesedow and Gibbs (*285*) demonstrated that oxygen was taken up as a direct result of ultraviolet irradiation, and its role in sensitization of the polymer to thermal breakdown was proposed. It was suggested that the free radicals produced in the polymer chain by ultraviolet irradiation react with oxygen, thus forming peroxide groups. The latter activate the chlorine atoms, thereby giving rise to a weak spot in the polymer. The observed catalysis of HCl evolution by the presence of oxygen can be satisfactorily accounted for by this mechanism.

From a mechanistic viewpoint, the rate of oxygen consumption is more meaningful than the rate of evolution of gases. The latter measurements cannot be made until the reactions are carried so far that secondary processes complicate interpretation of the initial stages of photochemical degradation. Grassie and Weir (*286*) showed that the oxygen uptake technique could be used to investigate the mechanism of photochemical oxidation of polystyrene in the early stages. A detailed description of their apparatus was given (*66*). Studies were carried out on the dependence of the rate of oxygen absorption on light intensity and oxygen pressure. These were compared with the kinetics of photochemical degradation in vacuum in the absence of oxygen. The reaction rate was determined by the diffusion rate of oxygen into the film. An explanation for the mechanism of photochemical oxidation at 254 nm was given and was shown to be different than the mechanism at 365 nm. The rate of oxygen absorption of polystyrene was also measured by Tsang (*287*) who used a manometric oxygen uptake technique for investigating the photochemical degradation of polystyrene and its model compounds. The results were correlated with infrared analysis of the photooxidation products.

REFERENCES

1. R. C. Hirt and N. Z. Searle, in *Weatherability of Plastic Materials* (M. R. Kamal, ed.), Wiley-Interscience, New York, 1967, pp. 61–83.
2. ASTM Designation D 795-65T, in *1969 Book of ASTM Standards*, Pt. 27, Am. Soc. for Testing Materials, Philadelphia, Penna., 1969, pp. 322–327.
3. R. A. Kinmonth, Jr., *Soc. Plastics Eng. Trans.*, **4,** 229 (1964).
4. C. A. Brighton, in *Weathering and Degradation of Plastics* (S. H. Pinner, ed.), Columbine Press, Manchester, England, 1966, Chap. 4.
5. N. Z. Searle, *Appl. Opt.*, **3,** 923 (1964).
6. R. B. Withrow and A. P. Withrow, in *Radiation Biology* (A. Hollaender, ed., with collaboration of S. B. Hendricks), Vol. III, McGraw-Hill, New York, 1956, p. 180.
7. P. H. Estes, *Mod. Plastics*, **37,** No. 1, 134 (1959).
8. R. A. McCarthy, *Plastics Technol.*, **4,** 640 (1958).
9. Westinghouse Lamp Div., Bloomfield, N.J.
10. B. G. Achhammer, F. W. Reinhart, and G. M. Kline, in *Polymer Degradation Mechanisms*, NBS Circular 525, Supt. of Documents, U.S. Govt. Printing Office, Washington D.C., 1953, Chap. 17; *J. Res. Natl. Bur. Std. (U.S.)*, **46,** 391 (1951).
11. C. S. Schollenberger and K. Dinsbergs, *Soc. Plastics Eng. Trans.*, **1,** 31 (1961).
12. J. P. Luongo, *J. Polymer Sci.*, **42,** 139 (1960); *J. Appl. Polymer Sci.*, **3,** 302 (1960).
13. F. M. Rugg, J. J. Smith, and R. C. Bacon, *J. Polymer Sci.*, **13,** 535 (1954).
14. J. F. Heacock, F. B. Mallory, and F. P. Gay, *ibid.*, Pt. A-1, **6,** 2921 (1968).
15. G. R. Cotten and W. Sacks, *ibid.*, Pt. A, **1,** 1345 (1963).
16. A. V. Karyakin, G. V. Grishin, and B. D. Kurykin, *Vysokomolekul Soedin*, **7,** 389 (1965) (in Russian); *Polymer Sci. U.S.S.R.* (English Transl.), **7,** 429 (1967).
17. E. Ya. Vaiman, *Zh. Prikl. Khimii*, **41,** 362 (1968) (in Russian); *J. Appl. Chem. U.S.S.R.* (English Transl.), **41,** 341 (1968).
18. K-P. S. Kwei, *J. Polymer Sci.*, Pt. A-1, **7,** 1075 (1969).
19. O. B. Zapolskii, *Vysokomolekul. Soedin.*, **3,** 376 (1961) (in Russian); *Polymer Sci. U.S.S.R.* (English Transl.), **3,** 114 (1961).
20. A. V. Uvarov and S. V. Yakobovich, *Lakokras Mat.*, No. 6, 49 (1961).
21. A. H. Reine and J. C. Arthur, Jr., *Textile Res. J.*, **40,** 90 (1970).
22. O. B. Zapolskii, *Vysokomolekul. Soedin.*, **7,** 615 (1965) (in Russian); *Polymer Sci. U.S.S.R.* (English Transl.), **7,** 678 (1965).
23. K. A. Leitman, *Vysokomolekul. Soedin.*, **A9,** 1565 (1967) (in Russian); *Polymer Sci. U.S.S.R.* (English Transl.), **9,** 1758 (1967).
24. E. I. Kirillova, *Plast. Massy*, No. 3, 10 (1964) (in Russian); *Soviet Plastics* (English Transl.), No. 3, 14 (1965).
25. A. V. Karyakin and A. I. Funtikova, *Vysokomolekul. Soedin.*, **7,** 1171 (1965) (in Russian); *Polymer Sci. U.S.S.R.* (English Transl.), **7,** 1294 (1965).
26. B. G. Achhammer, M. J. Reiney, and F. W. Reinhart, *J. Res. Nat. Bur. Stand. (U.S.)*, **47,** 116 (1951).
27. H. C. Beachell and L H. Smiley, *J. Polymer Sci.*, Pt. A-1, **5,** 1635 (1967).
28. N. Grassie, *J. Appl. Polymer Sci.*, **9,** 999 (1965).

29. L. A. Wall and M. Tryon, *J. Phys. Chem.*, **62,** 697 (1958).
30. B. Baum, *J. Appl. Polymer Sci.*, **2,** 281 (1959).
31. H. C. Beachell and G. W. Tarbet, *J. Polymer Sci.*, **45,** 451 (1960).
32. H. H. Jellinek, *ibid.*, **4,** 1 (1949).
33. C. F. Berch, M. R. Harvey, and B. G. Achhammer, *J. Res. Natl. Bur. Stand. (U.S.)*, **60,** 481 (1958).
34. J. Mitchell, Jr., and L. R. Perkins, in *Weatherability of Plastic Materials* (M. R. Kamal, ed.), Wiley-Interscience, New York, 1967, pp. 167–173.
35. N. Grassie and N. A. Weir, *J. Appl. Polymer Sci.*, **9,** 975 (1965).
36. J. W. Tamblyn, G. C. Newland, and M. T. Watson, *Plastics Technol.*, **4,** 427 (1958).
37. L. H. Cross, R. B. Richards, and H. A. Willis, *Discuss. Faraday Soc.*, No. 9, 235 (1950).
38. R. B. Seymour, H. E. Tsang, and D. Warren, *Polymer Eng. Sci.*, **7,** 55 (1967).
39. W. T. M. Johnson, *Offic. Dig.*, **32,** 1067 (1960).
40. J. Fahrenfort, *Spectry. Acta*, **17,** 698 (1961); N. J. Harrick, *Internal Reflection Spectroscopy*, Wiley-Interscience, New York, 1967.
41. N. J. Harrick, *Ann. N.Y. Acad. Sci.*, **101,** 928 (1963).
42. J. J. Gedemer, *Appl. Spectrosc.*, **19,** 141 (1965).
43. M. G. Chan and W. L. Hawkins, *Am. Chem. Soc., Div. Polymer Chem., Preprints*, **9,** No. 2, 1638 (1968).
44. F. B. Marcotte, *J. Polymer Sci.*, Pt. A-1, **5,** 481 (1967).
45. Wilks Scient., Norwalk, Conn.
46. Barnes Eng., Stamford, Conn.
47. J. K. Barr and P. A. Fluornoy, in *Physical Methods in Macromolecular Chemistry* (B. Carroll, ed.), Vol. 1, Dekker, New York, 1969, Chap. 2.
48. W. W. Coblentz and R. Stair, *J. Res. Nat. Bur. Stand. (U.S.)*, **3,** 629 (1929).
49. M. J. Reiney, M. Tryon, and B. G. Achhammer, *ibid.*, **51,** 155 (1953).
50. L. Scarbrough, W. L. Kellner, and P. W. Rizzo, in *Polymer Degradation Mechanisms*, NBS Circular 525, Supt. of Documents, U.S. Govt. Printing Office, Washington D.C., 1953, Chap. 6.
51. P. I. Selivanov, E. I. Kirillova, and V. A. Maksimov, *Vysokomolekul. Soedin.*, **8,** 1418 (1966) (in Russian); *Polymer Sci. U.S.S.R.* (English Transl.), **8,** 1561 (1966).
52. C. Savides, J. A. Stretanski, and L. R. Costello, in *Advances in Chemistry Series No. 85, Stabilization of Polymers and Stabilizer Processes* (R. F. Gould, ed., with collaboration of N. A. J. Platzer), Am. Chem. Soc., Washington D.C., 1968, pp. 287–306.
53. L. A. Matheson and R. F. Boyer, *Ind. Eng. Chem.*, **44,** 867 (1952).
54. L. A. Wall, M. R. Harvey, and M. Tryon, *J. Phys. Chem.*, **60,** 1306 (1956).
55. P. I. Selivanov, V. L. Maksimov, and E. I. Kirillova, *Vysokomolekul. Soedin.* **A11,** 482 (1969) (in Russian); *Polymer Sci. U.S.S.R.* (English Transl.), **11,** 542 (1969).
56. P. Williams, *Ark. Kemi*, **29,** 485 (1968).
57. G. Oster, G. K. Oster, and M. Kryszewski, *J. Polymer Sci.*, **57,** 937 (1962).
58. H. H. G. Jellinek and I. J. Bastien, *Can. J. Chem.*, **39,** 205 (1961).
59. R. B. Fox, *J. Polymer Sci.*, Pt. A, **2,** 2085 (1964).
60. B. G. Achhammer, *Nat. Bur. Stand. Rep. 7171*, U.S. Dept. Comm., NBS, Dec. 1, 1960 to May 31, 1961, p. 3.
61. J. Voigt, *Makromol. Chem.*, **27,** 80 (1958).

62. V. V. Rode, A. S. Yarov, and S. R. Rafikov, *Vysokomolekul. Soedin.*, **6,** 2168 (1964) (in Russian); *Polymer Sci. U.S.S.R.* (English Transl.), **6,** 2399 (1964).
63. R. C. Hirt, R. G. Schmitt, and W. L. Dutton, *Solar Energy*, **3,** 19 (1959).
64. L. V. Nevskii, O. G. Tarakanov, and V. K. Belyakov, *Plast. Massy* No. 7, 43 (1966) (in Russian); *Soviet Plastics* (English Transl.), No. 7, 47 (1967).
65. J. A. Parkins, *Paint Varnish Prod.*, **47,** No. 13, 42 (1957).
66. N. Grassie and N. A. Weir, *J. Appl. Polymer Sci.*, **9,** 963 (1965).
67. ASTM Designation D 1925-63T, in *1969 Book of ASTM Standards*, Pt. 27, Am. Soc. for Testing Materials, Philadelphia, Penna., 1969, pp. 666–669; ASTM Designation E 308–66, *ibid.*, Pt. 30, pp. 1026–1046.
68. J. M. Jordan, R. E. McIlroy, and E. M. Pearce, in *Weatherability of Plastic Materials* (M. R. Kamal, ed.), Wiley-Interscience, New York, 1967, pp. 205–218.
69. H. R. Davidson and L. W. Imm, *J. Opt. Soc. Amer.*, **39,** 942 (1949); E. W. Peterson, *ibid.*, **39,** 1055 (1949); F. W. Billmeyer, Jr., *ibid.*, **50,** 137 (1960).
70. ASTM Designation E 313-67, in *1969 Book of ASTM Standards*, Pt. 30, Am. Soc. for Testing Materials, Philadelphia, Penna., 1969, pp. 1063–1067.
71. IES-SPI-NEMA Subcomm. of Light Control and Equip. Design Comm., *Illum. Eng.*, **53,** 284 (1958); **55,** 450 (1960).
72. F. W. Billmeyer, Jr., *Mater. Res. Stand.*, **6,** 295 (1966).
73. P. A. Sanders, *Rubber Chem. Technol.*, **22,** 465 (1949).
74. H. E. Hill, *J. Paint Technol.*, **41,** 321 (1969).
75. V. E. Gray and J. R. Wright, *Proc. Conf. Plastics Bldg. Struct., London, June 14–16, 1965*, Conf. Suppl. No. 1, Plast. Inst. Trans. and J.
76. H. H. Frey, *Kunststoffe*, **53,** 103 (1963).
77. V. E. Gray and J. R. Wright, *J. Appl. Polymer Sci.*, **7,** 2161 (1963).
78. V. E. Gray and J. R. Wright, *ibid.*, **8,** 1505 (1964).
79. V. E. Gray and J. R. Wright, *Polymer Eng. Sci.*, **5,** No. 4, 284 (1965).
80. S. R. Palit and P. Ghosh, *J. Polymer Sci.*, **58,** 1225 (1962).
81. J. F. Heacock, *J. Appl. Polymer. Sci.*, **7,** 2319 (1963).
82. I. N. Ermolenko and G. N. Savastenko, *Zh. Anal. Khim.*, **21,** 98 (1966) (in Russian); *J. Anal. Chem. U.S.S.R.* (English Transl.), **21,** 81 (1966).
83. P. G. Campbell and J. R. Wright, *Ind. Eng. Chem., Prod. Res. Develop.*, **5,** No. 4, 319 (1966).
84. M. I. Eiss and P. Giesecke, *Anal. Chem.*, **31,** 1558 (1958).
85. F. Feigl, *Spot Tests in Organic Analysis*, 5th edition, Elsevier, Amsterdam, 1960, pp. 220, 250, 253.
86. I. Boustead and A. Charlesby, *Eur. Polymer J.*, **3,** 459 (1967).
87. R. H. Partridge, *J. Chem. Phys.*, **45,** 1679 (1966).
88. A. Charlesby and R. H. Partridge, *Proc. Roy. Soc.* (*London*), **A283,** 312 (1965).
89. L. J. Basile, *J. Chem. Phys.*, **36,** 2204 (1962).
90. S. S. Yanari, F. A. Bovey, and R. Lumry, *Nature*, **200,** 242 (1963).
91. M. Leibowitz and A. Weinreb, *J. Chem. Phys.*, **46,** 4652 (1967).
92. M. T. Vala, Jr., J. Haebig, and S. A. Rice, *ibid.*, **43,** 886 (1965).
93. K. F. Plitt and S. D. Toner, *J. Appl. Polymer Sci.*, **5,** 534 (1961).
94. J. Jortner, *J. Polymer Sci.*, **37,** 199 (1959).
95. J. G. Pacifici and J. M. Straley, *ibid.*, Pt. B, **7,** 7 (1969).
96. A. P. Pivovarov, Yu. A. Ershov, and A. F. Lukovnikov, *Plast. Massy*, No. 10, 7 (1966) (in Russian); *Soviet Plastics* (English Transl.), No. 10, 11 (1966).

97. A. P. Pivovarov and A. F. Lukovnikov, *Vysokomolekul. Soedin.*, **A9,** 2727 (1967) (in Russian); *Polymer Sci. U.S.S.R.* (English Transl.), **9,** 3087 (1967).
98. V. A. Gachkovskii, *Zh. Strukt. Khim.*, **4,** 424 (1963) (in Russian); *J. Struct. Chem. U.S.S.R.* (English Transl.), **4,** 386 (1963).
99. E. A. Silin, *Vysokomolekul. Soedin.*, **A10,** 1786 (1968) (in Russian); *Polymer Sci. U.S.S.R.* (English Transl.), **10,** 2071 (1968).
100. G. E. Ashby, *J. Polymer Sci.*, **50,** 99 (1961).
101. G. E. Ashby, *Soc. Plastics Eng. Trans.*, **2,** 161 (1962).
102. M. P. Schard, *Polymer Eng. Sci.*, **5,** 246 (1965).
103. M. P. Schard and C. A. Russell, *J. Appl. Polymer Sci.*, **8,** 985 (1964).
104. R. F. Vasilev and A. A. Vichutinskii, *Nature*, **194,** 1276 (1962).
105. R. S. Hunter and L. Boor, in *Testing of Polymers* (J. V. Schmitz, ed.), Vol. 2, Wiley-Interscience, New York, 1966, Chap. 8.
106. A. E. Lover and J. A. Rhys, in *The Properties and Testing of Plastics Materials*, 3rd edition, Chem. Rubber Publ. Co., Cleveland, Ohio, 1968, pp. 203, 204.
107. ASTM Designation D 523-67, in *1969 Book of ASTM Standards*, Pt. 27, Am. Soc. for Testing Materials, Philadelphia, Penna., 1969, pp. 144–148; ASTM Designation D 2457-65T, *ibid.*, pp. 758–763.
108. Hunter Assoc., Inc., Fairfax, Va. 22030.
109. R. R. Knittel, *Mater. Res. Stand.*, **2,** 180 (1962).
110. Gardner Labs, Inc., Bethesda, Md. 20014.
111. J. H. Colling, W. E. Cracker, and J. Dunderdale, *J. Oil Col. Chem. Assoc.*, **51,** 524 (1968).
112. ASTM Designation D 1746-62T, in *1969 Book of ASTM Standards*, Pt. 27, Am. Soc. for Testing Materials, Philadelphia, Penna., 1969, pp. 578–580.
113. C. B. Havens, in *Polymer Degradation Mechanisms*, NBS Circular 525, Supt. of Documents, U.S. Govt. Printing Office, Washington D.C., 1953, Chap. 7.
114. ASTM Designation D 1003-61, in *1969 Book of ASTM Standards*, Pt. 27, Am. Soc. for Testing Materials, Philadelphia, Penna., 1969, pp. 411–418.
115. G. A. Baum, in *Weatherability of Plastic Materials* (M. R. Kamal, ed.), Wiley-Interscience, New York, 1967, pp. 189–204.
116. ASTM Designation D 1738-66, in *1969 Book of ASTM Standards*, Pt. 21, Am. Soc. for Testing Materials, Philadelphia, Penna., 1969, pp. 394–400.
117. J. J. Gouza and W. F. Bartoe, *Mod. Plastics*, **33,** No. 9, 157 (1956).
118. S. B. Newman and I. Wolock, *J. Res. Nat. Bur. Stand.* (*U.S.*), **58,** 339 (1957).
119. D. A. Hurst, *Soc. Plastics Eng. J.*, **12,** No. 5, 18 (1956).
120. S. Leghissa and O. Salvatore, *Polymer Eng. Sci.*, **6,** 127 (1966).
121. Instron Corp., Canton, Mass.
122. Scott Testers, Inc., Providence, R.I.
123. ASTM Designation D 638-68, in *1969 Book of ASTM Standards*, Pt. 27, Am. Soc. for Testing Materials, Philadelphia, Penna., 1969, pp. 189–203.
124. ASTM Designation D 747-63, *ibid.*, pp. 273–278.
125. ASTM Designation D 790-66, *ibid.*, pp. 303–310.
126. ASTM Designation D 882-67, *ibid.*, pp. 365–372.
127. ASTM Designation D 1623-64, *ibid.*, Pt. 26, pp. 169–174.
128. R. L. Bergen, Jr., in *Testing of Polymers* (J. V. Schmitz, ed.), Vol. 2, Wiley-Interscience, New York, 1966, Chap. 1.
129. H. S. Loveless, *ibid.*, Chap. 10.

130. W. D. Harris, in *Testing of Polymers* (W. E. Brown, ed.), Vol. 4, Wiley-Interscience, New York, 1969, pp. 399–431.

131. C. V. Stephenson, B. C. Moses, and W. S. Wilcox, *J. Polymer Sci.*, **55**, 451 (1961).

132. J. W. Tamblyn, G. C. Newland, and M. T. Watson, *Plastics Technol.*, **4**, 427 (1958).

133. H. M. Quackenbos and H. Samuels, in *Weatherability of Plastic Materials* (M. R. Kamal, ed.), Wiley-Interscience, New York, 1967, pp. 155–165.

134. H. S. Lilley, *Paint Technol.*, **22**, 441 (1958).

135. H. S. Lilley, *ibid.*, **23**, 51 (1959).

136. N. J. Marshall, *Offic. Dig.*, No. 391, 792 (1957).

137. C. V. Stephenson and W. S. Wilcox, *J. Polymer Sci.*, Pt. A, **1**, 2741 (1963); W. S. Wilcox, *Am. Chem. Soc., Div. Org. Coatings Plastics Chem., Preprints*, **21** No. 2, 382 (1961).

138. P. G. Kelleher, *Soc. Plastics Eng. J.*, **25**, No. 2, 53 (1969).

139. F. S. Kaufman, Jr., in *Weatherability of Plastic Materials* (M. R. Kamal, ed.), Wiley-Interscience, New York, 1967, pp. 131–139.

140. ASTM Designation D 1693-66, in *1969 Book of ASTM Standards*, Pt. 26, Am. Soc. for Testing Materials, Philadelphia, Penna., 1969, pp. 228–233.

141. R. J. Martinovich and G. R. Hill, in *Weatherability of Plastic Materials* (M. R. Kamal, ed.), Wiley-Interscience, New York, 1967, pp. 141–154.

142. ASTM Designation D 412-68, in *1969 Book of ASTM Standards*, Pt. 27, Am. Soc. for Testing Materials, Philadelphia, Penna., 1969, pp. 114–125.

143. ASTM Designation D 256-56, *ibid.*, pp. 77–84.

144. ASTM Designation D 1709-67, *ibid.*, Pt. 26, pp. 240–247.

145. A. E. Lover and J. A. Rhys, in *The Properties and Testing of Plastics Materials*, 3rd edition, Chem. Rubber Publ. Co., Cleveland, Ohio, 1968, p. 71.

146. ASTM Designation D 746-64T, in *1969 Book of ASTM Standards*, Pt. 27, Am. Soc. for Testing Materials, Philadelphia, Penna., 1969, pp. 267–272.

147. ASTM Designation D 1790-62, *ibid.*, pp. 581–584.

148. V. T. Wallder, *Ind. Eng. Chem.*, **42**, 2320 (1950).

149. A. W. Pross and R. M. Black, *J. Soc. Chem. Ind.*, **69**, 113 (1950); C. S. Myers, *Ind. Eng. Chem.*, **44**, 1095 (1952).

150. ASTM Designation D 149-64, in *1969 Book of ASTM Standards*, Pt. 27, Am. Soc. for Testing Matérials, Philadelphia, Penna., 1969, pp. 20–28; ASTM Designation D 150-68, *ibid.*, pp. 29–54; ASTM Designation D 257-66, *ibid.*, pp. 85–107.

151. A. H. Sharbaugh, in *Testing of Polymers* (J. V. Schmitz, ed.), Vol. I, Wiley-Interscience, New York, 1965, Chap. 5.

152. A. H. Scott, *ibid.*, Chap. 6.

153. R. H. Tucker, *ibid.*, Chap. 7.

154. R. W. Warfield, *ibid.*, Chap. 8.

155. T. W. Dakin, *ibid.*, Chap. 9.

156. P. F. Flory, *Chem. Rev.*, **35**, 51 (1944).

157. A. Charlesby, *Proc. Roy. Soc.* (*London*), **A222**, 60, 542 (1954).

158. R. A. V. Raff and K. W. Doak, *Crystalline Olefin Polymers*, Wiley-Interscience, New York, 1965, p. 901.

159. ASTM Designation D 154-65, in *1969 Book of ASTM Standards*, Pt. 21, Am. Soc. for Testing Materials, Philadelphia, Penna., 1969, pp. 54–59.

160. ASTM Designation D 115-68, *ibid.*, pp. 45–53.

161. Brookfield Eng. Labs, Stoughten, Mass.
162. ASTM Designation D 445-65, in *1969 Book of ASTM Standards*, Pt. 29, Am. Soc. for Testing Materials, Philadelphia, Penna., 1969, pp. 318–324.
163. D. I. Marshall, *ASTM Bull.*, No. 204, 40 (1955).
164. W. Roth and S. R. Rich, *J. Appl. Phys.*, **24,** 940 (1950).
165. M. R. Cannon and M. R. Fenske, *Oil Gas J.*, **33,** 52 (1935); **34,** 45 (1936).
166. L. Ubbelohde, *Ind. Eng. Chem. Anal. Ed.*, **9,** 85 (1937).
167. P. F. Onyon, *J. Polymer Sci.*, **22,** 13 (1956).
168. W. R. Krigbaum and Q. A. Trementozzi, *ibid.*, **28,** 295 (1958).
169. G. B. Taylor, *J. Am. Chem. Soc.*, **69,** 635 (1947).
170. A. Conix, *Makromol. Chem.*, **26,** 226 (1958).
171. J. N. Sen, S. R. Chatterjee, and S. R. Palit, *J. Sci. Ind. Res.* (*India*), **11B,** 90 (1952).
172. M. Kolinsky, *J. Polymer Sci.*, Pt. C, No. 16, 485 (1967).
173. H. Staudinger, *Chem. Ber.*, **70,** 66 (1933).
174. H. H. G. Jellinek and W. A. Schlueter, *J. Appl. Polymer Sci.*, **3,** 206 (1960).
175. H. Jacobs and R. Steele, *ibid.*, **3,** 239 (1960).
176. K. Kato, *ibid.*, **13,** 599 (1969).
177. K.-P. S. Kwei, *ibid.*, **12,** 1543 (1968).
178. K. R. Osborn, *J. Polymer Sci.*, **38,** 357 (1959).
179. ASTM Designation D 1238-65T, in *1969 Book of ASTM Standards*, Pt. 27, Am. Soc. for Testing Materials, Philadelphia, Penna., 1969, pp. 455–466.
180. W. L. Peticolas and J. M. Watkins, *J. Am. Chem. Soc.*, **79,** 5083 (1957).
181. A. Charlesby and S. H. Pinner, *Proc. Roy. Soc.* (*London*), **A249,** 367 (1959).
182. C. V. Stephenson, *J. Polymer Sci.*, **55,** 465 (1961).
183. D. K. Taylor, *Plastics Inst.* (*London*), *Trans. J.*, **28,** 170 (1960).
184. A. Charlesby, *J. Polymer Sci.*, **11,** 513 (1953).
185. J. R. Schaefgen and P. J. Flory, *J. Am. Chem. Soc.*, **72,** 689 (1950).
186. P. J. Flory and J. Rehner, Jr., *J. Chem. Phys.*, **11,** 512 (1943).
187. P. J. Flory, *Principles of Polymer Chemistry*, Cornell Univ. Press, Ithaca, New York, 1953.
188. H. Determann, *Gel Chromatography*, Springer, Berlin, 1968; P. Flodin and J. Porath, in *Chromatography* (E. Heftman, ed.), Reinhold, New York, 1961, Chap. 13; J. Cazes, *J. Chem. Educ.*, **43,** A567, A625 (1966); K. A. Boni, *J. Polymer Sci.*, Pt. A-2, **6,** 1567, 1579 (1968).
189. M. F. Vaughan, *Nature*, **188,** 55 (1960); **195,** 801 (1962).
190. J. C. Moore, *J. Polymer Sci.*, Pt. A, **2,** 835 (1964).
191. T. Williams, *ibid.*, Pt. A-2, **6,** 1613, 1621, 1627 (1968).
192. A. R. Shultz, *J. Appl. Polymer Sci.*, **10,** 353 (1966).
193. K.-P. S. Kwei and T. K. Kwei, *ibid.*, **12,** 1551 (1968).
194. M. Kasha, *J. Opt. Soc. Am.*, **38,** 929 (1948); L. R. Koller, *Ultraviolet Radiation*, 2nd edition, Wiley, New York, 1965, pp. 168–176.
195. W. C. Warner and E. E. Gruber, *Ind. Eng. Chem.*, *Prod. Res. Devel.*, **5,** 219 (1966).
196. J. A. Melchore, *ibid.*, **1,** 232 (1962).
197. R. T. Dean and J. P. Manasia, *Mod. Plastics*, **33,** No. 2, 131 (1955).
198. M. I. Frolova and A. V. Ryabov, *Vysokomolekul. Soedin.*, **1,** 1453 (1959) (in Russian); *Polymer Sci. U.S.S.R.* (English Transl.), **2,** 1 (1960).
199. A. S. Kenyon, in *Polymer Degradation Mechanisms*, NBS Circular 525, Supt. of Documents, U.S. Govt. Printing Office, Washington, D.C., 1953, Chap. 5.

200. R. C. Hirt, N. Z. Searle, and R. G. Schmitt, *Soc. Plastics Eng. Trans.*, **1,** 21 (1961).

201. T. Hirashima, *Kogyo Kagaku Zasshi* (*J. Chem. Soc. Japan, Ind. Chem. Sect.*), **70,** 1538 (1967).

202. L. D. Johnson, W. C. Tincher, and H. C. Bach, *J. Appl. Polymer Sci.*, **13,** 1825 (1969).

203. H. A. Taylor, W. C. Tincher, and W. F. Hamner, *ibid.*, **14,** 141 (1970).

204. I. H. Leaver, *Textile Res. J.*, **38,** 729 (1968).

205. A. R. Shultz and S. M. Leahy, *J. Appl. Polymer Sci.*, **5,** 64 (1961).

206. N. Z. Searle and R. C. Hirt, *Am. Chem. Soc., Div. Polymer Chem., Preprints*, **3** (2), 56–61 (1962).

207. R. C. Hirt and N. Z. Searle, *Soc. Plastics Eng., Preprints*, Regional Tech. Conf. (RETEC) on Stability and Degrad. Polymers, Washington D.C., June 4, 1964, pp. 286–302.

208. R. C. Hirt, N. Z. Searle, and W. L. Dutton, Symposium on Molecular Fluorescence, 11th Pittsburgh Conf. on Anal. Chem. Appl. Spectrosc, Pittsburgh, Penna., March 4, 1960, Paper No. 164 (Abstract only printed).

209. Herbach and Rademan, Inc., Philadelphia, Penna., 19107.

210. J. E. Wilson, *J. Am. Chem. Soc.*, **72,** 2905 (1950).

211. J. Petit and G. Zaïtoun, *Compt. Rend.*, **256,** 2610 (1963).

212. B. Baum, *Soc. Plastics Eng. J.*, **17,** 71 (1961).

213. Yu. I. Temchim, *Plast. Massy*, No. 12, 60 (1968) (in Russian); *Soviet Plastics* (English Transl.), No. 12, 63 (1968).

214. B. G. Achhammer, *J. Polymer Sci.*, **8,** 555 (1952); B. G. Achhammer, in *Polymer Degradation Mechanisms*, NBS Circular 525, Supt. of Documents, U.S. Govt. Printing Office, Washington D.C., 1953, Chap. 14.

215. M. I. Frolova, L. V. Nevskii, and A. V. Ryabov, *Vysokomolekul. Soedin.*, **3,** 877 (1961) (in Russian); *Polymer Sci. U.S.S.R.* (English Transl.), **3,** 703 (1962).

216. J. H. Flynn and W. L. Morrow, *J. Polymer Sci.*, Pt. A, **2,** 91 (1964).

217. J. H. Flynn, W. K. Wilson, and W. L. Morrow, *J. Res. Nat. Bur. Stand.* (*U.S.*), **60,** 229 (1958).

218. J. H. Flynn and W. L. Morrow, *J. Polymer Sci.*, Pt. A, **2,** 81 (1964).

219. R. L. Desai and J. A. Shields, *Makromol. Chem.*, **122,** 134 (1969).

220. G. H. Hartley and J. E. Guillet, *Macromol.*, **1,** 165 (1968).

221. N. Grassie and J. R. MacCallum, *J. Polymer Sci.*, Pt. A, **2,** 983 (1964).

222. N. Grassie, *Pure Appl. Chem.*, **12,** 237 (1966).

223. C. V. Stephenson, J. C. Lacey, Jr., and W. S. Wilcox, *J. Polymer Sci.*, **55,** 477 (1961).

224. G. H. Hartley and J. E. Guillet, *Macromol.*, **1,** 413 (1968).

225. C. David, W. Demarteau, and G. Geuskens, *Polymer*, **8,** 497 (1967).

226. L. A. Wall and R. E. Florin, in *Analytical Chemistry of High Polymers* (G. M. Kline, ed.), Pt. II, Wiley-Interscience, New York, 1962, Chap. XII.

227. N. Grassie, *Pure Appl. Chem.*, **16,** 389 (1968).

228. E. E. Schneider, M. J. Day, and G. Stein, *Nature*, **168,** 645 (1951).

229. R. J. Abraham, *Trans. Faraday Soc.*, **54,** 1133 (1958).

230. N. M. Atherton, H. W. Melville, and D. W. Whiffen, *J. Polymer Sci.*, **34,** 199 (1959).

231. A. Charlesby and D. K. Thomas, *Proc. Roy. Soc.* (*London*), **A269,** 104 (1962).

232. P. Yu. Butyagin, I. V. Kolbanev, and V. A. Radtsig, *Soviet Phys.—Solid State*, **5**, 1642 (1964).
233. K. Hotta and R. S. Anderson, paper presented, 4th Internatl. Symp., Free Radicals, Washington D.C., 1959.
234. S. E. Bresler, E. N. Kazbekov, and E. M. Saminskij, *Vysokomolekul. Soedin.*, **1**, 132, 1374 (1959) (in Russian); *Polymer Sci. U.S.S.R.* (English Transl.), **1**, 55, 540 (1960).
235. L. H. Piette, *NMR and PMR Spectroscopy*, Pergamon, New York, 1960, p. 218.
236. I. D. Campbell and F. D. Looney, *Australian J. Chem.*, **15**, 642 (1942).
237. M. G. Ormerod and A. Charlesby, *Polymer*, **5**, 67 (1964).
238. A. T. Bullock and L. H. Sutcliffe, *Trans. Faraday Soc.*, **60**, 625 (1964).
239. M. C. R. Symons, *J. Chem. Soc.*, 1186 (1963).
240. D. W. Ovenall, *Nature*, **184**, 181 (1959).
241. C. H. Bamford, G. C. Eastmond, and Y. Sakai, *ibid.*, **200**, 1284 (1963).
242. H. Fischer, *J. Polymer Sci.*, Pt. B, **2**, 529 (1964).
243. P. Kourim and K. Vacek, *Tetrahedron Lett.*, **23**, 1051 (1962).
244. P. Kourim and K. Vacek, *Trans. Faraday Soc.*, **61**, 415 (1965).
245. R. E. Michel, F. W. Chapman, and T. J. Mao, *J. Chem. Phys.*, **45**, 4604 (1967).
246. H. W. Melville, *Z. Electrochem.*, **60**, 276 (1956).
247. L. A. Wall and R. E. Florin, in *Analytical Chemistry of High Polymers* (G. M. Kline, ed.), Pt. II, Wiley-Interscience, New York, 1962, pp. 537–539.
248. M. Iwasaki and Y. Sakai, *J. Polymer Sci.*, Pt. A-1, **7**, 1537 (1969).
249. I. Ungar, W. Gager, and R. Leininger, *ibid.*, **44**, 295 (1960).
250. J. K. Kircher, *J. Phys. Chem.*, **69**, 189 (1965).
251. J. Sohma, T. Komatsu, and H. Kashiwabara, *J. Polymer Sci.*, Pt. B, **3**, 287 (1965).
252. H. A. Atwater and B. A. McCue, *ibid.*, **7**, 477 (1969).
253. T. Komatsu, *ibid.*, Pt. C, No. 16, 535 (1967).
254. R. E. Michael, F. W. Chapman, and T. J. Mao, *ibid.*, Pt. A-1, **5**, 677 (1967).
255. D. J. E. Ingram, M. C. R. Symons, and M. G. Townsend, *Trans. Faraday Soc.*, **54**, 409 (1958).
256. Y. Hajimoto, N. Tamura, and S. Okamoto, *J. Polymer Sci.*, Pt. A, **3**, 255 (1965).
257. A. A. Berlin, *Vysokomolekul. Soedin.*, **9A**, 1677 (1967) (in Russian); *Polymer Sci. U.S.S.R.* (English Transl.), **9A**, 1888 (1967).
258. O. G. Tarakanov, L. V. Nevskji, and V. K. Belyakov, *J. Polymer Sci.*, Pt. C, No. 23, 193 (1968).
259. V. K. Belyakov, *Vysokomolekul. Soedin.*, **A10**, 599 (1968) (in Russian); *Polymer Sci. U.S.S.R.* (English Transl.), **10**, 700 (1968).
260. R. E. Florin, L. A. Wall, and D. W. Brown, *Trans. Faraday Soc.*, **56**, 1304 (1960).
261. R. E. Florin, L. A. Wall, and D. W. Brown, *J. Polymer Sci.*, Pt. A, **1**, 1521 (1963).
262. R. F. Cozzens, W. B. Moniz, and R. B. Fox, *J. Chem. Phys.*, **48**, 581 (1968).
263. R. J. Abraham and D. H. Whiffen, *Trans. Faraday Soc.*, **54**, 1291 (1958).
264. Iu. D. Tsvetkov, Iu. N. Molin, and V. V. Voyevodskii, *Vysokomolekul. Soedin.*, **1**, 1805 (1959) (in Russian); *J. Polymer Sci. U.S.S.R.* (English Transl.), **2**, 165 (1960).
265. S. Ohnishi, T. Tanci, and I. Nitta, *J. Chem. Phys.*, **37**, 2402 (1962).
266. H. Fischer, *Kolloid-Z.*, **180**, 64 (1962).
267. L. A. Wall and R. B. Ingallis, *J. Chem. Phys.*, **41**, 1112 (1964).

268. B. Rånby and P. Carstensen, in *Irradiation of Polymers* (R. F. Gould, ed., N. Platzer, Symp. Chairman), Advances in Chemistry Series 66, Am. Chem. Soc., 1967, Chap. 18.
269. H. Yoshida and B. Rånby, *J. Polymer Sci.*, Pt. B, **2,** 1155 (1964).
270. B. Rånby and H. Yoshida, *ibid.*, Pt. C, No. 12, 263 (1966).
271. A. A. Kachan, *Vysokomolekul. Soedin.*, **A10,** 471 (1968) (in Russian); *Polymer Sci. U.S.S.R.* (English Transl.), **10,** 546 (1968).
272. K. Tsuji and T. Seiki, *J. Polymer Sci.*, Pt. B, **7,** 839 (1969).
273. H. L. Browning, Jr., H. D. Ackerman, and H. W. Patton, *ibid.*, Pt. A-1, **4,** 1433 (1966).
274. O. R. Hughes and L. C. Coard, *ibid.*, **7,** 1861 (1969).
275. N. Grassie, *Chemistry of High Polymer Degradation Processes*, Wiley-Interscience, New York, 1956, Chap. 4.
276. N. Grassie, *ibid.*, pp. 240, 241.
277. A. Votinov, P. Kobeko, and F. Marei, *J. Phys. Chem. U.S.S.R.* (English Transl.), **16,** 106 (1942).
278. L. H. Buxbaum, *Angew. Chem.*, **7,** 182 (1968).
279. P. G. Kelleher, D. J. Boyle, and R. J. Miner, *Mod. Plastics*, **46,** No. 9, 188 (1969).
280. W. L. Hawkins, W. Matreyek, and F. H. Winslow, *J. Polymer Sci.*, **41,** 1 (1959).
281. F. H. Winslow, W. Matreyek, and S. M. Stills, *Am. Chem. Soc., Div. Polymer Chem., Preprints*, **7,** No. 1, 390 (1966).
282. J. Unterzaucher and K. Burger, *Chem. Ber.*, **70,** 1392 (1937).
283. F. H. Winslow and W. L. Hawkins, in *Weatherability of Plastic Materials* (M. R. Kamal, ed.), Wiley-Interscience, New York, 1967, pp. 29–39.
284. ASTM Designation D 1434-66, in *1969 Book of ASTM Standards*, Pt. 27, Am. Soc. for Testing Materials, Philadelphia, Penna., 1969, pp. 494–508.
285. D. Druesedow and C. F. Gibbs, in *Polymer Degradation Mechanisms*, NBS Circular 525, Supt. of Documents, U.S. Govt. Printing Office, Washington D.C., 1953, Chap. 4.
286. N. Grassie and N. A. Weir, *J. Appl. Polymer Sci.*, **9,** 987 (1965).
287. H. S. Tsang, "A Study of the Effect of Additives on the Stability of Polystyrene and its Model Compounds," Doctoral Dissertation, University of Houston, Texas, May 1969.

AUTHOR INDEX

Numbers in parentheses are reference numbers and indicate that an author's work is referred to although his name is not cited in the text. Numbers in italics show the page on which the complete reference is listed.

C

D

E

F

I

L

M

N

T

SUBJECT INDEX

Q

R